Wonderful and wonder-full! This splendidly illustrated book explores total solar eclipses and their effect on us through art, music and words.

Dame Jocelyn Bell Burnell DBE FRS FRSE FRAS FInstP,
Astrophysics, University of Oxford

A courtside seat to watch scientists, scholars, artists, and musicians toss ideas right (awesome!) and wrong (interesting!) back and forth over centuries of wonder. Nothing as real and completely out of human control or influence as the total solar eclipse has been so endlessly fascinating, provocative, and compelling; get your ticket here.

Michael O'Hare, Professor of the Graduate School at the University of California
and Goldman School of Public Policy

A total solar eclipse is a spectacle without equal. Henrike Christiane Lange and Tom McLeish study the human and cultural impact of totality. Every human culture has a mythology about solar eclipses. These stories should be told, and this book is an excellent survey of many cultures across the continents and throughout the centuries. I especially enjoyed the excerpts from Tom McLeish's travel diary from August 2017 which capture the thrill of the chase and the allure of the corona in the co-authored Introduction. Chapter 2 by my late friend Jay Pasachoff on the solar corona is a masterclass in science communication. I highly recommend *Eclipse & Revelation* to anyone interested in solar eclipses and their many interactions with humanity.

Michael Zeiler, cartographer and eclipse chaser

Superb! This book touched my soul! Fabulous stuff and the first of its kind! So rich in thought with delectable prose on the history, art and science of the ages that surround solar eclipses. Henrike Lange and Tom McLeish have done something extraordinary: From one momentary cosmic event they virtuously generated lasting inspiration to chase knowledge and wisdom. From the present, they manage to look back and forth through all collegiate fields of study. Even the most experienced eclipse chasers will feel enriched and further enlightened by an eclipse after delving into this book. In ancient Greece, every audience member knew the play's story as they took their olives and wine to the amphitheater. That's what this book does for the reader. A true tome on the total solar eclipse.

Mike Kentrianakis, American Astronomical Society 2017
Total Solar Eclipse Project Manager, Amateur Astronomers Association of New York,
IAU's Working Group on Solar Eclipses

As a very experienced eclipse follower, I was not quite expecting to be led down as many new pathways as those served up in *Eclipse & Revelation*. I guess the word "revelation" should have warned me. I enjoyed the book, as much for its expeditions from my normal comfort zone as its reconstruction of familiar places. At the end, I emerged with my over 30 eclipse experiences freshly anchored in mysticism, history, and legend while clothed in new garments of art, music and literature. For both the scientist and the artist, *Eclipse and Revelation* gives view of the "other side"—a forest of ideas that lead into unexpected clearings where a strange, new facet of the eclipse experience is revealed to the inquiring reader. The work becomes a tapestry that stitches history, passion, nature, weather, art, and music with a thread of mysticism and wonder. This ambitious volume welds so many dissimilar views of a shared experience: One chapter explains the physics behind the glow of the corona; another, Dante's eclipse muse. Ancient and medieval history with their embedded spiritualism and terror blend with the magnificence of art and poetry; animal and atmosphere respond in muted sympathy, all bound by the glory of the total eclipse. In *Eclipse & Revelation*, the reader, veteran eclipse chaser or novice, will discover that the eclipse world is much larger than just the visual experience.

Jay Anderson, Canadian meteorologist and eclipse chaser

Genius! Truly marvelous and relevant work, beautifully illustrated and delivered: an utterly brilliant new take on interdisciplinary collaborations between the arts, humanities, and sciences exploring a gripping natural phenomenon across human history. Unlike any other, this book includes fascinating perspectives and early science from ancient Asia, Assyria, Babylonia, India, China, Greece and Rome, the middle ages and the early modern era, the scientific revolution to the present, ritual, theology, and field reports, a cornucopia of art historical, theoretical, philosophical, literary, and musical content, and a wide range of environmental aspects such as the reactions to total solar eclipses of bees, mammals, birds, even reef fish, seals, and zooplankton–all topped off with the latest meteorological methods and a conclusion that creates a poetic awareness of the entire cosmos. This uniquely reader-friendly volume has been meticulously researched and didactically prepared with deep expert knowledge for a wide general audience. Its welcoming prose, warmth, clarity, and inclusive approach invite all readers to consider themselves future students in each potential new field and discipline. Lange and McLeish deliver a passionate defence of the liberal arts and a delightful account of the perpetual curiosity, excitement, joy, and enduring love of wisdom at the core of the scientific and scholarly life.

Andrew Stewart, Professor Emeritus in History of Art and Classics
at the University of California, Berkeley

Eclipse and Revelation

Eclipse and Revelation

Total Solar Eclipses in Science, History, Literature, and the Arts

Edited by

HENRIKE CHRISTIANE LANGE

University of California, Berkeley

and

TOM McLEISH

University of York

OXFORD
UNIVERSITY PRESS

Great Clarendon Street, Oxford, OX2 6DP,
United Kingdom

Oxford University Press is a department of the University of Oxford.
It furthers the University's objective of excellence in research, scholarship,
and education by publishing worldwide. Oxford is a registered trade mark of
Oxford University Press in the UK and in certain other countries

Published in the United States of America by Oxford University Press
198 Madison Avenue, New York, NY 10016, United States of America

British Library Cataloguing in Publication Data

Data available

Library of Congress Control Number: 2023945986

ISBN 9780192857996

DOI: 10.1093/oso/9780192857996.001.0001

Printed and bound by
Bell and Bain Ltd, Glasgow

Cover image: Paul Nash, *The Eclipse of the Sunflower* (detail), with kind permission of the
British Council Collection, London.

As flint sparks flint, so with friends:

In loving and grateful memory of

JAY MYRON PASACHOFF

(1943–2022)

and

THOMAS CHARLES BUCKLAND MCLEISH

(1962–2023)

'Love is strong as Death ... many waters cannot quench Love.' Song of Songs.
So if you love, and are loved, you have nothing to fear from death.
Nothing especially complicated about it.

Tom McLeish, 13 January 2023

Compound image from 65 originals from Andreas Möller at Piedras del Aguila made by Roman Vaňúr/Jay Pasachoff, including a coronal mass ejection (CME) on the left.
Source: Jay M. Pasachoff, Vojtech Rušin, and Roman Vaňúr.

Preface: 'Cosmos' is for Harmony

Henrike Christiane Lange and Tom McLeish

We knew what we would be doing for the next seven years: the schedule for this book was elegantly set by the cosmos itself, spanning the interim between the great American total solar eclipse of 2017 and that of 2024. Our interdisciplinary initiative as an editorial team began during a shared fellowship at the Notre Dame Institute for Advanced Study (NDIAS) in August 2017 under the motto 'VERUM— BONUM— PULCHRUM.' Ever since, we have been uniting in our conversation the fields of natural philosophy, art history and architecture, astrophysics, music, theology, soft matter physics, literature, medieval and early modern studies, and European history, along our parallel backgrounds in the arts, sciences, history, theology, and *Geistesgeschichte*. But most importantly, our collaboration in art and science is fuelled by our shared enthusiasm for knowledge as a united enterprise, for language and connection, for dimensions of music and faith—in short, for the mystery of existence itself. Those were the themes that gathered us in the late summer of 2017 as fellows at NDIAS in Indiana, Midwest USA, from where Tom was able to fulfil his lifelong wish and witness a solar eclipse in the path of totality, in Crofton, Kentucky (see the travel diary excerpts in the Introduction). And so we went down a rabbit hole the size of a universe.

At NDIAS, we hosted an initial panel on interdisciplinary perspectives on total solar eclipses with scholars from a wide spectrum of disciplines each contributing a reflection on how their fields have responded to the extraordinary phenomenon. The intellectual engagement vividly illustrated the demand for a multi-perspectival book on the total solar eclipse. So the work towards such a book would be our next step to explore further cross-disciplinary conversations—maybe even to create a model of disciplinary discourses in contact with each other, in a true liberal arts tradition that helps to heal the artificial divisions between the academic fields and disciplines. The time at NDIAS was as productive as it was mutually inspiring. Having published his book on a scientist's reading of the Biblical book of Job, *Faith and Wisdom in Science*, Tom McLeish was then finishing *The Poetry and Music of Science: Comparing Creativity in Science and Art* at NDIAS, and published soon thereafter *Soft Matter Physics: A Very Short Introduction*, while Henrike Lange finished her book manuscript for *Giotto's Arena Chapel and the Triumph of Humility*, and finalized research on Cimabue, Giotto, Dante, Donatello, Mantegna, Botticelli, and Raphael.[1] These publications opened the wide thematic field in which this long-standing collaboration came to flourish.

---

[1] See Tom McLeish, *Faith and Wisdom in Science*, Oxford, 2014; *The Poetry and Music of Science: Comparing Creativity in Science and Art*, Oxford, 2019 (new edition with an additional chapter on poetry, Oxford, 2022); and *Soft Matter Physics: A Very Short Introduction*, Oxford 2020; see Henrike Christiane Lange, 'Cimabue's True Crosses in Arezzo and Florence.' In *Material Christianity: Western Religion and*

As art historian Roberta J. M. Olson and astrophysicist Jay M. Pasachoff wrote in their groundbreaking and richly illustrated interdisciplinary study *Cosmos: The Art and Science of the Universe*, these two fields are already natural allies—they allow a collaboration 'from separate but extremely visual professions'[2] as the astronomer and the art historian are both used to looking at visual phenomena with a well-trained gaze, to cultivating a deep and interconnected visual memory, and to using visual evidence as a source of knowledge and, subsequently, as interpretative matter. Olson's and Pasachoff's *Cosmos* rightly appears under the Platonic motto 'Astronomy compels the soul to look upwards and leads us from this world to another' (Plato, *The Republic*, I, 342).[3] Continuing their decades-long interdisciplinary conversation, Roberta J. M. Olson and Jay M. Pasachoff contributed respective chapters on the total solar eclipse to this book (Pasachoff Chapter 2, Olson Chapter 10).

Colleagues and new friends joined our interdisciplinary journey over the years, helping to form the illuminating ideas represented in this volume by the full team—each author is an established expert in their own fields as well as a pioneering interdisciplinary thinker. Mike Frost—the Director of the historical section of the British Astronomical Association and a friend from Tom's time at Emmanuel College, Cambridge—kindly supported the work during the final editorial stages from late 2022 on all levels of production, but especially on that of scientific discourse and its translation for wider audiences of 'eclipse-chasers.' Authors in our networks incorporated feedback from both editors and in conversation with each other—under Tom McLeish's lead for the science chapters; under Henrike Lange's direction for the arts, humanities, and history chapters; and since Winter 2022 also with valuable feedback on all final drafts from Mike Frost. Uniquely, all authors have read all of the chapters in various drafts, commented upon each others' work at multiple stages, and included feedback from other authors, not just from the editors, into their final chapters. Cross-reading, and the subsequent emergence of those governing themes that only become visible when an object is illuminated from different directions, brought the authorial team together in years of dialogues that included the unexpected COVID-19 shelter-in-place lockdown and the global pandemic with all its known and unknown consequences.

Our teaching and service for the general public extends itself beyond the confines of departments, labs, archives, museums, and the campus. It includes a vast array of outreach to general audiences such as McLeish's collaborative work for schools

---

the Agency of Things, eds. Christopher Ocker and Susanna Elm, Cham, 2020, pp. 29–67; 'Portraiture, Projection, Perfection: The Multiple Effigies of Enrico Scrovegni in Giotto's Arena Chapel.' In *Picturing Death 1200–1600*, eds Stephen Perkinson and Noa Turel, Leiden, Boston, and Paderborn, 2020, pp. 36–48; 'Relief Effects in Donatello and Mantegna.' In *The Reinvention of Sculpture in Fifteenth-Century Italy*, eds Amy Bloch and Daniel M. Zolli, Cambridge, 2020, pp. 327–43; 'A Drawing of a Cherub's Head for Raphael's Disputa in the Stanza della Segnatura.' In *Master Drawings* 60, 2022, pp. 291–302; 'Giotto's Triumph: The Arena Chapel and the Metaphysics of Ancient Roman Triumphal Arches.' In *I Tatti Studies* 25, 2022, pp. 5–38; *Giotto's Arena Chapel and the Triumph of Humility*. Cambridge, 2023; and 'Ephemerality and Perspective in Dante's Marble Reliefs and Botticelli's Drawing for *Purgatorio* 10.' Forthcoming in *Reading Dante with Images: A Visual Lectura Dantis*, Vol. 2, ed. Matthew Collins.

[2] See Roberta J. M. Olson and Jay M. Pasachoff, *Cosmos: The Art and Science of the Universe*, London, 2019, p. 7.

[3] Olson and Pasachoff, *Cosmos*, p. 9.

across the UK and his work as Chair of the Royal Society's Education Committee, and Lange's work in public higher education at the University of California, Berkeley and within the wider UC system over the past decade. We conceptualized this book to open the door into our daily lives as scholars and researchers—to offer, for a wider readership from all fields, a rare inside view of how scientists and humanities scholars can work together when they gather around something exciting and mysterious, exemplifying a variety of disciplines by sharing their diverse perspectives on the same, universal phenomena.

We were casting our net as widely as possible when finalizing our invitations for the authors' team between 2018 and 2020—thinking about the full spread of historical and geographical areas and media that could potentially be represented in such a volume on total solar eclipses. In doing so, we found that, rather than sketching out a superficial coverage of a complete register for all times and places, the specificity of the approaches gathered here is most valuable for the book's interdisciplinary mission. Likewise valuable is the inclusion of different timbres for reporting significant personal and academic experiences, such as the very personal report of Mike Frost in his capacity as eclipse-chaser which serves as our pivot between those chapters with a mainly scientific approach and those coming from a history, arts, and humanities perspective, and David Bentley Hart's report from the 2017 eclipse.

Eclipse and Revelation therefore does not reach for any kind of completeness; on the contrary, it purposefully opens up avenues for future interdisciplinary work in exemplifying depth and details of each scholarly contribution in their collected variety. The years of working on this project in its many stages have proven that our fragmented disciplines have so much to give to each other when engaged in conversation around a central subject of human and material significance. We have already seen that this kind of openness inspires readers of all ages to find their own way into thinking about and seeing eclipses, through their initial interest in any single aspect of science, history, art, and theology in a rewarding way, while opening perspectives into previously unvisited fields of knowledge and aesthetic-creative endeavours in their lives.

Across the represented disciplines, the chapters gathered for the four thematic parts of this volume combine large overviews with very focused case studies—our contributing authors had absolute freedom in their method of choice. The greatest value of this enterprise, we hope, is going to be the combination of a purposefully wide opening into all possible disciplines and approaches. At the same time, it was an interesting lesson from the project that such cross-disciplinary integration is stimulated only by a drastic reduction of the discourse to *total* solar eclipses specifically and exclusively. In this combination of freedom and focus, the book is entirely new in the vast landscape of cosmos literature for expert and general readership. This prompt inspired the entire team over years to find the common ground of imagination, creativity, discovery, beauty, and awe in each field's learning and knowledge-production. As we will see in the chapters on artistic representations of total solar eclipses, some of the phenomenological challenges are the parameters of the inherent invisibilities of a total eclipse, but also of its fluidity, its motion, meaning, and transcendence— these and more set also the challenge of this volume for its writers and its readers, asking for an adjustment of the framework for each chapter's represented discipline (a short introductory note and a chapter glossary provide orientation and some core technical terms at each of those junctures).

Our plans since 2017 had always involved authors' conferences at the Royal Society in London and a series of in-person writing workshops at the University of California, Berkeley. The global pandemic cancelled these and many more plans, as we had to realize immediately at the very beginning of the shelter-in-place lockdown, starting for us with the cancellation of a long-anticipated panel on Grosseteste and Giotto, 'Visionaries of Vision,' for the Medieval Academy of America (planned for March 2020 in Berkeley). Cambridge and London were the centres of our editorial workshops following Notre Dame; we are grateful for the time and local support there, especially at the Royal Society in London and Emmanuel College at Cambridge, and for the most encouraging interest and kind support offered by the Medieval Academy of America as well as UC Berkeley's Departments of History of Art and Italian Studies, Berkeley's Medieval Studies Program, the Institute of European Studies, the Center for the Study of Religion, the Arts + Design Initiative at the Arts Research Center, and Berkeley's Designated Emphasis in Renaissance and Early Modern Studies for our conference and campus events planned for March 2020. We are therefore even more delighted that, in 2024, we will be able to recover some of these lost opportunities with our three celebratory exhibitions between the US West Coast and the City of Westminster— *Eclipse & Revelation: A Berkeley Astrophysics & Art History Exhibition* on the UC Berkeley campus in California, *Eclipse & Revelation: An Exhibition at Burlington House* (with items from the Royal Astronomical Society) at Burlington House on Piccadilly in London, and *Eclipse & Revelation: An Exhibition at Carlton House Terrace* (with eclipse materials from the Royal Society) at the Royal Society in London.

As the British astronomer William H. McCrea put it, 'The astronomer's view of the cosmos may be influenced to a great and almost unknowable extent by a combination of lucky and unlucky circumstances.'[4] Somehow, somewhere along the way from 2020 to 2023, we came to understand that this has been even more true in these years of unprecedented challenges and, regardless, grace. We had little to no control over the exact combination of lucky and unlucky circumstances, but we still were able to make the most of them, so that, much like the paradoxical nature of the eclipse-revelation, our formerly positive connotation of the beautiful word 'corona' still shines bright and clear into the face of this difficult time.

---

[4] W. H. McCrea, 'Astronomer's Luck.' In *Quarterly Journal of the Royal Astronomical Society* 13, 1972, pp. 506–19, at p. 506. McCrea explains this in a larger context, summarizing the 'haunting thought' of the combination of lucky and unlucky circumstances as follows: 'The state of astronomy at any epoch depends to a great extent upon the examples of various phenomena that are available for observational study, upon the instruments that are available for this study, upon the existence of relevant mathematical and physical theories and of serviceable accounts of them, upon the combination of workers who happen to be active and interactive at the time, and so on. These are all matters of "luck" in the sense in which the word is used here. There must also be converse cases of important phenomena that are not discovered because no example is conveniently located, no suitable instrument available, etc. Thus the astronomer's view of the cosmos may be influenced to a great and almost unknowable extent by a combination of lucky and unlucky circumstances. Moreover, even his own existence may be a matter of rare "luck," in a somewhat different sense.'

The moment of finishing a book for the 'design freeze' (Mike Frost) is already one of arrival and fulfilment. Despite the unexpected global hardship, together we have been able to bring this volume to completion. It exemplifies our lifelong enthusiasm for the observation of the stars, the Moon, and the Sun, their inner workings, and our shared joy in everything that can be methodologically seen, observed, discerned, studied, conceptualized, expressed and represented, said and communicated, multiplied, and, essentially, *intensified* by sharing—especially by sharing in rich and fruitful interdisciplinary perspectives on the experience of the world, the cosmos, and life itself. We celebrated the book contract in early 2020 in the Royal Society in London and then worked with even more determination through the years of the lockdown and pandemic, authors by then only gathering on videocalls, on the phone, and via email, with such generous ongoing commitment and determination, and brilliant contributions from friends and colleagues around the globe. Our gaze up to the nocturnal or diurnal sky, while reading and researching the *Divine Comedy* and the mystics all over again, truly turned from here to eternity. Far above us, Sun, Moon, and stars continued their course as predicted. They lit our way at day and night, no matter if we could see them or not. They still constantly connect all the places in which our authors, colleagues, and loved ones were kept during this phase across so many different time zones, seemingly separated only by the dimension of space while always united by the dimensions of time, thoughts, and ideas. In that, our interdisciplinary total solar eclipse-chasing was never constricted by the laws of chronological time and keeps generating new opportunities for discovery—the real astronomer's luck—across all dimensions of being.

Henrike Lange
Tom McLeish

Berkeley and York
February 2023

Paul Nash, *The Eclipse of the Sunflower* (oil painting), 1945.
Source: British Council Collection, London.

Acknowledgements

Looking back at almost seven years of eclipse-chasing and interdisciplinary thinking together, we have lots to be grateful for. Pursuing any intellectual project of this length and magnitude gives one a unique focus in life and scholarship. A venture that is so fascinatingly prismatic and universally applicable also provides perpetual inspiration and excitement. This endeavour was all at once interdisciplinary, multidisciplinary, creative, and truly generative in its many scientific, cross-historical, theological, artistic, and humanities-driven implications and associations. When, finally, an idea in truth has a deep and awe-inspiring cosmic dimension, every day under Sun, stars, and Moon is enriched with the thrills of lived discovery and a sense of meaningfulness and existential connection. In all this, our families, partners, friends, and colleagues were present and involved in many different ways: Thank you for sharing and enduring our enthusiasm and for supporting us with lasting love and care, feedback, patience, encouragement, and joy.

Many individuals and institutions have supported this book in their own ways throughout the extraordinarily long phases of its gestation, generation, and completion (remarkably before, during, and after a global pandemic lockdown). Among them, we want to thank especially the Pasachoff family and the Jay and Naomi Pasachoff Archive for their generous help in finalizing Jay's wishes for the publication of his chapter on the solar corona—and our families, especially Julie, Katie, Nick, Max, Rosie, Joel, and Phoebe Joy, and Annegret, Thomas, Christopher, Frank, Martin, Andreas, Ilse, Heinrich, Hildegard, and Sam. Some of our most cherished friends and colleagues have been particularly important from the very beginning of the project in 2017 at the Notre Dame Institute for Advanced Study: NDIAS Director Brad Gregory, NDIAS Associate Director Donald Stelluto, NDIAS's heart and soul Carolyn Sherman, the contributors and audience to our first experimental panel in 2017, the fellows of the NDIAS cohort of 2017–18, and especially Patrick Griffin, Kaya Şahin, and Bernie McGinn.

Special thanks to the five anonymous reviewers for their extremely helpful feedback, productive criticism, and much-appreciated support, and thanks to everyone at Oxford University Press handling the double-blind peer-review process as well as the production of the book with great professional care and kindness. Sonke Adlung and Giulia Lipparini, our wonderful editors, have given the book a home at Oxford and sustained our work throughout this entire time. We are also very grateful to OUP's licensing team and to Recorded Books Media for the inclusion of *Eclipse and Revelation* in RBmedia's audiobook programme. Dear colleagues in Durham and York supported Tom in many ways, especially throughout 2022–23, and Henrike in and at Berkeley—Mario Biagioli, Whitney Davis, Anne Derbes, Barbara Forbes, Christopher Hallett, Maureen Miller, Brenda Schildgen, Randolph Starn, and Andrew Stewart. Very special thanks to Nick James for reviewing the 'Eclipse-Chaser's Toolkit' for us.

Warm thanks to the Medieval Academy of America and to the John Templeton Foundation for a seed grant that acknowledged the interdisciplinary and theological dimensions of the topic, and many thanks to all image-granting institutions, especially to the UC Berkeley Historical Slide Library, the British Library, the British Museum, and to the British Council in London for our cover image of the *Eclipse of the Sunflower* by Paul Nash. Heartfelt thanks also to Leo Caldas for his generous contribution of the superb photo and video link showing birds flying into the total solar eclipse for Chapter 13, to Jocelyn Holland, and to Sir Tom Stoppard OM CBE FRSL HonFBA and the staff of United Artists for the quotes from *Arcadia* in Chapter 11.

Warm thanks also to Open Science Librarian Sam Teplitzky (liaison to the Earth & Planetary Sciences Department and Lawrence Berkeley National Laboratory) and all staff and friends involved with *Eclipse & Revelation: A Berkeley Astrophysics & Art History Exhibition* on the Berkeley campus in California.

Very special thanks in the UK to Keith Moore, Head of Collections at the Royal Society, for his friendship and brilliant work on *Eclipse & Revelation: An Exhibition at Carlton House Terrace* (our exhibition with items from the Royal Society) at the Royal Society in London, and to Sian Prosser, Librarian and Archivist for the Royal Astronomical Society, for all the inspired work on *Eclipse & Revelation: An Exhibition at Burlington House* (our book launch exhibition with items from the Royal Astronomical Society) at Burlington House on Piccadilly in London.

From the authors specifically and individually, we would like to include thanks: from Roberta J. M. Olson to Hugo Chapman of the British Museum, Tom Baione of the American Museum of Natural History, Nadine Orenstein and Elizabeth Zanis of the Metropolitan Museum of Art, Freyda Spira of the Yale University Art Gallery, and Alexandra Mazzitelli, for their kind assistance with images of the works of art; from Elaine Stratton Hild to Wayne Grim, for providing the image of his compositional score '233rd Day'; from Steven Portugal to Professor Craig White of Monash University; from Giles Harrison to Professor Tim Palmer (Oxford Physics) who kindly helped with Figure 14.2 (with calculations from the European Centre for Medium range Weather Forecasts' Ensemble Prediction System), to Professor James Milford (Reading Meteorology) who obtained the data in Figure 14.6c, to Dr Graeme Marlton (Met Office) who provided Figure 14.4, as well as to Professor Karen Aplin (Bristol Engineering) who provided review comments; and the following special thanks from Anna Marie Roos:

I would like to thank the University of Lincoln for giving me the time and space to research and write, and eclipse observers and chasers—past and present. I would also like to thank all contributors for their insights and collegiality and especially Tom McLeish, Henrike Lange, and Mike Frost for their exceptional stewardship of this publication.

Last but not least, the editors extend the warmest thanks to all of our contributing authors in our total solar eclipse 2017–24 dream team—Alison Cornish, Mike Frost, Giles E. M. Gasper, Giles Harrison, David Bentley Hart, Elaine Stratton Hild, C. Philipp E. Nothaft, Roberta J. M. Olson, Jay M. Pasachoff, Steven J. Portugal, Anna Marie Roos, and John Steele—and, once more and especially from our hearts, to our dear friend Mike Frost. Following Tom's diagnosis, Mike greatly reinforced the scientific half of the editorial team from Christmas 2022 onwards and, a true friend since the first days in Emmanuel College, Cambridge in 1981, supported Tom—who then knew that Henrike would not be left alone with the final tasks to see our book through to publication in 2024.

Henrike Lange
Tom McLeish

Berkeley and York
February 2023

Contents

Book Summary

Two questions guide this seven-year editorial project: First, how can we approach the phenomenon, representation, and interpretation of total solar eclipses? Second, how can we heal the historical divide between natural sciences on one hand, and humanities, arts, history, music, and theology on the other? This volume presents Tom McLeish's and Henrike Lange's answer to both questions: They invited a group of colleagues to reflect with them on these issues between the two great American eclipses of 2017 and 2024. The result is an exciting look behind the scenes—into labs and studies, archives and museums, around fieldwork in astronomy, meteorology, animal behaviour, and ecophysiology—in short, an invite to all kinds of total solar eclipse-chasing presented by experts and innovative thinkers across disciplines. All chapters have been read and commented upon by the editors and the entire team of authors multiple times, carefully prepared for readers and students from all backgrounds, and grouped into four parts: 'Cosmos,' 'History and Religion,' 'Arts and Literature,' and 'Animals, Weather, Environment.'

The book is unique in combining the analysis of a strictly narrow focus (on only *total* solar eclipses) with an unlimited set of methods and approaches, freely developed by the authors themselves in reflection upon their disciplines, their positions as public intellectuals and leading experts in their fields, and their practice as innovative interdisciplinary thinkers. Bound into one volume, these voices invite us to imagine a liberated mode of discovery, perception, creativity, and knowledge-production across the traditional academic divisions. A prismatic representation of total solar eclipses emerges, itself rising to a model of communal thinking, together, across the borders between more than two dozen disciplines.

A Preface, Introduction, and Conclusion by Lange and McLeish lay out their groundbreaking interdisciplinary work and collaboration in the sciences, natural philosophy, and the arts and humanities. This book is Tom McLeish's final project and scholarly testament; the volume is dedicated to his memory as well as to the memory of astrophysicist Jay M. Pasachoff, contributing author of a chapter about the solar corona which is also Pasachoff's final piece of writing.

Further thematic chapters are by Tom McLeish and Mike Frost (astronomy), C. Philipp E. Nothaft (medieval and early modern intellectual history, history of science, and history of astronomy), John Steele (history of the exact sciences in antiquity, Assyriology), Giles E. M. Gasper (medieval history), Anna Marie Roos (early modern science history), David Bentley Hart (theology and literature), Alison Cornish (Italian literature, Dante Studies), Roberta J. M. Olson (art history, cultural history, science history), Henrike Christiane Lange (history of art and architecture, history of literature), Elaine Stratton Hild (music history), Steven J. Portugal (animal behaviour and ecophysiology), and Giles Harrison (meteorology). The 'Eclipse-Chaser's Toolkit,' prepared by Mike Frost, completes the volume. This book is a friendly companion to the chase of knowledge, encouraging its readers to embark upon their own interdisciplinary journey of discovery.

List of Contributors

In the Order of Their Appearance in this Volume

The Editors

Tom McLeish FRS, Professor of Natural Philosophy (Emeritus) in the School of Physics, Engineering and Technology at the University of York, UK, also held positions at York's Centre for Medieval Studies and York's Humanities Research Centre. He held previous academic posts at the universities of Cambridge, Sheffield, Leeds, and Durham. Internationally recognized as theoretical physicist and '*penseur anglais*' (Emmanuel Macron, 'Ce en quoi je crois,' *L'Express*, 23 December 2021), McLeish has produced a wide range of interdisciplinary, pedagogical, philosophical, and theological work. His research in 'soft matter and biological physics' draws on interdisciplinary collaborations to study relationships between molecular structure and material properties. He led the UK 'Physics of Life' network and held a five-year research fellowship focusing on the physics of protein signalling and the self-assembly of silk fibres. Tom also pursued a programme of interdisciplinary research between the sciences and humanities, including the framing of science, theology, society and history, education, and philosophy, leading to the books *Faith and Wisdom in Science* (Oxford, 2014), *The Poetry and Music of Science* (Oxford, 2019), and *Soft Matter: A Very Short Introduction* (Oxford, 2020). He co-led the *Ordered Universe* project, a large interdisciplinary study of thirteenth-century science. From 2008 to 2014 Tom served as Pro-Vice-Chancellor for Research at Durham University and was from 2015–20 Chair of the Royal Society's Education Committee. A Reader in the Anglican Church, Tom was awarded the 2018 Archbishop of Canterbury's Lanfranc Award for Education and Scholarship. Tom's scientific-theological legacy includes the Christian leadership in the sciences initiative, ECLAS (Equipping Christian Leadership in an Age of Science). Having finalized with Henrike Lange all editorial comments on the chapters and the *Eclipse of the Sunflower* cover, Tom completed with Henrike in February 2023 the entire co-authored manuscript of *Eclipse and Revelation*.

Henrike Christiane Lange is Associate Professor of History of Art and Italian Studies at the University of California, Berkeley. The author of *Giotto's Arena Chapel and the Triumph of Humility* (Cambridge, 2023) earned her MA, MPhil, and PhD at Yale University after finishing her *Magister Artium* in art history as well as Italian language and literature at Universität Hamburg (Germany). Lange's highly interdisciplinary work engages social art history, architecture, and literature in the European medieval to early modern age—especially in Rome, Florence, Venice, and Padua around artists such as Cimabue, Giotto, Donatello, Mantegna, Botticelli, Michelangelo, and Raphael, and writers such as Augustine, Francis of Assisi, Dante, and Petrarch. Lange is the recipient of the 2020 Prytanean Faculty Award for research, teaching, and service. Her service as a role model for the young scholars of the Prytanean Women's Honor Society includes an active interdisciplinary arts, humanities, and science programme along with initiatives for women scientists within the College of Letters & Science at Berkeley. As Distinguished Fellow at the Notre Dame Institute for Advanced Study (2017–18), Lange expanded her lifelong interest in visual knowledge, music, and learning at the intersection of the arts, humanities, and natural sciences in a theological context. Her editorial collaboration with Professor of Natural Philosophy Tom McLeish FRS began around the great American Eclipse of 2017.

Contributing Authors

Mike Frost FRAS , FIET is the director of the historical section of the British Astronomical Association (BAA). His day job is systems engineering in the steel industry, but astronomy has always been a central part of his life. He has an MSc in astronomy, is a fellow of the Royal Astronomical Society, a founder member of the Society for the History of Astronomy, and has been a BAA member for 29 years and a council member for 14 years. He enjoys travel to view astronomical events such as aurorae and meteor showers, but his absolute favourite spectacle will always be total solar eclipses.

Jay M. Pasachoff, Field Memorial Professor of Astronomy at Williams College (Williamstown, MA) and Chair of the Working Group on Eclipses of the International Astronomical Union, was a veteran of 36 total solar eclipses and 75 solar eclipses of all kinds, as well as two transits of Venus and several transits of Mercury. He is coauthor with Roberta J. M. Olson of *Cosmos: The Art and Science of the Universe*; and with Alex Filippenko of *The Cosmos: Astronomy in the New Millennium*, and author of Peterson's *Field Guide to the Stars and Planets*.

C. Philipp E. Nothaft is a Fifty-Pound Fellow at All Souls College, Oxford. He has published widely on the history of mathematical astronomy in medieval Europe.

John Steele is Professor of the History of the Exact Sciences in Antiquity in the Department of Assyriology at Brown University, RI. His research focuses on the history of ancient astronomy, in particular the history of Babylonian astronomy and its scholarly and social context, the circulation of scientific knowledge in antiquity, and the reception of ancient astronomy in the early modern and modern periods. He is the author of several books including recently *The Babylonian Astronomical Compendium MUL.APIN* (coauthored with Hermann Hunger; Routledge, 2019), and is the editor of the forthcoming *Oxford Handbook of Ancient Astronomy* (Oxford University Press).

Giles E. M. Gasper was educated at the University of Oxford and the Pontifical Institute of Mediaeval Studies, Toronto. He has been in post at Durham University since 2004, in the History Department and with close links to Theology and Religion. His areas of interest are the religious, intellectual, and cultural histories of the European Middle Ages, and its inheritances from late antiquity and the early church. Creation and the natural world, the economy of salvation, and concepts of order form his main themes of interest. These are explored through monastic and scholastic theology and science, historical writing, literary texts, and craft and culinary manuals. Recent publications include a five-volume series with Oxford University Press, *The Scientific Works of Robert Grosseteste*, featuring new editions, English translations, and interdisciplinary commentary.

Anna Marie Roos is Professor of the History of Science and Medicine at the University of Lincoln (UK). Roos has been a visiting fellow at All Souls College, Oxford; a Fellow at the Huntington Library (San Marino, CA); a John Rylands Fellow at the University of Manchester; and a Beinecke Fellow at Yale. Roos was the recipient of the John C. Thackray Medal for her work on the history of natural history, and in 2023 she delivered the Gideon de Laune Medal Lecture at the Worshipful Society of Apothecaries of London. She is a fellow of the Linnean Society of London and the Society of Antiquaries of London. Roos is the Editor-in-Chief of *Notes and Records: the Royal Society Journal of the History of Science*. Author of ten books and editions, and several journal articles and book chapters, her latest book is *Martin Folkes (1690–1754): Newtonian, Antiquary, Connoisseur* (Oxford University Press, 2021).

David Bentley Hart is a Collaborative Researcher at the University of Notre Dame. His specialties are philosophical theology, systematics, patristics, classical and continental philosophy, philosophy of mind, and Asian religion. Much of his recent work has concerned the genealogy of classical and Christian metaphysics, ontology, and the philosophy of mind. He is the author of twenty books—including volumes of philosophy, cultural criticism, philosophical theology, fiction, and literary essays—as well as more than 750 articles in both scholarly and trade journals. His critical translation of the New Testament appeared from Yale University Press in 2017 (second edition 2023).

Alison Cornish is Professor of Italian Studies at New York University and President of the Dante Society of America. She is the author of *Reading Dante's Stars* (Yale, 2000), *Vernacular Translation in Dante's Italy: Illiterate Literature* (Cambridge, 2011), a commentary on Dante's *Paradiso*, translated by Stanley Lombardo (Hackett, 2017), and *Believing in Dante: Truth in Fiction* (Cambridge, 2022), as well as a number of essays on Dante, Petrarch, and Boccaccio. During the seventh centenary of the poet's death, she organized a crowd-sourced series of video conversations between members of the Dante Society of America, entitled 'Canto per Canto: Conversations with Dante in Our Time.'

Roberta J. M. Olson is a distinguished art historian and curator. Her theory that the Star of Bethlehem Giotto painted in the Adoration scene in Scrovegni Chapel was a portrait of Halley's Comet in 1301 (see *Scientific American*) led the European Space Agency to christen their satellite to that comet (1985–6) 'Giotto.' Olson's many award-winning books include: *Cosmos: The Art and Science of the Universe*, and *Fire in the Sky: Comets and Meteors, the Decisive Centuries, in British Art and Science* (coauthored with Jay M. Pasachoff); *Artist in Exile: The Visual Diary of the Baroness Hyde de Neuville*; *Making It Modern: The Folk Art Collection of Elie and Viola Nadelman*; *Audubon's Aviary: The Original Watercolors for 'The Birds of America'*; *The Florentine Tondo*; *Fire and Ice: A History of Comets in Art*; *Ottocento*; and *Italian Drawings 1780–1890*. Her interdisciplinary research concerns drawings, Italian Renaissance art, nineteenth-century art, and the interconnections between art and astronomy, focusing on comets, meteors, and solar eclipses.

Elaine Stratton Hild serves as an editor with *Corpus monodicum*, a long-term research project housed at the Universität Würzburg (Germany). Her responsibilities include publishing volumes of previously unedited plainchant, transcribed from medieval manuscripts. Dr Hild's work with medieval plainchant has been supported by the DAAD (Deutscher Akademischer Austauschdienst), the University of Notre Dame Institute for Advanced Study, and the Fulbright Foundation. Her edition of proper tropes (*Tropen zu den Antiphonen der Messe aus Quellen französischer Herkunft*) was published by Schwabe Verlag in 2017. Dr Hild's most recent book, *Music in Medieval Rituals for the End of Life* (Oxford University Press, 2024), presents the first modern edition of the chants sung at the bedsides of the sick and dying during the Middle Ages.

Steven Portugal is a comparative ecophysiologist. His research is located at the interface of the physiology, sensory ecology, and behaviour of vertebrates. Steve is a Reader in Animal Behaviour and Physiology at Royal Holloway University of London. Most of his work focuses on how animals adapt their behaviour and ecology to the challenges of their environment within the constraints of their own physiological and anatomical limitations. Such questions are particularly important in the light of global environmental change and exploitation of natural resources in the emerging field of conservation physiology.

Giles Harrison works on atmospheric electricity and atmospheric measurements in the Department of Meteorology at the University of Reading, UK, where he is a Professor of Atmospheric Physics. He has doctorates from Imperial College and Cambridge University, and is a member of the Academia Europaea. In 2016 he was awarded the Appleton Medal by the Institute of Physics, and in 2021 he received the Christiaan Huygens Medal of the European Geosciences Union. He conceived and led the National Eclipse Weather Experiment (NEWEx) in 2015, and an 'Eclipse Meteorology' themed issue of *Philosophical Transactions*. His postgraduate textbook on meteorological instruments is now also in a Chinese edition.

Introduction
Chasing the Total Solar Eclipse
On the Road and in the Archive

Henrike Christiane Lange and Tom McLeish

Flights of Imagination

Though the journey had been planned for a long time, not everything could be determined. The exact date and timings for the track across North America on August 17th 2017 had been calculable, indeed calculated, for centuries—the meeting of milepost with Moon's shadow known in advance to within a fraction of a second, its Saros Cycle set for millennia. Even so, although the placement of the planets is predictable, the accumulation of clouds is not. The distribution of clear skies within a day's drive of the summer research location in Indiana was only itself clear the day before. This was how a shorter route to a nearby but clouded Illinois was updated to a longer drive south and east to Kentucky—to Crofton, Kentucky, 37° 3' 30" N 87° 29' 16" W. A strange thought occurred at the wheel, while seeking out long, lonely, and traffic-free back-roads that led towards the track of totality: my personal 'world-line' and that of the moon converge by operations of nature today at this place, this time, in intersecting fulfilment of a personal dream of decades.

Excerpt from Tom McLeish's Travel Diary, August 2017

It does not matter whether the cool, darkened exhibition room belongs to the British Library in London, or to the John Rylands Library of the University of Manchester, or to any other collection that possesses medieval manuscripts illustrating total solar eclipses. In all of those places of curation and reading, encountering an eclipse on a manuscript page elicits a common sensual disruption—the seeing of something unseen. The 'seen' is a black Sun, the 'unseen' the passage of the vast body of the Moon passing in front of the vaster and proportionally more distant orb of the Sun, blocking out its light. Human observers have tracked the regularity of the Sun and Moon's apparent motions across the sky for at least as long as civilizations have kept astronomical records. Some drew the conclusion that the nearer moon must occasionally pass in front of the more distant sun, although the distances of neither from the Earth were known to the ancients. Many also realized that these occurrences are predictable, following complex and subtle series of near-periodicity.

A medieval manuscript's illuminations mark the appearance of a total eclipse either with the absence of gold, or the presence of strong highlights, or some perfect geometry tracing the shadow of the moon over the Earth. Depending on the function of the manuscript, its presentation can depend mostly on tables of lunar variations, gilded to the point of eclipse (Figure I.1), where the total eclipse leaves no gold showing,

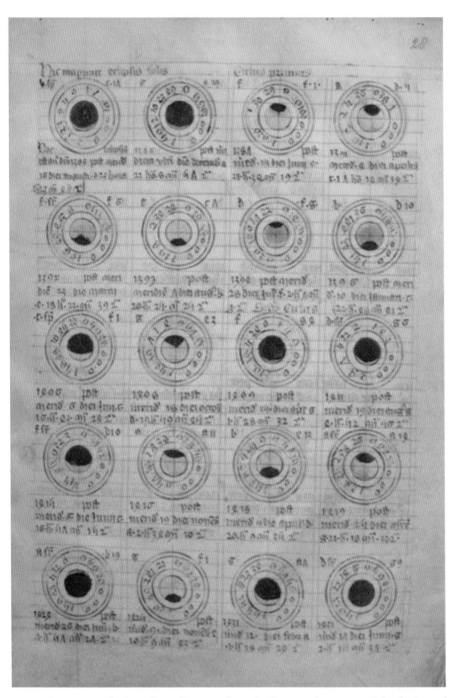

Figure I.1 A page of solar eclipse diagrams from the fourteenth-century *Kalendarium* of John Somer, and other astronomical texts, bound in British Library additional ms. 10628.

Source: Courtesy of the UC Berkeley Historical Slide Library (est. 1938–2018), Horn Collection, Doe Memorial Library in Berkeley, California.

as in the lowest quadrant of the page's right column. Or, as we will soon see, it can inspire an illuminator to draw dramatic views of its effects on the earth and its inhabitants, surfaces and spaces of reverberation and spectators witnessing a darkening of the sun during the day—since Biblical times a standard indicator of fundamental change.

The heat of the summer's day had sent us scurrying for the cool of any available shady tree, but as the noonday sun threatened to roast the large assembled crowd, its strength seemed to wane, though the sky was, as anticipated, cloudless. Spectators emerged into a subdued daytime illumination as they made images of the cosmic spectacle appear on makeshift canvases placed onto the ground. The dappled patches of sunlight on the ground beneath our broad-leafed tree of refuge took on strange crescent forms— blurred images in nature's own multiple camera obscura of the partially eclipsed sun (Figures I.2 and I.3).

(Excerpt from Tom McLeish's Travel Diary, August 2017)

A strange chill, and a stranger dimness descended on the field; the shadows assumed an extraordinary sharpness as the last few seconds of pared-down sun's edge cast them, before the final precipitous plunge into total eclipse. The most unfamiliar noonday darkness over us was fringed at the horizon by the burning glow from outside the umbra itself—yet no twilight bathed this landscape.

(Excerpt from Tom McLeish's Travel Diary, August 2017)

Figure I.2 Crofton Total Eclipse Consortium: Nature's own multiple camera obscura made visible during the total solar eclipse of 2017, seen from Kentucky.

Figure I.3 Crofton Total Eclipse Consortium: Projection of the eclipse in progress during the total solar eclipse of 2017, seen from Kentucky.

The *Cambrai Apocalypse* imagines the total eclipse as a cosmic disaster (Figure I.4). Supporting the function of a prophetic text, the twelfth-century illustration from northern France sets a black sun and a sickle-moon into a frill-bordered opening in the sky. In the vision of the prophet-apostle St John, shown standing next to

Figure I.4 A miniature from the Apocalypse of Cambrai depicting the opening of the seventh seal. Municipal Library of Cambrai, France.
Source: UC Berkeley Historical Slide Library, Bony Collection.

the disaster and, as visionary of the scene, in appropriately exaggerated hierarchical scale, the influence of the dark sun reaches the ground in the form of earthquakes, buildings collapsing and falling into some earthly matter, and the former inhabitants of the palaces and cities crawling out into the chaotic right corner of the image.

There, masses of rising earth, mudslides maybe, build themselves up under the indication of two small islands of ground, one of them still hosting an unearthed tree. The people—noblemen, kings and queens—and their subjects are all experiencing the same fate. But this effect, common to every man and every woman, does not only occur with the final great equalizer of the Apocalypse—any total eclipse in the here and now, casting darkness over the day, will haunt and capture the imagination of those experiencing it.

Now at last could my eyes be safely cast skywards to gaze on the long-sought object of the hunt—the silken luminosity of the solar corona (Figure I.5).
(Excerpt from Tom McLeish's Travel Diary, August 2017)

Immense yet delicate, the pearly filaments of the sun's distant atmosphere seemed to bundle into three filigreed tufts, reaching out in strange stillness to 4 or 5 solar

Figure I.5 Crofton Total Eclipse Consortium: The total solar eclipse of 2017, seen from Kentucky (exposed to show structure in the outer corona; the inner regions are overexposed).

radii. I imagined the vast magnetic fields whose presence was betrayed by the coronal filaments. Close inspection past the rim of the moon picked out the metallic-pink hydrogen-light glow of prominences—huge flames of glowing gas ejected from the sun's surface, another solar structure, normally invisible, made visible by the eclipse. Then, after the shortest 2 minutes 30 seconds ever experienced, the solar surface made a first reappearance through some nameless valley on the moon's trailing edge. Expecting the normal yellow-golden impression of sunlight, I was surprised by the searing whiteness of the 'diamond ring' (Figure I.6). Colours in the sky, as on Earth, are distorted under eclipse.

Excerpt from Tom McLeish's Travel Diary, August 2017

As Sir Patrick Moore put it, 'a total eclipse of the Sun is the grandest sight in all Nature.'[1] Although the eclipsed Sun occupies an angle of just one half a degree across in the sky—the width of a finger held at arm's length—the strange significance of the new 'hole in the sky' and its unworldly corona fixes the attention so that the entire spectacle seems to fill the beholder's field of view. It requires a lot of observational and imaginative skill to move an eclipse halfway out of the scene, and to bathe the foreground interior into a dully reflecting light, a fading echo of the actual brilliance in the sky, as we will find in an example from Italian Renaissance fresco painting.

[1] See Patrick Moore, *On the Moon.* London, 2001, p. 112.

Figure I.6 Crofton Total Eclipse Consortium: The diamond ring appearing during the total solar eclipse of 2017, seen from Kentucky (exposed to show structure in the outer corona; the inner regions are overexposed).

In one field embedded in the murals of Raphael's *Loggie* in the Vatican, painted by Raphael and his pupils (among them Giulio Romano, Gianfrancesco Penni, Vincenzo Tamagni, Perino del Vaga, Polidoro da Caravaggio, Giovanni da Udine, Tommaso Vincidor da Bologna, Pellegrino da Modena, and others), an enormous eclipsed sun appears or disappears in the dramatic *chiaroscuro* ceiling painting of Isaac and Rebekah as they are observed by King Abimelech (Figure I.7).[2] The spying king peers over the wall against which the lovers lean; prints and majolica reproducing the scene illustrate both, the voyeurism and the effects of the light, more clearly. Acts and ideas of cover, intimacy, observation, and revelation haunt the image. Architecture structures the scene, relating the lovers in their hidden corner of a palace to the cosmic event that casts its light on them. As nonchalantly as the painter has displaced the eclipse, he has also almost cut the compositional standard force of the majestic triumphal arch, fitting it ever so tightly into the scene's rectangular frame. Both forms, the eclipse and the arch, command their conventional visual significance regardless of the daringly decentred positioning, or even cropping, of their most salient

[2] See Roberta J. M. Olson and Jay M. Pasachoff, *Cosmos: The Art and Science of the Universe*. London, 2019, pp. 61–2. See also Roberta J. M. Olson and Jay M. Pasachoff, 'Blinded by the Light: Solar Eclipses in Art-Science, Symbolism, and Spectacle.' In *The Inspiration of Astronomical Phenomena VI*, ed. Enrico Maria Corsini. San Francisco, 2011, pp. 205–15, here esp. pp. 207–8. See also Marilyn Aronberg Lavin and Irving Lavin, *The Liturgy of Love: Images from the Song of Songs in the Art of Cimabue, Michelangelo, and Rembrandt*. Kansas, 2001, pp. 54–5 and 62.

Figure I.7 *Isaac and Rebekah Spied upon by Abimelech* (1518–19), fresco, school of Raphael. Loggia on the second floor, Palazzi Pontifici, Vatican.

Source: UC Berkeley Historical Slide Library, Baxandall and Partridge Collection.

geometrical features. Knowledge of the biblical story of Isaac and Rebekah (Genesis 26:8) underscores the brilliance of this interplay between hiding and revealing, between a fleeting moment in the sky and the massive, eternalizing palace architecture. The married lovers have to pretend to be siblings in a foreign land. The moment of eclipse, the displacement of light, and possibly even the idea that everyone else is looking at the sky in that precise instant, gives the couple a chance for one to move one leg over the other's, and to share a passionate kiss. The light from the corona falls into their little corner in the sharp shape of the Roman arch, complicating the idea of the architecture's triumphal theme and a happy ending for the lovers (though the majolica version warns, 'Nec Spe, Nec Metu,'—without hope, without fear). The artificially enlarged image of the darkened sun allows Raphael and his workshop not only to endow the scene with sinister foreboding as the kiss between Isaac and Rebekah was actually observed by the King Abimelech, but also to mirror the exposed details of their tryst in the intricate fibrillation of the corona itself. Comparison with the photograph of the 2017 eclipse (see Figure I.5) indicates that Renaissance artist knew of the rapid increase in the corona's brightness towards the solar surface, as well as its filamentous structure.

I recalled my previous attempt at witnessing totality—the 1999 eclipse visible from the extreme south-west peninsular of England. Completely clouded-out on a Devonshire clifftop, the experience was not nothing. We did not see the eclipsed sun, but were instead treated to the overwhelming encounter of the moon's shadow, not apparent under clear skies, as I now knew. A few minutes before totality, the western horizon blackened into a fearful darkness that rushed towards us at a thousand miles per hour until we were enveloped under its cloak. I experienced the extraordinary impression of hearing the rumble of some vast cosmic machinery responsible for this sweeping of the long conical arm of the moon's shadow over the surface of the earth. I received a visceral impression of the lever-arms and circular orbits of the vast solar system above our heads.

Excerpt from Tom McLeish's Travel Diary, August 2017

The idea of an external viewpoint of the imagination onto the eclipse is as old as the ancient Greek astronomies of Heraclitus, Aristotle, and Ptolemy. Medieval philosophers at the dawn of the first European universities adopted and developed their ideas in textbooks written for scholars of the 'quadrivium.' This curriculum of four mathematical arts that all twelfth- and thirteenth-century students would study included astronomy (alongside geometry, arithmetic, and music).

The diagram of Figure I.8 accompanies a late-thirteenth-century manuscript copy of the short treatise *De Sphera* (*On the spheres*) by the Oxford polymath and later bishop of Lincoln, Robert Grosseteste. Both text and diagram feature the full geometry of the Moon's shadow as a *cone*, its base sitting on the unlit half of the satellite's surface, its point reaching out into space directly opposite the sun, and sweeping the Earth's surface when the three bodies momentarily find themselves along a single line. Extraordinarily elegant, the diagram gains its striking dynamic appearance from the simple prolongation of two lines to the right, over the outmost yellow circumference and the subsequent addition of just a few calligraphed words in red and black respectively, subtly destabilizing the border between diagram and blank page. Its construction is also implicitly three-dimensional: the two innermost intersecting circles represent the two planes of the Moon's orbit, and that of the ecliptic (the plane of the Earth and Sun). The projective device represents the five-degree tilt between these two planes in space, projected onto the single plane of the diagram. A grasp of this feature of spatial astronomical geometry is essential to the reader of a text on eclipses, for the Sun, Moon, and Earth can only line up in the directions at which the two planes intersect.

The text itself, and its lavish diagrams, adopt the pre-Copernican cosmology that has the Earth central to the cosmos, and all other bodies in orbit around it. This 'geocentric' model for the universe was gradually replaced by the Sun-centred, or 'heliocentric' model following Copernicus's 1543 publication of *De Revolutionibus Orbium Coelestium* (*On the Revolutions of the Heavenly Spheres*). The insight that the shadows responsible for eclipses are giant cones in space survives the transformation, however, for this is a consequence of geometric optics, a science whose basic laws, such as the linear propagation of light, had been established at least since Euclid in the fourth century BC. The tradition of texts on cosmology and astronomy embodies

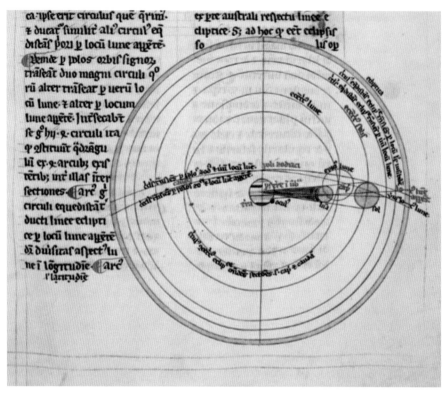

Figure I.8 Solar eclipse diagram within the text of Robert Grosseteste's *On the Sphere* (*De Sphera c.*1215) in British Library Harley ms. 3735 f. 82r.

Source: UC Berkeley Historical Slide Library, Horn Collection.

the longest tradition of the scientific imagination—the human imperative to switch from the perspective and perception of the eye itself, to that of the 'mind's eye.' As the eclipse bears down upon the impressions with its overwhelming transformation of the sky, it also inspires a recreation in the mind of the orbits, rays, and shadows of the cosmos.

A Gateway to Multiple Perspectives

These twin-tracked accounts of meetings with the eclipsed Sun map out the motivation and goal of this book. In the first encounter with the awe-inspiring phenomenon, experience was mediated by direct sight and sense, the place on the track of the moon's shadow across North America in August 2017, as well as by a moderate scientific understanding of what lay behind the visual impression. The second, guided by the discipline of literary studies and the history of art, took place at the quiet reading tables of British university libraries and in the Vatican's famous spaces of wall painting, generating a virtual tour of galleries and reading rooms. The artefacts and images found there testified, in turn, to a rich history of inspiration, imagination, and encounter with the eclipsed Sun. There are as many ways of responding to the unforgettable experience as there are human means to do so. Total solar eclipses appear in art, in prose and poetry, in music—in addition to the long history of mathematical astronomy that grapples with their understanding and prediction. This rich tradition of inspiration complements and contrasts the multiple sciences that respond to eclipse: not only solar astrophysics itself, but also atmospheric science, animal behaviour, and planetary astronomy are furnished with new data, when atmosphere and land are darkened and cooled in mid-day, and the veil of scattered blue sunlight we call 'sky' is momentarily swept aside.

There is a total eclipse of the sun somewhere on Earth every eighteen months or so. But because of the coincidentally close apparent diameters of sun and moon in the sky, a view of the completely obscured solar disk is restricted to a track swept out by the path of the moon's shadow, varying from only a few kilometres wide to a hundred or so. The track is much broader than that when the shadow strikes the Earth at a glancing angle—so when the eclipse is near the poles. Total eclipses are therefore extremely rare events for any particular place on the planet's surface—most people never experience one, and those who do have typically travelled deliberately to the 'track of totality'. Notwithstanding, their extreme impact in terms of visual display with the shocking darkness and starlight in the daytime and the glory of the solar corona combined with their ancient history of predictability and portent means that solar eclipses cast a huge cultural shadow and are present in the imagination of most. There is almost no human discipline that does not have a perspective onto this primal cosmic experience.

Many motivations bring all these perspectives together in this volume, but the first was the realization that, without their combined force, it is impossible to grasp the human meaning of the event as it has built up in art, story, and science over millennia. In our own time, however, we have taught ourselves to compartmentalize thinking and learning. Education treats the 'arts and humanities' and the 'sciences' separately, then subdividing further into the splintered specialisms of today's world of learning. A formation in either the humanities or the sciences typically avoids their extended history of communication, their common origins, and their current deep-lying connections. Even within their own domains, conversations that connect history, philosophy, and literature on one hand, and physics, chemistry, and biology on the other, are few and muted. Contributing to a renewed interdisciplinary conversation is therefore a second, and reciprocal, motivation for this book.

The contributors believe that the idea and experience of the eclipsed sun becomes a focal point, an attractor capable of drawing together once more disciplines that have become fragmented and separated in today's academic world. Their goal is therefore to use the touchstone of the total solar eclipse to open pathways for the volume's readers, as it has for its writers, into these multiple human endeavours of science and the arts and humanities. Such centripetal attraction of concrete objects, ideas or phenomena has, after all, proved of the greatest effectiveness in refocusing divergent disciplines to a single conversation once more. Archaeology, for example, reknits the scientific analysis of material remains with historical knowledge of their creators, a geographical grasp of their territory, the social science of their communities, narrative scholarship of their religious and cultural thought. The history of eclipses nests, likewise, within the history of science, which at its best attracts to a single table not only historians, but mathematicians, scientists, philologists, theologians, and philosophers. Such radical dialogue is also always reciprocal, so when science revisits its own past as a guest of history or literature, it not only informs those disciplines, but can receive new imaginative energy. A recent collaborative examination of a thirteenth-century treatise on the rainbow, for example, inspired research between psychology and visual neuroscience into a new coordinate system for the abstract space of colours.[3] As we have already noticed, our particular example beckons to artists, art historians, poets, and playwrights as well.

The Two Cultures: A Common Origin Eclipsed

For this larger project of communication across special fields, it is worth reflecting for a moment on the background to the increasingly narrow and disconnected education in the colonial, postcolonial, and globalized West of the last two centuries. At the beginning of the nineteenth century, the world of science and literature looked much more connected. English poet William Wordsworth wrote in 1820 that:

> The remotest discoveries of the Chemist, the Botanist, or Mineralogist, will be as proper objects of the Poet's art as any upon which it can be employed, if the time should ever come when these things shall be familiar to us, and the relations under which they are contemplated by the followers of these respective sciences shall be manifestly and palpably material to us as enjoying and suffering beings.

Yet science has only rarely qualified as 'proper object of the poet's art', despite the rich experiences of natural phenomena that inspire and form the scientific imagination such as the cheerful daffodils, to revisit one of Wordsworth's own most brilliant poetic passions in which he combines the notions of a wandering cloud, the breeze, daffodils, and the stars that 'twinkle on the milky way.'[4] Instead, the sciences, arts, and humanities since Wordsworth's time have been driven apart through a cultural narrative, itself known by various names. The mid-twentieth century's 'Two Cultures' debacle was propelled by an infamous lecture and book of that title by British scientist and novelist C. P. Snow. Yet, nearly a century before Snow had crossed swords with literary critic F. R. Leavis through the ensuing bitter debate, Matthew Arnold and

---

[3] Hannah E. Smithson *et al.*, 'Color-coordinate system from a 13th-century account of rainbows.' *Journal of the Optical Society of America* 31, 2014, A341–A349.

[4] See William Wordsworth, 'I Wandered Lonely as a Cloud:' 'I wandered lonely as a cloud / That floats on high o'er vales and hills, / When all at once I saw a crowd, / A host, of golden daffodils; / Beside the lake, beneath the trees, / Fluttering and dancing in the breeze. // Continuous as the stars that shine / And twinkle on the milky way, / They stretched in never-ending line / Along the margin of a bay: / Ten thousand saw I at a glance, / Tossing their heads in sprightly dance. // The waves beside them danced; but they / Out-did the sparkling waves in glee: / A poet could not but be gay, / In such a jocund company: / I gazed—and gazed—but little thought / What wealth the show to me had brought: // For oft, when on my couch I lie / In vacant or in pensive mood, / They flash upon that inward eye / Which is the bliss of solitude; / And then my heart with pleasure fills, / And dances with the daffodils.'

T. H. Huxley had publicly, though less vociferously, voiced the separate pathways of arts and sciences that had become manifest by the end of the nineteenth century. In spite of the desire of liberal educational reforms shared by the two Victorian figures, and their lifelong personal friendship, they seem already to be condemned to divisive views. According to Arnold, the study of literature revealed 'human freedom and activity,' in contrast to science, which showed only 'human limitation and passivity.' Huxley's perception of the axis of freedom and captivity was already the opposite; for him the exclusive study of letters was 'a thrawldom of words.'[5]

Some of the seeds of separation were resown by the Romantic movement, poet John Keats complaining in the former medical doctor's long poem *Lamia* that '[natural] philosophy will clip an angel's wings . . ., unweave a rainbow.' This accusation that science eviscerated the world of the very mysteries on which art and poetry drew was made at the time of Wordsworth's expression of the opposite, and when his colleague Coleridge was both undertaking experiments and writing poetry with scientist Humphry Davy at the Royal Institution. Yet in the same century the Vice-President of the British government's Committee of Council on Education would recommend that a liberal education was too costly to the nation and that practically oriented 'science instruction' should occupy the curriculum in future. Romantic relations of science and art in the nineteenth century were turbulent indeed as a unified vision became clouded, then totally eclipsed, by the adoption of a complex of polarized views. The opposite conception of freedoms and limitations offered by science and literature, that emerged between Huxley and Arnold, became reflected in a mutual polarization of cultural damage.

As university curricula shifted from exclusively classical disciplines to include the new sciences, secular institutions of education were founded, and science became increasingly professionalized and institutionalized, a separation of ways and closing of communication marked not only the sciences and arts, but separated the study of science from its own history and philosophy as well. This aspect of the growing polarization of learning became enshrined in new language. William Whewell, Master of Trinity College Cambridge, suggested in 1834 that the term 'natural philosopher' be replaced by 'scientist.' In spite of the outright rejection by such towering figures as James Clark Maxwell and Michael Faraday, by the end of the nineteenth century the notion of a love of wisdom suggested by the older term (*philo-sophia*—'love of wisdom' in Greek) had given way to the knowledge-claim implied by a 'science' (*scire*—'to know' in Latin). The self-conscious and critical self-reflection that puts the humanities in constant dialogue with their own history and with the philosophical foundations of knowledge was 'parcelled off' in the case of science into disciplines of their own. The formation of most scientists focused on the establishment of a body of knowledge and set of hypotheses on nature's workings that, one must fear, leave not much room for the imagination and all-important bisociative processes. The narrative of a divorce of imagination from science echoes cries from before the anguished fragmentations of Keats, Wordsworth, and Whewell. William Blake, at the end of the eighteenth century, declared that his business was 'to imagine,' in contrast to the

[5] Arnold and Huxley quotes from Paul White, 'Ministers of Culture: Arnold, Huxley and Liberal Anglican Reform of Learning.' *History of Science* 43, 2005, pp. 115–38.

calculation of the early modern scientists and thinkers: 'in the grandeur of Inspiration to cast off Rational Demonstration . . . to cast off Bacon, Locke and Newton'; 'I will not Reason and Compare—my business is to Create.'[6]

The seeds of fragmentation were sown, it seems, right at the start of the modern period, from the foundational work of Francis Bacon, and the subsequent establishment of the Royal Society (and other professional scientific bodies internationally) which enshrined a marginalization of imagination as a partner to reason in truth-making, proposing reason as the sole permitted pathway. Thomas Sprat, the early historian of the Royal Society, in a text titled *History of the Royal Society* which was arguably more of a manifesto, urged its followers to 'separate the knowledge of *Nature*, from the colours of *Rhetoric*, the devices of *Fancy*, or the delightful deceit of *Fables*.'[7] From the same tradition came the formal, sparse, and deliberately plain style of scientific prose that characterizes most of its publication to this day, and contributes to the impression that there is no human creativity in science, and no possible dialogue between art and literature on one hand, and the industry of scientific knowledge-creation on the other.

Yet throughout the modern period there comes a different testimony from many practitioners of science itself who draw on much earlier traditions of 'imagination' as an essential pathway to truth-telling as the data of the senses alone. They testify to a persistent, if quiet, narrative that also points to imagination as core and key to the scientific process itself. From Albert Einstein:

'I am enough of an artist to draw freely upon my imagination. Imagination is more important than knowledge. Knowledge is limited. Imagination encircles the world.'[8]

Much like Leonardo da Vinci centuries before him, Einstein refers to the great imaginative task of science that is so enormous in its scope, and hubris, that it is hardly mentioned today. For it is not possible to deduce the structure and dynamics of the world from observations, however technically advanced and adroit. Gravity, atoms, genetic information, the delicate operation of brain tissue's synapses, the glorious passage of the whirling Moon in front of the Sun—all these have to be imagined before that creation of inner vision can be formally compared with observations. This first, and most vital, stage of the true method of science has been quietly suppressed for generations.

The dubious form, and even doubtful existence of a 'scientific method' has notwithstanding contributed to a centrifugal force that has propelled sciences and humanities apart in public imagination. Yet the application of 'method' in science is as restricted as it is in the arts. The expert layering of oil paint, or the compositional structure of a fugue, may be supported by methodological approaches, but they do not exhaust the creative process of a painting or a piece of music. Similarly, if the trial of hypotheses by 'experimental method' may be to some extent made methodical, as

[6] William Blake, *Milton*, London, 1804, book 2, pl. 41; *Jerusalem*, ch. 1, pl. 10.

[7] See Thomas Sprat, *The History of the Royal Society of London, for the Improving of Natural Knowledge*. London, 1667, p. 62.

[8] Albert Einstein and Leopold Infeld, *The Evolution of Physics*. Cambridge, 1938.

has been discussed at length by philosophers of science following Karl Popper and others, then the formulation of those hypotheses in the first place most assuredly cannot be. Science reimagines the universe—its creative acts are in those conceptions, and these in turn must surely feed from the entire gamut of human experience.

Understanding the historical forces that splintered our conceptualization of disciplines suggests that we may mend them with words. Mending happens, for instance, by way of reinstating the former dialogue between disciplines recognized by Wordsworth as well as by way of celebrating the imaginative energy within science identified by Einstein. The raw experiences of the natural world that diverge into both poetry and science, but that can also converge into poetry-inspired science and science-inspired poetry, are brought into the sharpest relief by this oldest and most terrifying of cosmic phenomena. The reflections of Samuel Taylor Coleridge (to which we have already alluded through his partnership with scientist Humphry Davy) bear witness to the deeply analogous relationship between science and poetry that lies beneath the other connections of this volume. Coleridge writes of his project with Wordsworth, the *Lyrical Ballads*, the object:

> To give the charm of novelty to things of every day, and to excite a feeling analogous to the supernatural, by awakening the mind's attention to the lethargy of custom, and directing it to the loveliness and the wonders of the world before us; an inexhaustible treasure, but for which, in consequence of the film of familiarity and selfish solitude, we have eyes, yet see not, ears that hear not, and hearts that neither feel nor understand.[9]

We will return to giving 'the charm of novelty to things of every day' in Chapter 11 with its references to Tom Stoppard's *Arcadia*. Responding to such a poetic move to open eyes to what lies behind the perceptual surface of the world, contemporary poet and Coleridge scholar Malcolm Guite comments, 'The poet, as much as the philosopher or the scientist, is concerned with helping us look beyond surfaces at what is really there.'[10] Poetry, like science, opens eyes of understanding to new perspectives, creates connections, and sets loose new narratives, even shocking ones. Even the process of poetry itself, of shaping imaginative energy through the creative constraint of form, constitutes a powerful, and also accurate, metaphor for the process of doing science—for what could call on greater imagination than the reconception of the cosmos? And what could constitute a tighter form for that imagination than the structure of that cosmos as we observe it?

Witnessing a total solar eclipse, the experience that drives the reflections of this book, takes the onlooker 'behind the surface' in this shared sense, and in the most powerful way, seeding the imagination in science, poetry, and art. The experience leaves an impression that nothing is unchanged—not the sky, nor the flora and fauna on the ground, nor the human being looking up into a dramatically different sky and feeling suddenly how the universe aligns around her, according to laws that have

---

[9] See Samuel Taylor Coleridge, *The Works of Samuel Taylor Coleridge, Prose and Verse: Complete in One Volume*. Philadelphia, PA, 1852, p. 308.
[10] Malcolm Guite, *Faith, Hope and Poetry*. London, 2012, p. 164.

always been there but only reveal themselves in the darkness of the eclipsed sun and the bright light of the corona around it. A diamond ring, the movement of the shadow over the sun generates brief emergences of sparkles around it, the animals grow silent, and the crickets start to sing. Afterwards, everything appears in a new and fresh light (Figure I.9).

A total solar eclipse appears in the Bible to mark the moment of Christ's death on the cross.[11] It is used metaphorically and mimetically in literature, the visual arts, music, and film to announce change, mystically and mysteriously alluding to some fundamental principle of life, a revelation, a new era. It is a constant in all exegetical arts, such as theology and philosophy, and appears in poetic, painterly, and even sculptural imagery. When a 'great eclipse' is announced, millions of people make their way into the path of totality. Sales of books on the astrophysical phenomenon skyrocket, as do those of special eclipse glasses. 'Total Eclipse of the Heart' plays all week; BBC, CNN, and all other stations report live from the heart of the ecliptic darkness.

Figure I.9 Crofton Total Eclipse Consortium: The Sun after the total solar eclipse of 2017, seen from Kentucky.

---

[11] The 'darkness over the whole land' is mentioned in Matthew 27:45 as well as in Mark 15:33 and Luke 23:44–45—not surprisingly, as the three synoptic gospels (unlike that of John) use information from the unknown Q-source and regularly display such correspondences. The happening functions in contrast to the bright star of the birth of Jesus Christ, 'Thus, Christ's birth and death are seen as cosmic events; the universe could not be indifferent to such happenings.' See Georg Luck, *Arcana Mundi: Magic and the Occult in the Greek and Roman Worlds*. Baltimore, MD, 1985, pp. 314–15.

Aiming to deliver a conceptual panorama almost as all-embracing as the 360-degree horizon produced by the total eclipse, this book opens up an inter-, cross-, and multi-disciplinary discussion. Chapter by chapter its writers walk through the different elements of what a total eclipse is, how it appears, how it has been and is experienced across time and across the diverse fields of knowledge and art, and how the sciences and humanities can begin a dialogue by sharing mutually illuminating visions of total eclipses. This book will do just that. Its authors convene from fields as diverse as ancient, medieval, and early modern history, astrophysics, history of science, atmospheric optics, literature, paleoastronomy, critical theory, plasma physics, cultural astronomy, art history, animal behaviour, planetary dynamics, music, engineering, physics, and theology, for there is no aspect of human experience over which the eclipse has not cast its shadow.

An eclipse consists of simultaneous acts of hiding and revealing, a bringing into light what had previously been obscured, and so the eclipse brings about an apocalypse (literally, the uncovering, from the Greek *apokalupsis*, from *apokaluptein* 'uncover, reveal'). Even the visual experience reflects this metaphor, for as the sun is darkened, so the veil of the sky is lifted. Blue becomes transparent black as the brightest stars can be perceived to be shining at that daytime hour. The delicate fronds and striations of the corona appear, and possibly the silvery pink of those solar flames called 'prominences' decorating the edge of the lunar disk. There is a strange timelessness to the short duration of the event, for it reveals in imagination those inherited connections between observers to their ancient ancestors who also stood transfixed under the black sun for those few terrifying minutes.

Existing Eclipse Literature and Differentiation

There are many books on the total solar eclipse. The vast majority are guides to observation: travel, the track of totality, tips on working with the weather, what to look out for, safety, and so on. Some open a glimpse into historical and artistic ways in a typically abbreviated, bite-sized way (for instance, *Total Eclipses: Science, Observations, Myths and Legends* edited by Pierre Guillermier and Serge Koutchmy includes a section on 'myths and legends' associated with eclipses added to the presentation of science and experience).[12] Excellent recent literature for a general audience has emerged around the 2017 eclipse such as Mark Littmann and Fred Espenak's *Totality: The Great American Eclipses of 2017 and 2024* and David Baron's *American Eclipse: A Nation's Epic Race to Catch the Shadow of the Moon and Win the Glory of the World*.[13] While Bryan Brewer's *Eclipse: History, Science, Awe* does have three pages on 'eclipses in literature,' it focuses, after a short historical introduction, on lavish illustrations.[14] Frank Close's *Eclipse: Journeys to the Dark Side of the Moon* is a personal story from

---

[12] See Pierre Guillermier and Serge Koutchmy, *Total Eclipses: Science, Observations, Myths and Legends*. Chichester, 1998.

[13] See Mark Littmann and Fred Espenak, *Totality: The Great American Eclipses of 2017 and 2024*. Oxford, 2017. See also David Baron, *American Eclipse: A Nation's Epic Race to Catch the Shadow of the Moon and Win the Glory of the World*. New York, 2017.

[14] See Bryan Brewer, *Eclipse: History, Science, Awe*. Seattle, WA, 2017.

an articulate astronomer of the fascination of eclipses; although there is mention of history and culture, the focus is on the diarized experiences of eclipse-viewing.[15]

Lots of recent general literature explores eclipses, historically like Armitage's *The Shadow of the Moon: British Solar Eclipse Mapping in the Eighteenth Century*; maths-historically like Montelle's *Chasing Shadows: Mathematics, Astronomy, and the Early History of Eclipse Reckoning*; with a focus on celestial shadows and eclipses in general like Westfall's and Sheehan's *Celestial Shadows: Eclipses, Transits, and Occultations*; with a discussion of Sun, Moon, and Earth like Nordgren's *Sun Moon Earth: The History of Solar Eclipses from Omens of Doom to Einstein and Exoplanets*; or focusing on the Sun or Moon individually like Golub and Pasachoff's *Nearest Star: The Surprising Science of our Sun*, Golub and Pasachoff's *The Sun*, and Leatherbarrow's *The Moon*.[16] New printing technologies brought forth an entire genre of exciting 3D books on space such as Brian May's and David J. Eicher's *Cosmic Clouds 3-D: Where Stars Are Born*.[17]

The science project of total solar eclipse-chasing is also highly personal and emotional. As Guillermier and Koutchmy put it, '[first] contact is an emotional moment. After all the preparations, to be in the right place at the right time and to have provided for the moment is a source of great satisfaction.'[18] Kate Russo, coming from a background in psychology, aims at a more holistic vision and delivers a great example for eclipse-chasing and writing that merges with her professional knowledge and language in her 2012 book *Total Addiction: The Life of an Eclipse Chaser*.[19] Interwoven with her own story Russo delivers a personal, psychological, oral-history, and sociological look at the 'eclipse-chasers' who make annual trips to each total solar eclipse.

A formidable human sciences approach to astronomy in general that also became an important companion for this project is developed in *Cosmos: The Art and Science of the Universe* by Roberta J. M. Olson and Jay M. Pasachoff.[20] Building on their decades-long collaboration, Olson and Pasachoff (both also contributing authors to this volume) provide comprehensive topical chapters in the tradition of Alexander von Humboldt's *Kosmos* (1845 and 1862), recognizing how 'every aspect of the universe is connected in a web of life, on Earth and in the heavens.' *Cosmos* is the kind of

[15] See Frank Close, *Eclipse: Journeys to the Dark Side of the Moon*. Oxford, 2017.

[16] See also, among others, Geoff Armitage, *The Shadow of the Moon: British Solar Eclipse Mapping in the Eighteenth Century*. Tring, Hertfordshire, 1997; Clemency Montelle, *Chasing Shadows: Mathematics, Astronomy, and the Early History of Eclipse Reckoning*. Baltimore, MD, 2011; John Westfall and William Sheehan, *Celestial Shadows: Eclipses, Transits, and Occultations*. New York, 2015; Tyler Nordgren, *Sun Moon Earth: The History of Solar Eclipses from Omens of Doom to Einstein and Exoplanets*. New York, 2016. See also Leon Golub and Jay M. Pasachoff, *Nearest Star: The Surprising Science of our Sun*. Cambridge, 2014, Leon Golub and Jay M. Pasachoff, *The Sun*. London, 2017, and Bill Leatherbarrow, *The Moon*. London, 2018.

[17] See Brian May and David J. Eicher, *Cosmic Clouds 3-D: Where Stars Are Born*. Cambridge, MA 2020. Brian May notably has also produced a Bohemian Rhapsody celebration with *Queen in 3-D: Bohemian Rhapsody Edition*, Cambridge, MA, 2018.

[18] See Guillermier and Koutchmy, p. 149.

[19] See Kate Russo, *Total Addiction: The Life of an Eclipse Chaser*. Berlin, 2012.

[20] See Olson and Pasachoff, *Cosmos*. See also Roberta J. M. Olson and Jay M. Pasachoff: 'Astronomy: Art of the eclipse.' *Nature* 508, 2014, pp. 314–15. See also Roberta J. M. Olson and Jay M. Pasachoff, 'Depictions of the Moon in Western Visual Culture.' *Oxford Research Encyclopedias, Planetary Science*, published online: 25 June 2019, https://doi.org/10.1093/acrefore/9780190647926.013.55.

book that helps readers from the science side into history and the arts and humanities, and vice versa. The authors offer an ideally balanced and stunningly illustrated historical and interdisciplinary view for specialists and students alike.

While this volume has benefitted from the many books we have read over our lifetimes, and while it is itself deeply engaged in conversation with many publications across all our fields, there is nothing like it available yet—no other publication narrows down the question to just one specific phenomenon exclusively (*only* the *total solar* eclipse) while bringing the question to so many different, leading experts internationally in such a variety of fields. The combination of entire chapters devoted to eclipses in music and literature as well as their record in ancient, medieval, and more contemporary civilizations, readable accounts of how they are predicted, and the astrophysics of the corona, make this book unique for a reason: It is an ever-open invitation; its main intent is to encourage its readers to further discovery.

Approaching the Eclipse with this Volume

The chapters of this book can be thought of as doors along a corridor that can be opened in any order to access different approaches to the total eclipse. The corridor leads along different disciplines and expertises and, by giving access to them, it unites the diverting fields of the sciences, arts, humanities, theology, and social sciences (needless to say the corridor of eclipse knowledge is much longer and more varied than the many perspectives gathered in this book, and we hope that our contribution will spark even more varied critical and intercultural engagement). Following this introduction, the reader is invited to enter from this corridor different areas of scholarly, scientific, and lived experiences in order to see how a vast array of disciplines relate to total solar eclipses. A cosmic happening thereby becomes decipherable in the terms of astrophysics, measurable in historical shifts from ancient and medieval to modern times, approachable as the image of incommensurate theological–mystical virtues, graspable as a theme of literature across languages, visible in the mirror of the visual arts. It reverberates in ancient myths around the world, and resounds in musical interpretations, from medieval to modern. Its effects become visible in the reactions of animals as well as in the atmospheric optics as described by human beings. Its appeal is contagious in the account of an eclipse-chaser who, as many others, follows the paths of Sun and Moon around the world to experience and continuously re-experience the darkening revelation and the lifting of the darkness.

The volume's chapters are grouped into four parts: 'Cosmos,' 'History and Religion,' 'Arts and Literature,' and 'Animals, Weather, Environment.' The first four chapters introduce some basic science: Tom McLeish and Mike Frost (British Astronomical Association) lay out how the astronomical regularities of eclipses arise from the orbits of Earth and Moon in 'The Cosmic Clockwork: the How and When of Total Solar Eclipses,' followed by Jay Paschoff's (Williams College) account of the principle phenomenon revealed, 'The Unveiling of the Corona,' the glorious and extremely high-temperature high atmosphere of the Sun that appears in luminous striation around the dark Moon in totality. A series of historical perspectives bridges Parts I and II, beginning with C. Philipp E. Nothaft (All Souls College Oxford), whose chapter,

'Pre-Modern Astronomies of Eclipses in the Near East and Europe,' surveys the historical development of the theories and techniques that put ancient and medieval astronomers in a position to predict the occurrence and appearance of eclipses. Geographically and chronologically, ranging from the ancient Near East to Renaissance Europe, Nothaft's study focuses on the status of eclipses in Hellenistic Greek and later Arabic astronomy and the transmission of the relevant knowledge and tools to Latin Europe during the Middle Ages. The part's final chapter delivers a captivating account from Mike Frost, 'From Science to Story: Testimony of an Eclipse-Chaser.' As a self-confessed inveterate 'eclipse-chaser,' Frost writes a personal account of the appeal and the sheer aesthetic variety of eclipses from one year to the next, describing the tension between the worldly business of getting oneself to the right place at the right time and the other-worldly experience of the eclipsed Sun.

The core of historical perspectives is gathered in Part II on 'History and Religion': John Steele (Brown University) explores in 'Solar Eclipses Across Early Asia' ancient Asian traditions of eclipse prediction, recording, and interpretation. Giles Gasper (Durham University) focuses on the Middle Ages in Europe and detects the emotional reactions to, and political interpretations of, total eclipses in late medieval texts and chronicles in '"The Face of the World was Wretched, Horrifying, Black, Remarkable": Solar Eclipses in the Middle Ages.' Anna Marie Roos (University of Lincoln) investigates societal perceptions of the total solar eclipse in early modern England, with a focus upon the total solar eclipse of 29 March 1652 entitled 'Black Munday' in almanacs and broadside ballads, as well as in more elite texts, demonstrating an interplay between the changing worlds of belief in 'signs and wonders' and the development of the 'new science,' in her chapter '*Annus Tenebrosus*: Black Monday, Faith, and Political Fervour in Early Modern England.' Finally, David Bentley Hart (author of *The Beauty of the Infinite: The Aesthetics of Christian Truth*[21]) conceptualizes 'Signs and Portents: Reflections on the History of Solar Eclipses' and traces solar and lunar eclipses through the myths, histories, and world literature, and in an incalculable variety of ways. Bentley Hart probes in his cross-cultural discussion how eclipses have tended to be regarded, at most times and in all places, up until very recently, as carrying or announcing some kind of curse or misfortune (from ancient China to Rome, to Shakespeare, Milton, Descartes, and Heidegger, among others).

From this venture into theological and literary perspectives on total solar eclipses we move into Part III, focusing specifically on literature, the visual arts, and music. In chronological order of their main examples, the first chapter of this section shows President of the Dante Society of America Alison Cornish (New York University) discuss 'Dante's Total Eclipses.' The author of *Reading Dante's Stars*[22] explores the exalted status of astronomy in conversation with the science, philosophy, morality, and poetry of the Duecento to Trecento age in Italy and brings the poetic eclipse-chasing in Dante's *Paradiso* into a crucial literary focal point between the late Middle Ages and the early modern age. Art historian Roberta J. M. Olson (Wheaton College, Massachusetts, and the New York Historical Society Museum & Library)—the

---

[21] David Bentley Hart, *The Beauty of the Infinite: The Aesthetics of Christian Truth*. Grand Rapids, MI, 2004.

[22] See Alison Cornish, *Reading Dante's Stars*. New Haven and London, 2000.

co-author, with Jay Pasachoff, of *Cosmos*[23]—presents new material on the visual tradition of eclipse representations in the nineteenth century with 'Eclipsed? The Nineteenth-Century Quest to Capture Solar Eclipses in Art, Science, and Technology.' Olson's chapter includes the discoveries of scientists, thinkers, and practicing artists such as Audubon, Constable, Caspar David Friedrich, Goethe, Halley, Linnell, Runge, Turner, and Wordsworth. Olson's research and her collaborations with Jay Pasachoff have been among the first serious considerations of artistic representations of total solar eclipses within their scientific and cultural context. Henrike Christiane Lange (University of California, Berkeley) continues Cornish's and Olson's lines of inquiry in art, literature, experience, image theory, and representation in 'Total Eclipse of the Art: Vision, Occlusion, Representation.' Lange covers a wide spread of materials, ranging from early modern painting to early photography and film, political iconography, contemporary digital image processing, and nineteenth-century painting and poetry. As she works along the break between Romanticism and the rational age, Lange takes cues from Tom Stoppard's science play *Arcadia* (1993) that structure the chapter's different points of view ('Looking Up,' 'Looking Down,' 'Looking Around,' and 'Looking Inwards'). Closing the book's part on artistic inspiration and representation with exciting insights from centuries of music history, Elaine Stratton Hild (University of Würzburg) listens to the reactions of musicians who have referenced the phenomenon of the total solar eclipse in their compositions—from Handel to Pink Floyd. Her survey examines musical responses to the eclipse of the Sun in European and North American pieces: the eclipse appears as a metaphor in sung texts, as a marketing strategy in song titles and album cover art, and as an inspiration for musical structures and textures. The chapter, 'When Words Fail: Eclipse, Music, and Sound,' also considers the changing compositional practices that have led to differing musical portrayals of the natural phenomenon.

The final part, 'Animals, Weather, Environment,' returns to a scientific perspective—but now contemplating the earthly effects of the eclipse. Steven Portugal (Royal Holloway) reports in 'Animal Behaviour and Eclipse' on how non-human animals react to the sudden darkness and chill—some awakening, others falling silent—and on the fascinating ways in which the entire cosmos becomes a mobile, outdoor lab, and how new technologies of observation and animal tracking devices might improve our future understanding of animals' experiences. Portugal follows species from the size of zooplankton and fireflies to that of elephants to see how the shared experience impacts their various behavioural patterns. Giles Harrison (Department of Meteorology, University of Reading) looks through the lens of the Earth's atmosphere, discussing the changes that the sweeping shadow makes to the dynamics that drive weather in 'Weather and the Solar Eclipse: Nature's Meteorological Experiment.' Beyond the obvious impact of weather on visibility, Harrison argues, total solar eclipse can be characterised as providing a well-defined 'cause' with the weather's less predictable response the resulting 'effect.' This allows for interpretation of an eclipse as a natural meteorological experiment within which long reported anecdotal eclipse-generated phenomena can be investigated and understood.

---

[23] See Olson and Pasachoff, *Cosmos*.

Finally, the editors (Lange and McLeish) draw out from the collaborating community what we have learned through the sharing of our perspectives, and those of the people through time and across space whose contributions and questions inform our own, opening further doors and windows with a reflective conclusion, 'The Moon and the Sun in the Afternoon.' To apply the metaphor once more, the rooms opening from the now light and spacious gallery-corridor of this book exist not just in the context of the scientific imagination, or of scholarly discourses, but open themselves potentially back onto all of human imagination. Ultimately, a full perception, a full comprehension of the total eclipse might only be attempted in the totality of rooms and passages connecting them. The goal is then to merge disciplines and perspectives into an idea of the eclipse which is greater than the sum of its parts.

Let's now open some of these doors.

I
COSMOS

1

The Cosmic Clockwork

The How and When of Total Solar Eclipses

Mike Frost and Tom McLeish

Astronomy, Astrophysics, Soft Matter Physics, Natural Philosophy, Theology of Science

This first chapter tackles the underlying dynamic geometry of eclipses from a scientific point of view—when and why do total solar eclipses occur? The cosmic clockwork that creates total solar eclipses essentially consists in Euclidean geometry enhanced with planetary motion: The motion of the orbital dynamics of Earth around the Sun, and that of the Moon around Earth (yet strongly affected by the force of the Sun). The phenomenon in motion provides the glossary with surely its favourite word—'syzygy'— meaning a lining up of three bodies. A second ingredient to the geometry is constituted by the need to match the periods of cycles of Earth and Moon so that all align, giving rise to the saros cycle known to the ancient world and the more modern concept of the inex cycle.

annular eclipse A solar eclipse in which the Moon is too small to cover the Sun completely and a ring (annulus) of the Sun remains uncovered.

antumbra The area of the shadow in which an annular eclipse can be seen.

apse (or apsis) the nearest or furthest point in the orbit of a planetary body from the primary body it is orbiting around.

inex A period of just under 29 years after which eclipses often repeat (though not as close a repeat as after a saros).

lunar eclipse An eclipse in which the Moon passes into the shadow of the Earth.

partial eclipse An eclipse in which only part of the eclipsed body is hidden.

penumbra The area of the shadow in which a partial eclipse can be seen.

precession A periodic change in the orientation of a rotating body. Orbital precession is a periodic change in the orientation of the orbit.

saros A period of just over 18 years after which eclipses often repeat.

solar eclipse An eclipse in which the Moon passes in front of the Sun.

syzygy When three solar system bodies lie in a straight line.

total eclipse An eclipse in which the whole of the eclipsed body is hidden.

totality the period of time when the eclipse is total from a given location (for a solar eclipse this is brief, minutes at most; for a lunar eclipse it can be longer).

track of totality The (relatively small) portion of the Earth's surface in which a total solar eclipse can be seen. The track of totality generally sweeps from west to east across the Earth's surface.

umbra The area of the shadow in which a total eclipse can be seen.

A practical note: *It is well worth reading this chapter alongside a simple household experiment with a light bulb in a darkened room. Cast the shadow of a small sphere (such as a ping-pong ball), representing the Moon, onto a larger sphere representing the Earth, by a diffuse light source representing the Sun.*

In preparation for the science, history, and art of total solar eclipses, this chapter follows the ancient precedent of an exercise in geometry, followed by one in arithmetic. Enshrined in Euclid's geometry, the most widely received textual legacy of the ancient Greek world, and in the medieval curriculum of the 'quadrivium,' geometry remains the starting point for an understanding of eclipses. Diagrams such as the medieval Figure I.8, and its modern counterparts in this chapter, constitute one aspect of the way to understanding how total solar eclipses arise. The sister subject to geometry within the quadrivium, arithmetic, provides the other aspect to an understanding of eclipse astronomy, the key to when they appear. Intricate numerical patterns to the series of dates on which eclipses happen, identified in the ancient world, are now understood in terms of the orbits of Moon and Earth. The chapter concludes with a comparative overview of solar eclipses generated by the moons of the solar system's other planets.

Geometries of the Total Solar Eclipse and the Shadow of the Moon

The principal geometric structure of the total eclipse is that of the shadow of the Moon, cast by the Sun. The shadow takes the form of a cone, whose base is the circumference of the Moon, and whose point is directed directly away from the Sun. Figure I.8 in the Introduction, illustrating the thirteenth-century astronomical treatise *De Sphera*, by Robert Grosseteste, contains this construction. At every point within this conical volume, the Sun is completely eclipsed—alternatively an observer there has no direct line of sight to any point on the Sun's surface. The reason for the finite conical shape is that the Sun is larger than the Moon, so that the two limiting lines of sight just grazing the edges of both Sun and Moon converge (at an angle very close to that subtended by the solar disk when observed from the Moon). The accompanying text from the treatise covers the logic in the case of the shadow of the Earth (responsible for Lunar eclipses when it falls upon the Moon), but the reasoning is identical:[1]

> For a lunar eclipse happens because the moon travels through the shadow of the earth, which is always projected opposite the sun. For since the sun is a luminous body, and the earth is a shadowy body, and rays are straight, and the sun is larger than the earth, it is necessary that the sun projects a shadow in the shape of a cone, and that the apex of the shadow terminates directly on the point on the ecliptic opposite the sun. So, since the sun is always beneath the ecliptic, so the apex of the shadow of the earth is always beneath the ecliptic.

[1] Robert Grosseteste, *De Sphera*, ed. C. Panti, trans. S. Sønnsyn, in Giles E. M. Gasper *et al.*, *Mapping the Universe: Robert Grosseteste's De sphera—On the Sphere*. Oxford, 2023, §57.

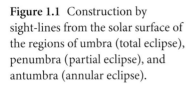

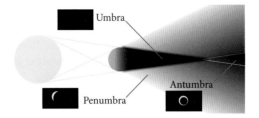

Figure 1.1 Construction by sight-lines from the solar surface of the regions of umbra (total eclipse), penumbra (partial eclipse), and antumbra (annular eclipse).

The connection between the conical shadow and the various types of eclipse of the Sun is demonstrated in Figure 1.1.

The conical region of space directed from the Earth, directly opposite the Sun, is termed the *umbra*. All points in this volume have the property, by construction, that there is no direct sight-line to any point on the Sun's surface. When any point on the Earth's surface passes though the umbra, a total solar eclipse is experienced there.

Immediately surrounding the umbra is a larger region, also conical, but this time a divergent cone (with the umbra-cone subtracted from it), rather than a convergent one, within which all points have some, but not all, of the Sun's surface obscured by the Earth. This region is the *penumbra*. From any point of the Earth's surface located in the penumbra, a partial eclipse is therefore visible. The closer to the boundary with the umbra, the larger is the eclipsed region of the Sun's surface, and the more noticeable is the partial eclipse. As the Earth moving through the Moon's shadow during an eclipse must pass through the penumbra before and after any interval that finds it within the umbra, any place experiencing a total eclipse must observe a partial eclipse before and after (unless the Sun rises or sets from that location during the total eclipse). Whilst a total eclipse is in progress at one location, observers at other nearby locations will see a partial eclipse—this is an example of parallax, the Moon being closer to the observer's eyes than the Sun.

The final qualitatively distinct region of the Moon's shadow lies directly beyond the point of the umbra's cone. This diverging cone, symmetric with respect to the umbra, is the *antumbra*. From the construction of sight lines to the solar surface from this region, it is evident that from points in the antumbra, there is a central region of the Sun completely obscured, but an annulus of solar surface near the limit of the apparent solar disk from which light is directly visible. This produces the bright ring-like phenomenon of the annular eclipse. Because both the orbits of Earth about the Sun, and Moon about the Earth, are elliptical (see the next section for more detail), there are occasions when the Moon passes directly between the Sun and Moon at a distance that means that the Earth's surface exists within the antumbra, rather than the umbra.

The observation of a solar eclipse is therefore equivalent to the point of observation on the Earth's surface passing through umbra, antumbra, or penumbra. This can clearly occur only when the Sun, Moon, and Earth are all very close to a single geometric line (the condition known as 'syzygy'). For this to occur it is at least necessary that the phase of the Moon is new—though more needs to be true, as at most new Moons the lunar shadow passes either north or south of the Earth. The details of orbital geometries and periodicities that determine when this is true, we review in the next section. Before we do so, it is worth focusing in a little more detail on the time-dependence of the falling of umbra and penumbra upon the Earth. For the lunar shadow moves relative to the Earth's surface, leaving a track of connected eclipse events, rather than a single event at a single location.

There are two important motions that determine the eclipse track. First, there is the orbital motion of the Moon, from west to east in the sky, at a speed that traverses a distance of the Earth's diameter in about seven hours. Second, the rotation of the Earth, also from west to east, carries points on the Earth's surface in approximately the same direction to the movement of the lunar shadow though space. The speed of this rotational motion depends on the latitude of an observing site, reducing to zero at the poles. At the equator, when it is a maximum, it carries a point from one side of the Earth (as seen from the Moon) to the other in 12 hours, so is not very much slower, on average, than the speed of the lunar shadow, and so contributes to lengthening the duration of totality. The combination of the two motions results in an 'eclipse track' that moves relative to the surface of the Earth (in most cases) predominantly from west to east. Since the axis of the Earth is tilted by roughly 23.5° to its orbit, the tracks typically sweep with a component in latitude as well.

Figure 1.2 shows examples of tracks of total and annular eclipses during the period 2021–40. When close to the equator or temperate latitudes, note that the tracks are predominantly west to east, with some change of latitude as well. The latitude change

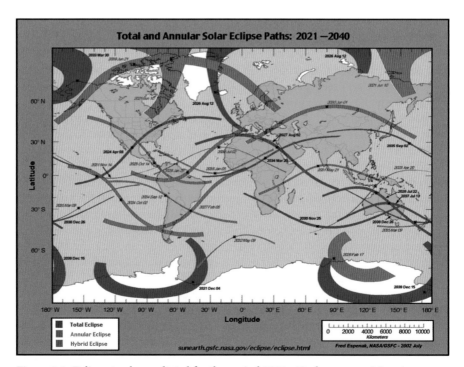

Figure 1.2 Eclipse tracks predicted for the period 2021–40, shown on a Mercator projection of the Earth.

may reverse between the early and late phases of the eclipse as the fall of the Moon's shadow changes from a glancing angle on sites towards sunrise, through a more perpendicular angle near midday, then again to a more tangential fall at the final sites appearing around the eastern rim of the Earth as it turns into the shadow, during their early morning.

Some tracks at high latitude display strange features. They can be very wide—this is due to the Sun appearing low on the horizon in the Arctic and Antarctic, so that both the Sun's rays and the umbra at mid-eclipse strike the Earth's surface at a near-tangential angle. The umbra is spread over a wide latitude. Portions of such tracks may also run, anomalously, from east to west. In these cases, the corresponding pole of the Earth is tilted towards the Sun (when that hemisphere is enjoying summer), so that the eclipse occurs between the pole and the extremity of the Earth visible from the direction of Moon and Sun.

A final feature of some eclipse tracks arises from their three-dimensional form on the Earth's surface. Because of the Earth's spherical curvature, those sites on the track near sunrise or sunset are further from the Moon (by up to as much as the radius of the Earth) than sites experiencing eclipse at local midday. If the relative positions of Moon and Earth are such that the point of the conical umbra falls between these two distances, it is possible for the eclipse to switch, during its track, from annular to total, and back again as the eclipse sites move from the antumbra into the umbra, and back again (an account of one such eclipse can be found in Chapter 4). In the limiting case where the cone just touches the Earth's surface, the assumption that the Moon's profile is exactly circular breaks down—the exact shape of the shadow cast by the uneven limb of the Moon becomes essential. The result can be a 'broken ring' eclipse, where the annulus is partially broken by the higher mountains on the Moon's limb.

While not the central topic of this volume, very similar considerations of orbital periodicity and geometry also regulate a completely different type of eclipse which is visible from the Earth's surface—a lunar eclipse. The Earth also casts a shadow, with umbra, penumbra, and antumbra; larger in size than the Moon's shadow. The Moon can pass into the penumbra and/or umbra of the Earth's shadow, causing a lunar eclipse, visible from the entire night side of the Earth. From the Moon's surface, a solar eclipse is visible. Some illumination of the Moon can still occur, by light which is bent around the Earth by refraction in the Earth's atmosphere. Thus, during a total lunar eclipse, the Moon appears a dullish red, illuminated at once by all the sunsets on Earth.

Ellipticity and Inclination of Lunar Orbit

The Earth's orbit around the Sun, and the Moon's orbit around the Earth, are both elliptical (although, as we will see later in the chapter, the Moon's orbit is surprisingly complicated). When the Moon is at its closest to the Earth (perigee), its apparent size in the sky is greatest—full Moons close to perigee are the 'supermoons' so beloved of the media in recent years. For this reason, an eclipse with the Moon close to perigee is more likely to be total. Conversely, when the Moon is furthest from the Earth, at apogee, its apparent size is smaller, and so the closer the Moon is to apogee the more likely central eclipses are to be annular than total. The eccentricity[2] of the Moon's orbit around the Earth varies but has a mean value of 0.055, and the Moon appears around 14% larger by diameter at perigee than it does at apogee.

The Earth's orbit around the Sun is also elliptical, but with an eccentricity of only 0.017 (zero gives a circular orbit). When the Sun is at its furthest from the Earth, aphelion, its apparent diameter in the sky is only 3% smaller than when the Sun is closest, perihelion. So, central eclipses are more likely to be total at aphelion, and more likely to be annular at perihelion, although the greater ellipticity of the Moon's orbit means that its distance from Earth is the larger effect.

Perihelion, the closest approach of the Sun to the Earth, currently occurs in early January. One surprising corollary to this is that, in the current epoch, total solar eclipses are more likely to happen in the northern hemisphere than in the southern hemisphere. This is because solar eclipses happen in the daytime, and each hemisphere sees more daytime during its summer. So, the southern hemisphere summer occurs when the Sun appears largest in the sky and is less likely to be covered completely by the Moon, because the Moon needs to be closer to its perigee to cover the Sun completely than for northern summer eclipses. On average, a random location in the northern hemisphere sees a total eclipse every 330 years, a random location in the southern hemisphere sees a total eclipse every 540 years. Because of the precession of the equinox, the slow circling of the Earth's axis around the ecliptic pole, this imbalance between hemispheres waxes and wanes over a period of 22,000 years.

Initial reflection on the condition for an eclipse—that the Sun, Moon, and Earth are temporarily in a straight line—might lead to the conclusion we visited above, that an eclipse should occur every month at new Moon. This would happen if the Sun, Moon, and Earth all orbited in the same plane. However, this is not the case. The Moon's orbit is tilted at five degrees to the Earth's orbit around the Sun (see Figure 1.3). It is an interesting question as to why this should be, and hints at the Moon's origin by collision rather than forming at the same time as the Earth (when it might orbit in the Earth's equatorial plane, or in the same plane as the Earth's orbit around the Sun).

Twice in each orbit, the Moon crosses the plane of the Earth's orbit around the Sun, the ecliptic—once 'up,' heading towards the north, and once 'down,' towards the south. We call these the ascending and descending nodes, and an eclipse can only happen when the Moon passes close to a node at the same time as it is in the same direction as the Sun. The two near-circles that are the Moon's true, inclined, orbit, and the projection of this orbit onto the plane of the ecliptic, are represented in the medieval diagram by Robert Grosseteste of Figure I.8. The variable distance of the Moon from the Earth is also represented in that Harley manuscript, by way of the 'epicycle' of the Moon (both lunar inclination and variable distance were known then, but needed to be accounted for within a geocentric model with circular orbits, rather than the heliocentric system with elliptical orbits of today's astronomy).

---

[2] The eccentricity, e, of an elliptic orbit is a measure of how far it departs from circular, and is defined as the difference between nearest and furthest distances, divided by their sum. A value $e = 0$ corresponds to a circle, and $e = 1$ the limiting case of a parabola, the curve of which just fails to close.

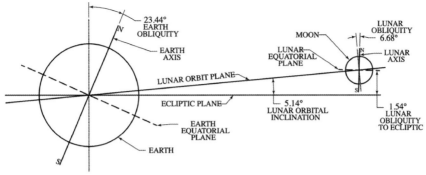

Figure 1.3 Diagram of the inclination of the Moon's orbit to the plane of the Earth's orbit around the Sun (ecliptic).

Source: Wikimedia Commons.

Long-term Patterns of Total Solar Eclipses: the Saros and Inex Cycles

How long does it take for the Moon to orbit around the Earth? This simple question requires a complex answer: From the perspective of the distant stars, one orbit is completed in 27.32 days. This is known as the sidereal month. This is not, however, the same as the period between new Moons, because the Earth is also in orbit around the Sun. The Moon has to progress about a twelfth further around its orbit in order to compensate for the Earth progressing by about a twelfth of its orbit around the Sun. The Earth's speed around the Sun varies during its orbit, so the time between new Moons, the synodic month, or lunation, varies between 29.18 and 29.93 days, with an average of 29.53 days.

Eclipses, as we discussed above, can only happen when Sun, Moon, and Earth lie in a straight line—syzygy. For this to happen, the Moon must be crossing the plane of Earth's orbit at the time of syzygy, either ascending or descending. The time between successive node-crossings of the same type is known as the draconic (draconitic) month. This is of length 27.21 days. One final periodicity requires mention, the time between successive perigees (the closest approach of the Moon to the Earth). This is the anomalistic month of 27.55 days.

Why should the Moon's orbit around the Earth be so surprisingly complicated? The laws of planetary motion, as expounded by Johannes Kepler, are straightforward—planets orbit the Sun in elliptical orbits. To a first approximation, this is true for the Moon's movement around the Earth, as was first pointed out by Jeremiah Horrocks[3] from a careful examination of the Moon's angular size through its monthly cycle. However, there is an important difference: The orbit of the Moon around the Earth cannot be considered in isolation. The gravitational attraction of the Sun must also be

---

[3] In *Principia Mathematica* Book III (Isaac Newton, London, 1687), Newton commends 'our country-man Horrox [sic] was the first who advanced the theory of the moon's moving in an ellipse about the earth placed at its lower focus.'

taken into account. Surprisingly, the Sun attracts the Moon about twice as strongly as the Earth does: it may be 400 times as distant, but the solar mass is a third of a million times that of the Earth. These two differences nearly compensate.[4] Sun, Earth, and Moon are perhaps the strongest example of a three-body system in the solar system; nowhere else are the mutual attractions so similar. The effect of this extra gravitational pull causes the structural features of the Moon's elliptical orbit around the Earth to precess.[5]

These three definitions of a month affect eclipses in different ways. Solar eclipses, by definition, can only occur at new Moon, so the intervals between eclipses must be close to multiples of the average synodic month. The same is true for lunar eclipses. Eclipses can also only occur when the Moon is close to an ascending or descending node, so eclipses must also occur at intervals close to multiples of draconic month (with, perhaps, an additional half if eclipses switch from ascending to descending node or vice versa). It is therefore sensible to look for multiples of these months which coincide. Periods that are also very close to multiples of the anomalistic month, though not essential for an eclipse to occur, will result in a similar apparent size of the Moon in the sky (recall that near apogee the Moon's disk is too small to fully cover the Sun). For example, it has been known for thousands of years that, just over 18 years after a lunar eclipse, there is usually another one. This discovery is usually ascribed to the Babylonians, though it is likely to have been noted by other cultures who kept records of sky events. The actual period between such eclipses is 6,585.32 +/-0.07 days. 6585 days is usually 18 years and 11 days, although it can be 18 years 10 days or 18 years 12 days, depending on how the intervening leap years fall. Edmond Halley named this period a saros.[6]

The saros corresponds to   223 synodic months
                                241.999 draconic months
                                238.992 anomalistic months

Solar eclipses are also often separated by a saros period. This, however, would have been much more difficult to notice, as partial eclipses are difficult to observe with the naked eye and so can easily be missed, and total eclipses are only visible on a narrow track, whereas lunar eclipses are visible across an entire hemisphere.

Solar eclipses occur at new Moons. So, 6,585.32 days after a new Moon, there is another new Moon (the 223rd such new Moon). Because it is very close to 242 draconic months, it is quite likely that the Moon will, once again, be passing through the node (ascending or descending) to give another eclipse. And it is also close to 239 anomalistic months, so the Moon is at almost at the same point in its orbit, and the

---

[4] The reader may easily convince themselves of this by recalling Newton's law for the gravitational force F between two bodies of mass M_1 and M_2 separated by a distance r: $F = GM_1M_2/r^2$, where G is the gravitational constant.

[5] Both apses and nodes precess, at different rates (8.88 years and 18.61 years respectively).

[6] E. Halley, 'Emendationes & Notae in Tria Loca Vitiose Edita in Textu Vulgato Naturalis Historiae C. Plinii.' *Philosophical Transactions of the Royal Society of London* 16, pp. 535–40, at 537–8. From Greek *sáros*, Akkadian *šāru* (meaning 3,600; it is not clear why Halley chose this).

same apparent size in the sky. As a result, 18 years and 11 1/3 days after a solar eclipse, there is likely to be another similar one.

However, the 0.32 days beyond an exact whole number means that the Earth has spun through nearly 120 degrees between the two eclipses. Two solar eclipses separated by a saros will then have tracks translated westwards by a third of a rotation. Because a sequence of 223 synodic months is so close to 242 draconic months, a whole series of eclipse takes place, each one separated by a saros interval. However, the very slight difference between the saros interval and an exact whole number of draconic months means that each eclipse in the series is at a slightly different point of the Moon's orbit relative to the node. Saros series last for between twelve and fifteen centuries, starting with a partial eclipse close to a pole. Every saros interval, there is an eclipse whose track gets closer and closer to the equator, eventually becoming a central eclipse (total, annular, or hybrid). After crossing the equator, the tracks progress towards the other pole, concluding with a series of non-central partial eclipses (there can be exceptions to this progression). A saros series consists of between 69 and 87 eclipses (but usually 71 to 73). For convenience, each series is allocated a saros number.

Some saros series are better than others: The April 2024 eclipse is in saros series 139, which produces a series of longer-duration total eclipses (they occur near to lunar perigee). The preceding eclipse in the series, 29 March 2006, produced an eclipse across North Africa, Turkey, and Russia, whose totality lasted a maximum of 4 minutes 7 seconds. 18 years and 10 days later, 8 April 2024 will produce an eclipse with a maximum of 4 minutes 28 seconds totality, and 20 April 2042 will give us up to 4 minutes 51 seconds of totality across Indonesia, Malaysia, Brunei, and the Philippines. After three successive saros intervals, a period named the exeligmos, there have been three shifts of 120 degrees of longitude and so the track of totality returns to the same terrestrial longitude, though shifted slightly to the north or south. So, the eclipse of 11 May 2078 also crosses Mexico and the continental United States, in this case the south-eastern states.

Another interesting period is the inex, of 358 lunations, 10,571.95 days, which is usually 20 days short of 29 years. This period was studied (and named) by George Van den Bergh in the mid-twentieth century.

> The inex period corresponds to   358 lunations
>   388.500 draconic months
>   383.674 anomalistic months

Because the inex is so close to half a lunation, an inex series is very long-lasting, in excess of thirty thousand years; over a thousand eclipses each separated by an inex period. Once again, an inex series starts with partial eclipses at high latitudes, but this time eclipses alternate between northern and southern latitudes, progressing slowly towards and then away from the equator. However, the lack of a numeric coincidence for anomalistic months means that successive eclipses are not similar (although there are similarities for every third eclipse, the triad period). This progressive pattern of eclipses within saros cycles lends itself to imaginative and attractive diagramming, such as in Figure 1.4, which displays information on all saros cycles actively producing

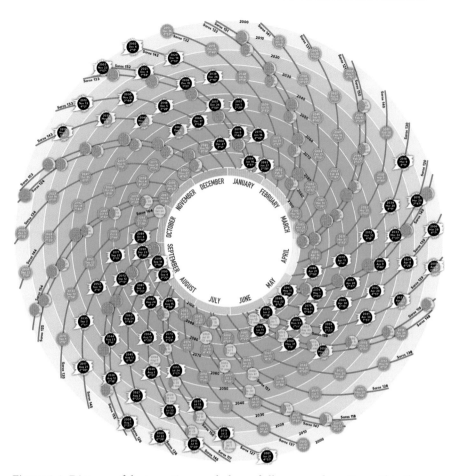

Figure 1.4 Diagram of the type, time, and place of all saros cycles active within the twenty-first century.

eclipses at the time of this book's publication (see the 'Eclipse-Chaser's Toolkit' at the end of this book—enjoy!).

As 223 and 358 are mutually prime,[7] each solar eclipse is in a unique saros and inex series. Van de Bergh produced a chart of eclipses indexed vertically by saros series and horizontally by inex series. More recently, Luca Quaglia and John Tilley generated a much more extensive chart of eclipses, covering 26,000 years, 13,000 into the past and 13,000 into the future, beyond which uncertainties in the calculation make further predictions unreliable.[8] There are other surprising relationships between saros and inex series; for example, an eclipse from saros series I is usually followed one inex later by an eclipse from saros series I+1, unless saros series I+1 has completed.

---

[7] That is to say, they share none of their prime-number factors: 223 is itself prime, and 358 = 2 × 179.
[8] See https://eclipse.gsfc.nasa.gov/SEsaros/SEpanorama.html.

There are other sequences. For example, the Metonic cycle of 235 lunations is a few hours over 19 years in duration, so both lunar and solar eclipses repeat every 19 years for several cycles. This period was also known to Babylonian, Greek, and Chinese astronomers, in particular Meton of Athens, after whom it is named. Metonic series do not persist for anywhere near as long as saros or inex series, but are useful for eclipse predictions. For example, consider the famous Antikythera mechanism (discussed by C. Philipp E. Nothaft in Chapter 5), a clockwork device dating from antiquity, found in 1901 in a shipwreck. It has gears to mimic both the saros and the Metonic series, allowing it to predict lunar and solar eclipses over a period of centuries with remarkable accuracy.

Eclipses Elsewhere in the Solar System

Eclipses are visible elsewhere in the solar system, although the extraordinary coincidence in apparent size between Sun and Moon does not occur anywhere else. The closest we get is on Mars, where the largest Moon, Phobos, can partially eclipse the Sun, as seen from the Martian surface—covering around 30% of it at maximum. Phobos is much closer to Mars than our Moon is to the Earth, and moves much faster around the planet, so eclipses are short in duration, no more than thirty seconds. Nonetheless, eclipses have been observed by the Opportunity and Curiosity Mars rovers (see Figure 1.5). The rovers have also observed the shadow of Phobos crossing the Martian landscape. Deimos, the smaller and outermost moon of Mars, can also cross the Sun, though its apparent size is so small that it is generally thought of as a transit rather than an eclipse.

On Jupiter, by contrast, the apparent sizes of the four Galilean moons—Io, Europa, Ganymede, and Callisto—are bigger than the apparent size of the Sun from the top of the Jovian atmosphere, so total eclipses of the Sun can be observed from there (assuming anyone is there to see them). Additionally, the smaller inner satellite Amalthea can also totally eclipse the Sun.

Jupiter's axial tilt is small, and the inner moons orbit in Jupiter's equatorial plane (perhaps indicating that that was where they formed), so eclipses occur frequently on

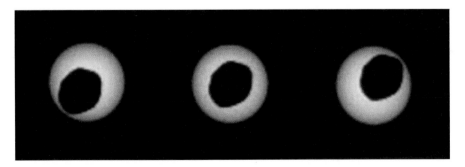

Figure 1.5 Phobos partially eclipsing the Sun, as captured by the NASA Mars rover Curiosity.

the surface of the planet. Because the Earth is not usually in line with Sun and Jupiter, we can observe these eclipses from Earth as shadow transits, where the shadow of a Galilean moon moves in lockstep with the moon across the surface of Jupiter. Careful timings of the onset of these shadow transits, particularly for Io, revealed that transits began later than expected when Jupiter was furthest from Earth. Ole Rømer (1644–1710), of Århus in Denmark, realized in 1676 that this was because light had a finite speed—an impressive discovery for seventeenth-century science.

From Saturn, the apparent size of the Sun is smaller still, so the major moons Mimas, Enceladus, Tethys, Dione, Rhea, and Titan, and the small inner moon Janus, can all cause a total eclipse to be seen from the top of Saturn's atmosphere. All these moons are also in equatorial orbits; however Saturn's axial tilt of 24 degrees means that eclipses are rare.

Likewise, several moons of Uranus can cause total eclipses on Uranus, but the extreme axial tilt of the planet (97.8°) means that eclipses can only take place in a short interval every 42 years, around the Uranian equinoxes. Neptune's axial tilt is not so extreme, but it still means that the equatorial moons can only produce eclipses around the equinoxes, 73 years apart. Triton, Neptune's largest moon, can produce eclipses, but these are also rare, because the moon's orbit is tilted to the equator, and of short duration, because Triton orbits Neptune in a retrograde direction. Finally, the moons of Pluto can produce eclipses, particularly Charon. Charon's shadow covers almost the whole of Pluto, so almost the whole planet will see a total eclipse.

Eclipses of the Sun are common throughout the solar system. However, the closest match between apparent size of Sun and moon occurs on our own planet: Only on Earth is the match close enough to allow us to see the inner atmosphere of the Sun during a total eclipse, and only this fine constraint, coupled with the inclination and strong precessions of the Moon's orbit, gives rise to the intricate numerology of eclipse cycles.

2

The Unveiling of the Corona

Jay M. Pasachoff

Astrophysics

The solar corona is visually the most striking feature of the eclipse. Governing many aesthetic representations, it also has a special place in the history of science. When the Sun is eclipsed, the occultation renders visible things that otherwise cannot be seen, leading to a better understanding of the properties and structure of the Sun's atmosphere that is usually obscured by scattered sunlight. Studies of the solar corona have led to advances in atomic physics and a better understanding of "space weather," which originates in the solar atmosphere and interacts with the Earth's magnetic field, to affect our lives here on Earth. There are still fundamental unanswered questions about the solar corona, which might be explored by space probes to the Sun and innovative solar telescopes.

Alfvén waves Magnetohydrodynamic (MHD) waves seen in plasma, for example in the solar corona (named after their discoverer, Hannes Alfvén).

Balmer series Transitions to the second-lowest electron energy levels in the hydrogen atom, each associated with a particular wavelength.

chromosphere The lowest part of the solar atmosphere.

coronagraph A scientific instrument which creates an artificial eclipse by blocking light from the solar corona with an occulting disk.

coronal mass ejection a large-scale eruption of plasma which passes through the corona.

coronal streamers Large, flame-shaped structures in the solar corona, streaming away from the solar surface.

H-alpha (hydrogen alpha) A transition between electron energy levels in the Balmer series for hydrogen. It produces the rose-red colour seen in prominences and the chromosphere.

Lyman series Transitions to the lowest electron energy levels in the hydrogen atom (each associated with a particular wavelength).

occulting disk The part of the coronagraph which blocks out the solar photosphere.

photosphere The bright, visible disk of the Sun.

plasma One of the four fundamental states of matter in the universe (along with solid, liquid, and gas), principally comprising charged particles (ions or electrons). The Sun is in the plasma state.

solar corona The outermost part of the solar atmosphere.

solar cycle An approximate 11-year cycle, first seen in sunspot activity, associated with the Sun's magnetic field. When magnetic polarity is considered, it is seen that the cycle is actually 22 years, swapping polarity every eleven years.

solar granulation A convection cell in the solar photosphere, typically 1,500 km across.

solar prominence An eruption from the solar disk, following magnetic field lines.

space weather Interactions of the solar wind with the solar system (in particular, with the Earth's atmosphere).

spectroscope An instrument to split light into its constituent wavelengths.

spicule A jet of plasma in the Sun's chromosphere, typically 300 km in diameter.

This chapter focuses on the most striking visual feature of a total solar eclipse: the solar corona, the outermost part of the solar atmosphere. Although it is as bright as the full Moon, the solar atmosphere is hidden behind the terrestrial blue sky and so is impossible to observe under normal conditions of sunlight. Extending several solar diameters away from the Sun's disk, the pearly, radial structure of the corona has streamers that take a different form at each eclipse, since the Sun's atmosphere is much more dynamic and changeable than the Earth's and varies with the 11-year solar-activity cycle, which should be reasonably high at the time of the 2024 total solar eclipse. Responding to the ever-changing solar magnetic field, the corona's charged particles are heated to over 2 million degrees Celsius, far hotter than the visible surface of the Sun, so how energy is injected into the corona and how it stays there is under study by scientists at eclipses and with some ground-based telescopes.

This chapter covers the physics of the corona as well as the solar chromosphere and the prominences—these are features closer to the Sun's visible surface which glow with the deep-red emission of heated hydrogen. It also discusses space weather and the impact on Earth of coronal mass ejections and the expanding solar corona.

Historical Presence, and Absence, of the Corona

Before the early modern period, the corona was not even mentioned. Accounts of eclipses in historical documents[1] almost exclusively discuss timing and location rather than any specific visual appearances of the event.[2] One famous example is the 1560 total solar eclipse observed by Danish astronomer, astrologer, and alchemist Tycho Brahe when he was still a student: An error by a day of the predicted time inspired him to collect the then best data about sky motions, especially of Mars, which in turn inspired Johannes Kepler to work out his three laws of planetary motion (1609 and 1619). Kepler's laws, it turns out as a consequence of Newtonian gravity, are applicable to all orbiting objects; not only bodies such as comets, which were known to Kepler, but also exoplanets and satellites, which were not.

Even the first map showing the passage of the zone of totality across the Earth (the 'eclipse track' discussed in Chapter 1)—prepared by Edmond Halley for the total solar eclipse of 1715, with a plea for the general public (now in the twenty-first century known as 'citizen scientists') to observe and correct the predictions (see Figure 2.1)—was devoted to timing rather than appearance. After all, as repeated

---

[1] F. R. Stephenson, *Historical Eclipses and Earth's Rotation*. Oxford, 2003.

[2] But, as Henrike Lange notes in Chapter 11, in 1571 Antoine Caron painted what looks very much like a corona. On Caron, see also Roberta J. M. Olson and Jay M. Pasachoff, *Cosmos*. London, 2019, pp. 62–4.

Figure 2.1 Halley map, collection of Jay and Naomi Pasachoff, on deposit at the Chapin Library, Williams College, MA.

historical and contemporary witness in this volume testifies, the sky darkening by a factor of about a million is a dramatic event. Less attention was paid to the actual appearance of the eclipse. It has been interpreted that the Chinese oracle-bones saying translated as 'three flames ate up the Sun, and a great star was visible' from 1307 BC could have been a mention of the Sun's atmosphere,[3] perhaps the corona or maybe a phenomenon known as 'prominences,' which extend above the solar limb.

Prominences (see Figure 2.2) are cooler gas than the corona; they match the 10,000 °C temperature of the solar chromosphere. At that temperature, the strongest colour is from the spectral line known as H-alpha, a bright red (see Figure 2.3). In addition to the relatively spiky but uniform-height chromosphere that surrounds the Sun, with hundreds of thousands of 'spicules' rising and falling with 15-minute periods, the prominences are more stable and can last weeks, held aloft by the Sun's small-scale magnetic-field enhancements.

The origin of the term 'corona' is unclear, but the word's Latin origin (first recorded in 1555–65 from *corona*, garland, crown) makes the term unusual. Other parts of the Sun, named later, have a Greek origin: 'photosphere' for the everyday surface (first recorded in 1655–65) and 'chromosphere' (first recorded in 1865–70) for the reddish rim visible during an eclipse just before the corona appears and just after it disappears.[4]

A significant early mention of the corona is thought to be that of Johannes Kepler's *Paralipomena to Witelo & Optical Part of Astronomy*, known as the *Optics* (1604, translated into English by Donahue, 2001, 2021). But though a ring of brightness around the Moon is noted, no shape was commented upon; at that time positions of celestial objects were noted but there was less interest in their physical characteristics.

Figure 2.2a A photograph, using the red H-alpha radiation from hydrogen, well known by solar astronomers, of a prominence that lasted stably for over a solar rotation. Its delicate structure reveals the magnetic field. (Big Bear Solar Observatory, then California Institute of Technology, now of the New Jersey Institute of Technology.)

---

[3] See https://eclipse.gsfc.nasa.gov/SEhistory/SEhistory.html; https://www.cam.ac.uk/research/news/oldest-recorded-solar-eclipse-helps-date-the-egyptian-pharaohs.

[4] See Leon Golub and Jay M. Pasachoff, *The Sun*. London, 2017.

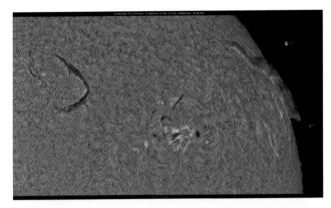

Figure 2.2b This image, falsely coloured in the reddish H-alpha from its original black-and-white imaging, captures a filament on 23 June 2022 just as some of it is rotated over the limb, making it show as a prominence. Note that the brightness of the filament/prominence itself as a single entity remains the same on disk (silhouetted against the photosphere) as it is over the solar limb (silhouetted against black sky). *Source:* Image by Martin Wise.

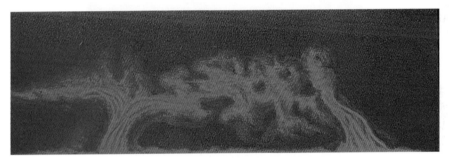

Figure 2.2c A prominence from *Die Sonne* (1872) by Fr. Angelo Secchi.

It was not known if the corona was part of the Sun or of the Moon, with many people assuming the latter. It was only in the nineteenth century that observations of the corona were simultaneously made from locations far enough separated on the European continent to allow parallax to indicate that the corona is a solar rather than a lunar feature. It took the invention of the portable spectroscope, taken by the French astronomer Jules Janssen to the 1868 total eclipse in India, to enable the discovery that the chromosphere and prominences were bright enough to be seen at other times as well as during total eclipses. Only with the 1871 eclipse was the corona definitively recognized as a solar phenomenon.[5]

[5] Megan Briers, Mixie Billina, and Deborah Kent, 'Chasing Change: The Lasting Legacy of India's 1871 Eclipse.' *Astronomy and Geophysics* 53, issue 1, February 2022, pp. 1.24–1.29.

Figure 2.3 Prominences at the 2015 total solar eclipse, observed from the Williams College Expedition to Svalbard in a spectral line of hydrogen. The red line that is the strongest in hydrogen's Balmer Series, representing the brightest colour that appears in the visible part of the spectrum. (An even stronger line appears in the Lyman series, but it and its sister spectral lines appear in the ultraviolet, at wavelengths that do not come through Earth's atmosphere, and can be studied only from spacecraft.)

Spectroscopy and the Million-Degree-Hot Corona

To the naked eye, the corona visible during a solar eclipse is pearly white. But to understand its physical structure, it is necessary to examine the light carefully across its spectrum. Isaac Newton, in the late seventeenth century, realized that sunlight could be decomposed by a prism into a rainbow of colour and then reassembled by a second prism. The nineteenth century saw tremendous advances in our spectral understanding of light and optics. The optician Joseph Fraunhofer in 1814–15 (published 1817) had improved on earlier spectroscopy by using a vertical, narrow slit to allow just a line of sunlight to enter his equipment. He discovered what we still call the Fraunhofer spectrum, with 574 parallel dark lines interrupting the solar rainbow (see Figure 2.4). He published his result in 1817, with the labels A through to H for

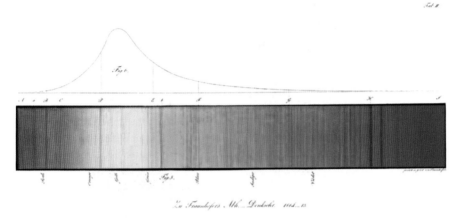

Figure 2.4 Fraunhofer's 1817 publication (colourised) of his 1814 spectrum of the Sun through a narrow slit. The intensity profile of the solar radiation also shows, notably peaking in the yellow, a part of the visible spectrum often assigned a filter labelled V for 'visible', along with RGB for red green blue, part of the ROY G BIV set of visible colors, where that V stands for violet.

Source: Collection of Jay and Naomi Pasachoff, on deposit at the Chapin Library, Williams College, MA.

the strongest (darkest) of the spectral lines, and with I for the end of his drawing. (The notation K was added decades later to label the obvious H and K pair.)[6]

Fraunhofer's discovery was a diagram, not a set of equations. In the 1880s, the Swiss mathematician and teacher Johann-Jakob Balmer realized that the frequencies (the inverse wavelengths) of a handful of the strongest spectral lines could be linked by a very simple formula: ($1/2^2-1/3^2$, $1/2^2-1/4^2$, $1/2^2-1/5^2$, and $1/2^2-1/6^2$) revealing an underlying simplicity in the physics, extending spectroscopy beyond mere mapping of spectral lines. In 1913, Niels Bohr realized that we can conceive of the hydrogen atom's nucleus (a single proton) surrounded by electrons on energy levels, and that, just as you can't hover between stairs on a staircase, the electrons can jump to other energy levels but cannot take intermediate values. Atomic systems with more nuclear particles and/or more electrons have more complex spectra.

By the time of the total solar eclipse of 1868, spectroscopic devices were much improved, so that the solar atmosphere could be studied during an eclipse. Jules Janssen, as mentioned,[7] took his new spectrograph to Guntoor, India to view the eclipse. During totality, when the Fraunhofer lines disappeared because the photo-sphere was covered, he was able to see several bright chromospheric lines shining out—so-called 'emission lines'—and realized that the brightest yellow line did not

---

[6] Jay M. Pasachoff and Terry-Ann Suer, 'The Origin and Diffusion of the H and K Notation.' *Journal of Astronomical History and Heritage* 13(2), 2010, pp. 121–7.

[7] See F. Launay, 'Observations of the total solar eclipse of 18 August 1868 carried out by Jules Janssen at Guntoor, India.' *Journal of Astronomical History and Heritage* 21, 2021, pp. 114–24. See also Roberta Olson's Chapter 10.

quite coincide with the location of the pair of yellow lines known to come from sodium (and which you can see if you toss salt—NaCl—into a flame).[8] Further, he realized that the emission lines were so bright that they could potentially be seen outside of eclipse, which he did the next day.

The English scientist Norman Lockyer had been ill and so had not travelled to the eclipse, but when his own spectrograph was delivered a few months later, it empowered him to collaborate with the chemist Edward Franklin to observe the emission lines from the chromosphere, the lowest layer of the solar atmosphere.[9] Lockyer and Franklin reported that the brightest-yellow line came from a new element, which they termed 'helium' since it appeared only on the Sun (see Figure 2.5). Decades would pass until, in 1895, chemists isolated the element helium on Earth. This was in spite of the simplicity of helium as next in the periodic table to hydrogen—the atomic nucleus of helium contains two protons to hydrogen's one. The topic is well discussed by Biman B. Nath,[10] who emphasizes the great importance of the discovery of helium. In the first three minutes after the Big Bang some 13.5 billion years ago, only hydrogen (including a trace of deuterium—heavy hydrogen) and a little helium and lithium

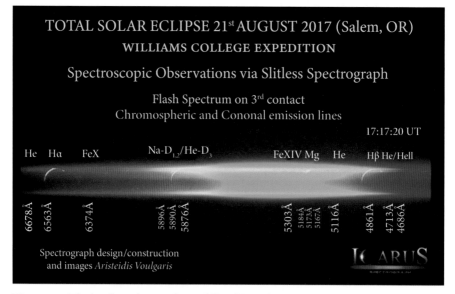

Figure 2.5 The brightest line in the middle of the solar spectrum at the end of totality is from helium and is brighter than the nearby pair of sodium lines; a sequence of basic hydrogen lines is also bright.

---

[8] Recent research indicates that the English astronomer Norman Pogson observed and commented on the same lines. B. B. Nath and W. Orchiston, 2021. 'Norman Robert Pogson and Observations of the Total Solar Eclipse of 1868 from Masulipatam, India.' *Journal of Astronomical History and Heritage* 24, pp. 629–51.
[9] See https://www.youtube.com/watch?v=RZfCqWZ8EAY.
[10] Biman B. Nath, *The Story of Helium and the Birth of Astrophysics*. 2013.

were formed; the bulk of the helium was formed from nuclear fusion inside stars.[11] Understanding that helium is a primordial element alongside hydrogen is at the base of our understanding today of the evolution of the elements and the evolution of stars.

On August 7, 1869, a total solar eclipse crossed the United States.[12] Charles Augustus Young from Dartmouth College and William Harkness from the US Naval Corps observed a green spectral line (see Figure 2.6). By analogy to Lockyer's and Franklin's helium, Young and Harkness speculated that it came from yet another undiscovered element, 'coronium.' But though helium indeed turned out to be a chemical element, coronium did not, as Mendeleev's periodic table got filled in.[13]

It took until 1940 to identify the elemental source of Young's and Harkness's 'coronium' lines, when the Swedish physicist Bengt Edlén made a series of observations of chemical elements increasing in their basic mass parameter but with the same number of outer electrons leading to the spectrum. He could extrapolate along the 'isoelectronic sequence' to see that the 'coronal green line' came from 13-times-ionized iron. Neutral iron is known as Fe I, so 13-times-ionized iron is Fe XIV. Iron is normally of 26 nuclear particles and so, when neutral, has 26 electrons, so retains 13 electrons when 13 escape. Magnesium, normally of 25 nuclear particles also retains 13 when 12 are taken away, but can be experimented with optically more readily than iron. It was easier to use lower-mass elements in spectral analysis, with the same number of retained electrons in their outer shells, as for the more highly ionized

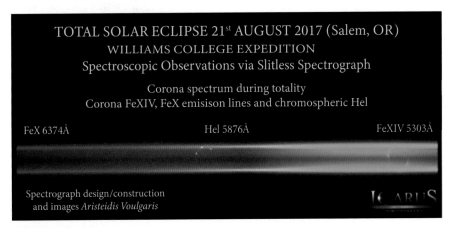

Figure 2.6 My team obtained spectra from the top of the Cerro Tololo Inter-American Observatory of the 2017 eclipse, using a spectrograph designed and operated with us by Aristeidis Voulgaris, Icarus Optomechanics. This display is cropped on either side of the coronal red and green lines; other spectral lines also show.

---

[11] Jay M. Pasachoff and William A. Fowler, 'Deuterium in the Universe.' *Scientific American* 23(5) (May 1974), pp. 108–18; reprinted 1988, *Particle Physics in the Cosmos*, with a new introduction.

[12] Thomas Hockey, 'Iowa in Eclipse,' *Astronomy and Geophysics*, October 2017, p. 5.10; Thomas Hockey, *America's First Eclipse Chasers*. With foreword by Jay M. Pasachoff, 2023.

[13] The Russian chemist Dmitri Ivanovich Mendeleev devised a two-dimensional way of displaying the chemical elements in which those with different chemical properties had their own column; there were originally some gaps, but they got filled in with no room for 'coronium.'

iron. The twentieth-century scientists, especially Bengt Edlén, moved from element to element experimentally in this way, extrapolating to the spectrum of iron XIV. Indeed, that transition is not favoured in atomic physics (that is, has a low 'transition probability') and is therefore called 'forbidden'; with square brackets to indicate forbidden lines, we say that the coronal green line is from [Fe XIV]. The unavoidable consequence of the presence of such a highly ionized species in sufficient quantity to produce the bright spectral line was an extremely high coronal temperature. For to ionize a gas it must be heated, and to ionize it several times it must be heated even more. So to reach iron-XIV, the gas must be at millions of kelvins![14] Further evidence of the extremely high temperature of the corona was supplied at that time by the realization by W. Grotrian that the absence of Fraunhofer spectral lines indicated high Doppler shifts in the corona, also corresponding to million-degree temperature. H. Peter and Bhola N. Dwivedi concluded in 2014 that much of the credit for realizing that the temperature is so high should be given to Hannes Alfvén rather than Grotrian and Edlén, who although they had many of the facts in hand, did not reach the conclusion.[15]

We do not have space here for a full explanation of atomic spectra and their history, but it is helpful to cite a similarly misleading case where the propinquity of hundreds of lines from iron in the Sun's Fraunhofer spectrum misled most astronomers to assume that the Sun was therefore mainly made of iron. When young Cecilia Payne (later Payne-Gaposchkin) in 1929 produced her Radcliffe thesis (at Radcliffe College because the President of Harvard refused to allow a female to be awarded a Harvard PhD) purporting to show that the Sun was almost entirely made of hydrogen, she was doubted by almost everyone, and most publicly by Henry Norris Russell of Princeton. But within a couple of years, and some measurements by Donald Menzel at Harvard, Payne was shown to be correct. Her result is now universally accepted.[16] Subsequently, a coronal red line was seen at an eclipse, and found to come from iron-X (as above 'iron-ten'—an iron atom that has lost ten of its electrons). Roughly, Fe X comes from gas at about a million kelvins, while Fe XIV comes from gas at about 1.4 million kelvins.

To conclude this section, we should return to consider the most immediately visible feature of the solar corona, its beautiful pearly white colour. With the hindsight of modern physics, we can explain this by two mechanisms. First, the electrons which have been stripped from the ionized atoms in the corona are free to move within the corona. Sunlight from the photosphere scattering off these electrons produces light of all colours, combining to form a white-light spectrum. We call this the 'K-corona' (K standing for 'kontinuierlich,' German for 'continuous'). Another contribution, further from the Sun, comes from sunlight which illuminates particles of the solar wind within the solar corona; again, no colour is favoured, and so this 'F-corona' (F for 'Fraunhofer') is also white in appearance. The F-corona extends a long way from the

---

[14] We can ignore the 273 °C difference between the kelvin temperature scale and the Celsius temperature scale for our purposes.
[15] See https://www.frontiersin.org/articles/10.3389/fspas.2014.00002/full.
[16] See Donovan Moore, *What Stars Are Made Of: The Life of Cecilia Payne-Gaposchkin*, 2020. See also Jay M. Pasachoff, 'On the stature of Cecilia Payne-Gaposchkin.' *Physics Today* 73(11), 2020.

Sun, eventually becoming the 'zodiacal light' which is visible close to the Sun just before dawn and just after sunset, given clear skies and no light pollution.

The Coronal Heating Problem and Coronal Structure

The measurement of the temperature of the gas is quite different from theoretically finding out what causes the coronal heating. After all, the everyday surface of the Sun, the 'photosphere' that we see when we look at the Sun and that gives off the light and heat that the Earth receives 93 million miles (150 million kilometres) away at its surface, possesses a temperature of 'only' 6000 °C. So, unlike the situation on Earth of moving away from a fire and getting cooler, as we move away from the solar photosphere into the corona, the temperature rises. (Note, however, that the density of the gas in the corona is very low, so the actual amount of stored energy is not great. It is now being sampled directly by NASA's Parker Solar Probe and the European Space Agency's Solar Orbiter spacecraft, both of which are closer to the photosphere than any previous probe.) This finding of a very high-temperature corona is a surprise, and the consequent 'coronal heating problem' remains a major one in astrophysics and applies not only to our Sun's corona but also to the corona around billions upon billions of other stars.[17]

Observations of the solar corona at an eclipse usually record 'coronal streamers' emerging from low latitudes on the Sun. These are flame-shaped structures of the high-temperature plasma, with wide bases against the solar edge, narrowing as they ascend. A second class of structure frequently observed (when they are not blocked by streamers) are 'plumes' ascending from the north pole and descending from the south pole, in a similar way to the pattern of iron filings near a bar magnet. These structures represent coronal gas held in shape by the Sun's magnetic field (another quantity measurable through careful observation of spectral lines—in this case their subtle 'splitting'). Charged particles, including ionized nuclei, can move easily along the magnetic field lines but cannot cross them, so our view of streamers at total solar eclipses reveals the pattern of the solar magnetic field outside the solar limb.

The average magnetic field on the solar surface is about the same strength as that of the Earth, the field that makes compasses point north–south. However, the solar magnetic field is typically a thousand times stronger in the regions that make up the darker photospheric 'sunspots' readily observed by small telescopes, and the larger ones even by the (suitably shaded) naked eye. The magnetic field that comes through the solar surface (photosphere) to make sunspots is generated in tubes that circulate underneath, and sometimes become pinched in a way that pokes the magnetic field through the surface, with one polarity while going up and the opposite while going back inside (N vs S, or + vs −).

The number of sunspots, and the magnetic field on the Sun's surface and in the corona, vary over a 'sunspot cycle,' more generally a 'solar cycle,' that lasts

---

[17] Jay M. Pasachoff, Jeffrey L. Linsky, Bernhard M. Haisch, and Albert Boggess, 'IUE and the Search for a Lukewarm Corona.' *Sky and Telescope* 57(5) (May 1979), pp. 438–43, https://www.researchgate.net/publication/4716590_IUE_and_the_search_for_a_lukewarm_corona.

approximately 11 years.[18] Early in each cycle, sunspots form at higher solar latitudes, then as their number and average size increases, they move to lower latitudes. Strictly, the solar cycle is 22 years, as the magnetic polarity of the common sunspot pairs switches between each 11-year cycle.[19]

The eclipse expeditions of the present author often look for vibrations on narrow magnetic loops of gas visible low in the solar corona. These magnetohydrodynamic waves (the 'hydro' part came from the basic physics that was learned in the nineteenth century in studying the flow of water) have long been looked at as a possibility for coronal heating. Known as Alfvén waves (after the 1970 Nobel prize winner Hannes Alfvén for his work on magnetohydrodynamics), they have vibrational periods of 30 seconds or so. My own team and students have been pursuing, with some success, 'surface Alfvén waves,' with sub-second periods. But there are over a dozen alternative explanations of coronal heating that have been proposed, such as 'nanoflares,' explosive events with higher frequency but less energy per event than ordinary solar flares, which reach temperatures of tens of millions of degrees—and whose frequency is linked to that of the solar-activity cycle.

Janssen discovered that the shape of the solar corona changed with the sunspot cycle[20]—that is, at the peak of sunspots, the streamers emerge from so many solar latitudes that the solar corona appears approximately circular. But at the trough of the sunspot cycle, most coronal streamers have origins within ±30° or so of the solar equator, so the corona is flattened by perhaps 30°.[21] Examples of both typical structures are given in Figure 2.10.

My team now monitors the shape of the corona at all the total solar eclipses (see Figure 2.7), most recently in 2017 in the United States, in 2019 in Chile/Argentina, in 2020 again in Chile/Argentina, and on 4 December 2021, from an aeroplane flying near Antarctica, east from Punta Arenas, Chile, with a coordinated team on the ice at Union Glacier, Antarctica (see Figure 2.8).

The Future of Coronal Observation

As we continue our ascent from the 2018–19 solar sunspot minimum, when most days showed no sunspots, to the 2024–5 sunspot maximum, the positions of coronal streamers viewed during totality will continue to distribute themselves evenly around the solar silhouette. In 2017, though, closer to sunspot minimum, the streamers were concentrated towards the solar equator (see Figures 2.7 and 2.10).[22] The study of 'space weather' has practical consequences (see http://spaceweather.com—new

---

[18] See graphs at https://www.sidc.be/SILSO/home.

[19] See Jay M. Pasachoff, Daniel B. Seaton, and Kevin P. Reardon 'Quick Study: Sunspots and Their Cycle.' *Physics Today* 76(2), 2023, pp. 54–5.

[20] https://www.sidc.be/SILSO/home.

[21] Pasachoff and Rušin 2022 updated a graph of the so-called Ludendorff flattening coefficient to include eclipses through the 2017 Great American Eclipse. See Jay M. Pasachoff and Vojtech Rušin, 'White-Light Coronal Imaging at the 21 August 2017 Total Solar Eclipse.' *Solar Physics* 297(3), 2022, article id.28. See also Jennifer Birriel and Joseph Teitloff, 'Solar Coronal Flattening during the Total Solar Eclipse of August 2017 from CATE Data.' *AAVSO Journal* 50, 2022, p. 252, at https://app.aavso.org/media/jaavso/3840_tWxdsiS.pdf.

[22] From Jay M. Pasachoff and Vojtech Rušin, 2022, 'White-Light Coronal Imaging at the 21 August 2017 Total Solar Eclipse.' *Solar Physics* 297(3), article id.28.

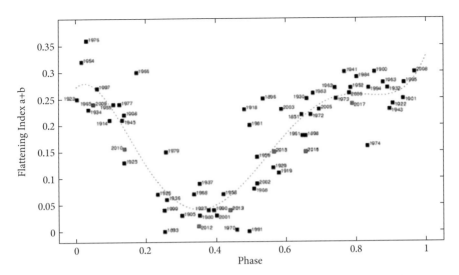

Figure 2.7 The Ludendorff flattening coefficient over the last hundred years, showing the deviation from round of the coronal overall shape at a distance of a solar radius above the solar limb, with the latest observations in red.

Figure 2.8 My photo out of the window of a charter flight due east from Punta Arenas, Chile, on 4 December 2021. The dark umbral shadow and bright atmospheric areas outside of totality are both visible. See Jay M. Pasachoff, 'Totality From Above the Clouds.' *Sky and Telescope*, 2021/22, online.

every day to see the sunspots and other solar-related situations), since the number of coronal mass ejections of material is linked to the solar-activity cycle. A coronal mass ejection hitting the Earth could severely damage many of the satellites in orbit and even lead to outage on power lines, for which there was a precedent of several hours of outage in Quebec in 1989 with subsidiary problems over the border in the United States.[23]

The US National Solar Observatory has built the Daniel K. Inouye Solar Telescope (DKIST) in Hawaii at the 10,000-foot (3,000-metre) altitude of the top of Haleakala Crater.[24] It began science observations in February 2022. With its four-metre off-axis main mirror, it uses sensors through the whole visible range and into the infrared, covering through the 28 μm infrared limit that is also the limit for the James Webb Space Telescope. It has an occulting disk (to be used together with a 'Lyot stop,' after the French scientist who, about 100 years ago, invented the 'coronagraph' that enabled the solar corona to be seen without an eclipse, using ultra-clean lenses instead of mirrors) that can be placed in the beam to hide the solar photosphere in order to observe the solar corona. Regular cleaning of DKIST's mirrors with carbon dioxide and water limits scattering by dust on the mirrors. The high resolution obtainable with its four-metre mirror is used to see the finest detail that shows the motion and evolution of magnetic fields in the inner corona, a resolution of only 70 km or 0.1 arc second, obtainable at 1.5 μm in the near-infrared spectrum.

Preliminary images from the 2021 engineering test phase of the DKIST are shown in Figure 2.9. A sunspot (a), with its dark umbra surrounded by its filamentary penumbra, is set upon a background of solar granulation. An extreme closeup of solar granulation (b) reveals a convective (boiling) phenomenon, with hot globules rising and showing as Texas-sized bright regions with cooled gas falling back down and showing as dark inter-granular lanes.

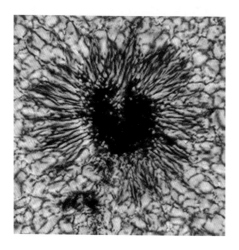

Figure 2.9a High-resolution images from the new Daniel K. Inouye Solar Telescope. (a) Sunspot with umbra and penumbra.

[23] https://www.nasa.gov/topics/earth/features/sun_darkness.html.
[24] See https://nso.edu/telescopes-3/dki-solar-telescope/.

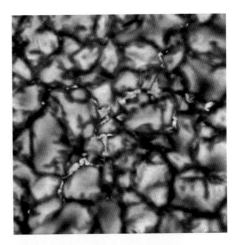

Figure 2.9b (b) Close-up of surface granulation.

Figure 2.10a White-light corona typical of a solar minimum (21 August 2017) by the Williams College Eclipse Expedition at Salem, Oregon. Processed by R. (Vanür) Hubčík working with V. Rušin.

The Parker Solar Probe will make its closest approach to the Sun in 2024, but even at the time of writing, it is making discoveries that increase our knowledge of the solar corona, detailed in the mission blog on the NASA website.[25] It made the first crossing by a spacecraft of the Alfvén critical surface, where material escapes from the Sun to become the solar wind, and found that it was not smooth but had spikes and valleys. It detected the first hints of a predicted dust-free zone, close to the Sun, where any dust particles will sublimate. One unexpected discovery was that of 'switchbacks' in

[25] See https://blogs.nasa.gov/parkersolarprobe/2022/04/29/amazing-achievements-from-parker-solar-probe/.

Figure 2.10b White-light corona typical of a solar maximum (11 August 1999).
Processed by Wendy Carlos.
Source: Jay. M. Pasachoff.

the solar magnetic field, aligned with magnetic funnels in the solar surface, which
emerge between the solar granules. More discoveries are anticipated.

Coronal studies: A Forever Legacy of Total Solar Eclipses

In this chapter, we have seen how scientific focus has changed over history from the
prediction and timing of eclipses to the astrophysics that they can reveal, principally
looking at the corona. We have further seen how this emphasis on astrophysics initially created complete ignorance of how many billions or trillions of other stars have
coronas or how our own Sun's—and presumably other stars'—corona is heated to
millions of degrees. There have been a dozen or more theories proposed to explain
how that surprisingly high temperature, discovered only about 80 years ago, might
be generated. Many fundamental questions remain but might become answerable
eventually from further detailed observations of total solar eclipses.

  The approaching sequence of total solar eclipses is expected to lead to significant advances in scientific astrophysical problems such as the heating of the solar
corona. Solar eclipse observations over the next few years can complement the coronal observations being made with the Parker Solar Probe and Solar Orbiter in orbit,
and the mountain-based Daniel K. Inouye Solar Telescope. Further, aligning spacecraft to a high precision, with the occulter sufficiently far from the cameras to keep it
in infinity-focus, constitutes part of the European Space Agency's PROBA-3 ASPI-
ICS (the Association of Spacecraft for Polarimetric and Imaging Investigation of

the Corona of the Sun) programme from 2023. The coronal image will be occulted only above $1.057\,R_{\odot}$ (1.057 solar radii) and be unvignetted above $1.17\,R_{\odot}$, providing imaging of the lowest corona, far below the $1.7\,R_{\odot}$ that has been available from the externally occulted C2 and C3 coronagraphs by the Large Angle Spectrometric Coronagraph (LASCO) systems of the US Naval Observatory on board the European Space Agency's Solar and Heliospheric Observatory, launched in 1995. Such advanced terrestrial and space imaging could even diminish the scientific need for coronal imaging at eclipses. However, the desirability of personal solar eclipse observation for scientists and the public will remain;[26] total solar eclipses will continue to provide inspiration to students and artists of all ages.

[26] S. V. Shestov, A. N. Zhukov, B. Inhester, L. Dolla, and M. Mierla, 'Expected Performances of the PROBA-3/ASPIICS Solar Coronagraph: Simulated Data.' *Astronomy and Astrophysics* 652, 2021, A4.

3

Pre-Modern Astronomies of Eclipses in the Near East and Europe

C. Philipp E. Nothaft

European Intellectual History (Medieval and Early Modern); History of Science (Medieval and Early Modern); History of Astronomy, Chronology, Time-Reckoning

While the study of the solar corona is barely two centuries old, and the study of its physics even more recent, the prediction of eclipses goes back much further in time. A wide variety of theories and techniques put ancient and medieval astronomers in a position to foretell the occurrence and appearance of total solar eclipses. Geographically and chronologically, the following chapter ranges from the ancient Near East to Renaissance Europe, placing a particular focus on the status of eclipses in Hellenistic Greek and later Arabic astronomy and the transmission of the relevant knowledge and tools to Latin Europe during the Middle Ages.

Antikythera Mechanism A geared mechanism for predicting astronomical events, dating from antiquity, found in a shipwreck off the Greek island of Antikythera in 1901.

apsides The plural of apsis (apse), the nearest or furthest point in the orbit of a planetary body from the primary body it is orbiting around. The apsides of the Moon's orbit around Earth are its perigee and apogee.

Archimedes of Syracuse (*c*.287–*c*.212 BC) Ancient Greek mathematician and physicist.

Aristotle (384–322 BC) Ancient Greek philosopher and polymath.

Byzantine empire The Christian Roman empire during late antiquity and the Middle Ages, ruled from Constantinople.

camera obscura Latin for a 'dark chamber,' into which the outside world (e.g. a solar eclipse) is projected through a pinhole, similar to a pinhole camera.

conjunction when the Moon has the same elongation as the Sun. This is when a solar eclipse can occur.

draconic/draconitic month The time between successive passages of the Moon through one of its nodes.

epicycle A mathematical construction to explain (approximately) the apparent motion of planets as a set of circular motions.

evection The largest inequality (variation of longitude) in the Moon's orbit, caused by the gravitational attraction from the Sun.

Newton, Sir Isaac (1642–1727) English mathematician and physicist.

nodes The two points in the Moon's orbit where the Moon crosses the ecliptic. Eclipses can only occur when the Moon is at a node.

opposition When the Moon has the opposite elongation to the Sun (i.e. it is on the opposite side of the celestial sphere). This is when a lunar eclipse can occur.

parallax The apparent difference in position of an object observed from two different locations.

Ptolemy (*c.*100–*c.*170) Ancient Roman mathematician, astronomer, and geographer.

quadrature When the Moon's elongation is perpendicular to the Sun (popularly known as first or last quarter of the Moon).

saros A period of just over 18 years, after which eclipses often repeat.

volvelle A wheel chart, a paper construction with rotating parts.

In October of 1084, an embassy of 150 Pecheneg warriors arrived at a Byzantine army camp at Lardea, south-eastern Bulgaria, where the young Emperor Alexios I Komnenos (*c.*1048–1118) was busy concentrating troops for a military campaign against their tribe. News of this diplomatic mission came at an inconvenient time for Alexios, who was determined to drive the Pechenegs back across the Danube and thereby restore control over his empire's northernmost provinces. Unwilling to negotiate a truce, Alexios was looking for an excuse to refuse the envoys. According to his daughter Anna, who gives a detailed account of the campaign in her *Alexiad* (*c.*1148), the saving idea came from one of Alexios's under-secretaries, a man named Nikolaos. We are told that he 'approached the emperor and gently whispered in his ear, "Just about this time, Sir, you can expect an eclipse of the sun".' The emperor was quick to seize on this information, telling the Pechenegs that he needed divine reassurance.

> 'The decision,' he said, 'I leave to God. If some sign should clearly be given in the sky within a few hours, then you will know for sure that I have good reason to reject your embassy as suspect, because your leaders are not really negotiating for peace; if there is no sign, then I shall be proved wrong in my suspicions.'

Anna tells us that the Pechenegs reacted with amazement when, within the next two hours, 'the whole disc of the sun was blotted out as the moon passed before it.'[1] Her description is not free of hyperbole considering that the solar eclipse of 2 October 1084,[2] as seen from Lardea, would have reached a magnitude of only around 0.85. Depending on the circumstances, such as the weather and time of day, this could have been insufficient to make an impression. At any rate, the emperor's ruse did nothing to save the Byzantine army, which suffered a crushing defeat in the subsequent battle.[3] This unsuccessful outcome aside, Anna Komnene's tale of her father's

---

[1] Anna Komnene, *The Alexiad*, VII.2, trans. E. R. A. Sewter. Harmondsworth, 1969, p. 221.

[2] The date is established in Konradin Ferrari-d'Occhieppo, 'Zur Identifizierung der Sonnenfinsternis während des Petschenegenkrieges Alexios' I. Komnenos (1084).' *Jahrbuch der österreichischen Byzantinistik* 23, 1974, pp. 179–84.

[3] Anna Komnene, *The Alexiad*, VII.3.

use of astronomy to manipulate his barbarian opponents is noteworthy for the message of cultural superiority it conveys. What a Pecheneg warrior from the Eurasian steppe may have viewed as an awe-inspiring or frightful celestial portent, an educated person from Constantinople could appreciate as a regularly occurring natural phenomenon. Byzantine astronomers not only understood the basic mechanics of solar eclipses, but they could even accurately predict their occurrence.[4] It was an ability that signalled high civilizational attainment, even besides the strategic advantages that might be drawn from it.

To know the date and time of a solar eclipse, not to mention other circumstances such as its duration and magnitude, ahead of its actual appearance is quite possibly the most complex scientific problem that pre-modern cultures found successful ways to address. The solution to this problem did not originate in the Byzantine Empire, nor did it present itself overnight. Rather, it was the result of work carried out by individuals and entire groups of thinkers in different societies existing centuries and millennia apart.[5] The present chapter traces some select aspects of this long history, by focusing on the contributions of astronomers in Mesopotamia and Europe—especially Greece—towards our scientific understanding of solar eclipses and the patterns of their recurrence (for which see Chapter 1). It also deals with the transmission of their insights to the Latin-speaking world of Western Europe during the Middle Ages, which will set the stage for some of the themes explored in later chapters (see esp. Chapters 6 and 8–10). This is not to detract from the important history of human engagement with eclipses outside the immediate European context, which will be explored by John Steele in Chapter 5.

In Babylon and other areas of the ancient Near East, the study of astronomy was wedded to celestial divination, as members of a priestly class were tasked with watching the skies for omens in order to forestall or mitigate their harmful effects. Eclipses played a key role in this activity, given their status as heralds of doom that could portend events as drastic as the death of a reigning king.[6] This belief in the ominous nature of eclipses in turn provided an impetus for keeping records of their appearance. Over time, habits of observation and record-keeping fostered attempts to formulate rules for their prediction. We find some early examples of such rules in the extensive omen collection known as *Enūma Anu Enlil*, parts of which date from the

---

[4] Detailed predictive calculations for solar eclipses have been preserved from the fourteenth century. See Barlaam of Seminara, *Traités sur les éclipses de soleil de 1333 et 1337*, ed. Joseph Mogenet and Anne Tihon. Leuven, 1977; Nicephoras Gregoras, *Calcul de l'éclipse du soleil du 16 juillet 1330*, ed. Joseph Mogenet et al. Amsterdam, 1983.

[5] The two foremost publications on this subject are John M. Steele, *Observations and Predictions of Eclipse Times by Early Astronomers*. Dordrecht, 2000, and Clemency Montelle, *Chasing Shadows: Mathematics, Astronomy, and the Early History of Eclipse Reckoning*. Baltimore, MD (2011). I shall draw on them heavily in what follows. See also Johannes Thomann, 'Sonnenfinsternisse in der Geschichte der Wissenschaft und Technik.' In Johannes Thomann and Matthias Vogel, *Schattenspur: Sonnenfinsternisse in Wissenschaft, Kunst und Mythos*, Basel, 1999, pp. 9–62.

[6] See, for example, Francesca Rochberg-Halton, *Aspects of Babylonian Celestial Divination: The Lunar Eclipse Tablets of Enūma Anu Enlil*. Horn, 1988, and 'Ina lumun attali Sîn: On Evil and Lunar Eclipses.' In *Sources of Evil: Studies in Mesopotamian Exorcistic Lore*, ed. Greta Van Buylaere, Mikko Luukko, Daniel Schwemer, and Avigail Mertens-Wagschal, Leiden, 2018, pp. 287–315; Paul-Alain Beaulieu and John P. Britton, 'Rituals for an Eclipse Possibility in the 8th Year of Cyrus.' *Journal of Cuneiform Studies* 46, 1994, pp. 73–86.

Old Babylonian Period (*c.*2000–1600 BC). In the extract quoted below, atmospheric conditions observed at the start of a particular lunar month are used to predict the occurrence of a solar eclipse on one of the last three days of a given lunar month.

> If the sun is red like a torch when it becomes visible on the first of Nisannu, and a white cloud moves about in front of it [. . .] and the east wind blows: in Nisannu [. . .] on the 28th, 29th or 30th an eclipse of the sun will take place and during that eclipse [. . .] in that month the king will die and his son will seize the throne [. . .].[7]

The suggestion, made in this source, that solar eclipses can occur on the 28th, 29th, or 30th day is appropriate for a lunar calendar that makes the month begin with the first visibility of the new-moon crescent, which tends to be observed a day or more after the last conjunction. Accordingly, the next conjunction, and with it the potential next solar eclipse, will occur at the end of the month.

A major early breakthrough towards a predictively successful theory of such phenomena was achieved once Babylonian astronomers had noted patterns in the intervals separating individual eclipses of the Sun and Moon (see Chapters 1 and 5). In the vast majority of cases, the observed time between consecutive lunar eclipses will be six months or some multiple thereof, yet there are occasional exceptions where the gap is shortened by one month. Further observations as well as mathematical considerations eventually alerted these astronomers to the existence of a larger pattern of 38 eclipses spread over 223 consecutive months or approximately 18 years, 11 days, and 8 hours (6,585 1/3 days). Babylonian use of this eclipse period, which is known as the 'saros' to modern historians and astronomers, may go back to the earliest systematically recorded eclipse observations, which commence in the mid-eighth century BC.[8] It was certainly in use by the sixth century BC, as seen from a cuneiform list of lunar eclipses (LBAT *1420) covering the first 29 years of the reign of Nebuchadnezzar II (604–576 BC). In the case of solar eclipses, saros-based methods of prediction were limited to pinpointing *possibilities* of eclipses, as Babylonian astronomers lacked the means of accounting for the geographical location of the observer as well as for the effects of parallax, which made the apparent position of the Sun or Moon contingent on latitude.[9] Without controlling for these and other components, it remained impossible to predict reliably whether a solar eclipse was actually going to be visible from a specific locality such as Babylon.

The Babylonian practice of predicting eclipses or eclipse possibilities based on the saros continued with relatively little modification until at least the first century AD. A spectacular witness to its inculturation into a Greek context is the famous Antikythera mechanism, a gear-driven astronomical calculator whose remains were recovered from a shipwreck in the Aegean Sea in 1901. Created at some point between the end of the third and the middle of the first century BC, the boxed mechanism drove a set of dials with revolving pointers that tracked astronomical and calendrical phenomena.

[7] Wilfred H. Van Soldt, *Solar Omens of Enuma Anu Enlil: Tablets 23(24)–29(30).* Istanbul, 1995, p. 5.

[8] On these observational records, see Peter J. Huber and Salvo De Meis, *Babylonian Eclipse Observations from 750 BC to 1 BC.* Milan, 2004. For a discussion of the saros cycle, see Chapter 1.

[9] For a discussion of parallax, see Chapter 1.

The lower half of the backplate was taken up by a spiral-shaped dial representing a full saros of 223 lunar months (Figure 3.1). To each of these months was assigned a separate cell, which in the case of an eclipse month carried an inscription composed of small glyphs and an index letter. The glyphs were there to differentiate between a lunar and a solar eclipse as well as to indicate its predicted hour (of day or night), while the index letter referred users to an inscription on the backplate that offered information about the eclipse's colour, magnitude, and direction of occultation.[10]

Figure 3.1 Computer reconstruction of the saros and exeligmos dials on the backplate of the Antikythera mechanism.

Source: © Tony Freeth. Originally published as Figure 11 in Tony Freeth, 'Eclipse Prediction on the Ancient Greek Astronomical Calculating Machine Known as the Antikythera Mechanism,' *PLoS ONE* 9(7): e103275.

---

[10] On the mechanism's eclipse scheme, see Paul Iversen and Alexander Jones, 'The Back Plate Inscription and Eclipse Scheme of the Antikythera Mechanism Revisited.' *Archive for History of Exact Sciences* (2019),

Placed within the main 'saros dial' was a much smaller dial, whose presence was predicated on an understanding that the length of a saros cycle does not contain an integer number of days, but instead approximates 6,583 1/3 days. By tripling this amount, one arrives at a larger eclipse cycle of 669 synodic months or just over 54 years, which was known to Greek astronomers as *exeligmos*. The small dial takes advantage of this cycle, reminding users to add eight hours to the inscribed times after one iteration of the saros and 16 hours after another iteration before everything returns to the start. Rather than relying simply on a conventional saros pattern, the precise eclipse times recorded within the cells of the main dial accounted for the variable velocity of the Sun and Moon, apparently on the basis of arithmetical schemes that were already known to Babylonian astronomers.[11]

The Antikythera eclipse dial as reconstructed from the remaining fragments recorded around ten solar eclipses fewer than the 38 possibilities contained in a conventional saros, as its creator(s) eliminated cases where the latitude of the Moon was too far south of the ecliptic. This reduction of eclipse possibilities can be interpreted as a conscious reaction to one of the effects of parallax, which at northern latitudes will make the Moon appear lower in the sky. For solar eclipses, this phenomenon will create instances where the eclipse will not be visible from a location on the northern hemisphere even though the lunar latitude remains within theoretically acceptable limits.[12] It has been suggested, moreover, that the computation of lunar latitudes carried out by the same creator(s) relied on a period relation equating 5,458 synodic months with 5,923 draconic months, which is more accurate than that implicit in the saros (223 synodic months = 242 draconic months). It, too, originated in late Babylonian astronomy.[13]

An aspect of the mechanism that seems distinctly non-Babylonian is the idea of factoring parallax into eclipse predictions, which rests on an understanding of eclipses as involving the interplay of three spherical bodies (Sun, Earth, Moon). Such an understanding was already well entrenched in Greek natural philosophy by the time Aristotle wrote his *On the Heavens*, in which he cites the curvature of the Earth's shadow during a lunar eclipse as proof of the Earth's spherical shape.[14] This spherical model paved the way towards a more profound geometrical analysis of eclipses, as it is in evidence in Aristarchus of Samos's *On the Sizes and Distances of the Sun and the Moon*. This text from the third century BC used eclipse shadows in an early effort to determine geometrically the relative sizes of the Earth, Sun, and Moon. Although Aristarchus's results were not very accurate owing to poor input data (his value for the apparent size of the Moon was about four times too large), his method constituted a major theoretical breakthrough, not least for the future of eclipse reckoning. One of Aristarchus's contemporaries, the famous Archimedes of Syracuse (286–212 BC),

pp. 469–511 (with references to further literature). See also Tony Freeth, 'Revising the Eclipse Prediction Scheme in the Antikythera Mechanism.' *Palgrave Communications* 5, 2019, article 7.

[11] Christián C. Carman and James Evans, 'On the Epoch of the Antikythera Mechanism and Its Eclipse Predictor.' *Archive for History of Exact Sciences* 68, 2014, pp. 693–774, at 722–52.

[12] Carman and Evans, 'On the Epoch,' pp. 707–19; Iversen and Jones, 'The Back Plate Inscription,' pp. 496--508.

[13] Iversen and Jones, 'The Back Plate Inscription,' pp. 508–9. The term 'draconic month' denotes the Moon's period of return to the same orbital node (see Chapter 1).

[14] See Aristotle, *On the Heavens* II.14, 297b, ed. and trans. W. K. C. Guthrie, London, 1939, pp. 252–3.

subjected the solar parallax to its earliest known mathematical analysis in a treatise known as *The Sand Reckoner*.

The most significant astronomer in the century after Aristarchus and Archimedes was Hipparchus of Rhodes (*c*.190–125 BC), whose writings have been lost for the most part. From the testimony provided by later authors, it can be inferred that he made pioneering steps towards combining Greek geometry with Babylonian parameters and period relations. Noticing the unequal durations of the four seasons, Hipparchus concluded that the Sun went around the Earth on an eccentric path, which caused its apparent speed to vary over the course of a year. He drew similar conclusions for the Moon, whose observed anomaly could be explained by having it travel on an epicycle that was itself carried by a deferent circle centred on the Earth. His specific contributions to eclipse theory include measurements of the size and distance of the Moon as well as attempts to assess the resulting parallax. In one documented case, he used two independent observations of a solar eclipse, made from different locations (Alexandria and the Hellespontine region), as the basis for establishing that the Moon's least distance from the Earth is 71 times the Earth's radius.[15] A further significant achievement on Hipparchus's part was to refine the eclipse intervals the Greeks had inherited from the Babylonians. He realized that solar eclipses could be spaced by seven-month and one-month intervals, not just by the five-month and six-month intervals assumed in the conventional saros cycle.

Hipparchus's insights and methods were developed further in the second century AD by the Alexandrian astronomer Ptolemy (*c*.100–175), whose *Mathēmatikē Syntaxis* ('The Mathematical Composition'), nowadays better known as the *Almagest* (*c*.150), in many ways represents the apex of Hellenistic mathematical astronomy.[16] One of Ptolemy's more drastic innovations concerned Hipparchus's kinematic model for the Moon, to which he added several new components. After discovering a significant discrepancy between the observed and predicted lunar position at quadrature, Ptolemy introduced an eccentric deferent with a rotating line of apsides (i.e. the diameter connecting the deferent's apogee and perigee). The resulting 'crank' mechanism (Figure 3.2) had the effect of bringing the Moon closest to the Earth when its elongation from the Sun reached 90° and 270°, while at conjunction and opposition (0° and 180°) the epicyclic centre was always found at its farthest position from the Earth. Ptolemy's discovery of a second anomaly of the Moon is linked to what is known as 'evection' in modern lunar theory, which, similar to the motion of the line of apsides and the epicyclic centre in Ptolemy's revised model, is a function of the Moon's elongation from the Sun. At conjunction and opposition, the new model effectively reduced to Hipparchus's simpler epicyclic model, such that its consequences for the theory of eclipse prediction remained limited in scope.

---

[15] Leslie V. Morrison, F. Richard Stephenson, and Catherine Y. Hohenkerk, 'On the Eclipse of Hipparchus.' *Journal for the History of Astronomy* 50 (2019), pp. 3–15; Christián C. Carman, 'On the Distances of the Sun and Moon According to Hipparchus.' In *Instruments—Observations—Theories: Studies in the History of Astronomy in Honor of James Evans*, ed. Alexander Jones and Christián Carman, online publication, 2020, pp. 177–203. DOI: 10.5281/zenodo.3928498.

[16] See *Ptolemy's Almagest*, translated by G. J. Toomer, revised edition, Princeton, NJ, 1998; Olaf Pedersen, *A Survey of the Almagest*, revised edition, New York, 2011.

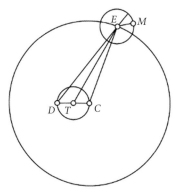

Figure 3.2 Schematic depiction of Ptolemy's final lunar model, with the centre of the Earth at *T*, the centre of the deferent at *C*, the centre of the epicycle at *E*, and the Moon at *M*. Point *D*, which is located opposite *C* with respect to *T*, serves as a reference point in computing the Moon's position on the epicycle and has the effect of introducing an additional non-uniformity (*prosneusis* or 'inclination') into its motion.

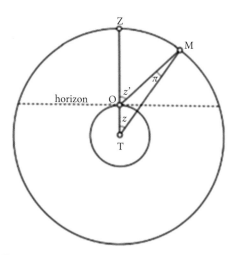

Figure 3.3 The parallax (π) of the Moon (*M*) shown as a function of the difference between its zenith distance (*z*) as seen from the centre of the Earth (*T*) and its zenith distance (*z'*) at the position of the observer (*O*).

Of far greater consequence in this area, especially when it came to understanding solar eclipses, were Ptolemy's advancements in the treatment of parallax. Building on the writings of his predecessor Hipparchus, he defined the parallax, π, of a given celestial object as the difference between its zenith distance as seen from the centre of the Earth and its zenith distance as seen from a given location on the Earth's surface (Figure 3.3). The Moon, owing to its close proximity to the Earth, was subject to a far more noticeable parallax than the Sun or any of the other planets. To measure its value empirically, Ptolemy designed what he called his 'parallactic instrument' (known to later astronomers as the parallactic 'rulers' or *triquetrum*), which made it possible to measure zenith distances to a high degree of precision. He subsequently used one of his results, an exaggerated lunar parallax of $\pi = 1;7°$ for a measured zenith distance of $50;55°$, to derive geometrically the mean lunar (i.e. Earth–Moon) distance. His result of 59 terrestrial radii came remarkably close to the modern value, which is around 60 terrestrial radii.

While the parallax of the Sun was too small to be measured directly, Ptolemy sought to derive a value on theoretical grounds from the calculated Earth–Sun distance. His estimate for this distance, which he had derived geometrically from observed eclipses, was approximately 19 times too small. It is therefore not surprising that the horizontal parallax of $0;2,51°$ Ptolemy inferred for the Sun on this basis is about 20 times the modern value. More significant than these errors were Ptolemy's efforts to make the solar and lunar parallax into easily computable quantities, by constructing dedicated numerical tables. The parallax table included in the *Almagest* featured no fewer than seven columns devoted to the Moon, which was a consequence of the complications wrought by the Moon's rapidly changing geocentric distance. This meant that the lunar parallax was not merely a function of the zenith distance, but in addition depended on the Moon's position on the epicycle and its elongation from the Sun. Importantly for the calculation of eclipses, Ptolemy also used his mastery of spherical trigonometry to show how the total parallax of the Moon could be divided into two separate terms for correcting the computed values of its ecliptic longitude and latitude.

Longitudinal parallax found a crucial application in predicting the midpoint of a solar eclipse, for which purpose the carefully computed 'true' time of conjunction (i.e. the moment of conjunction as seen from the centre of the Earth) had to be converted into an 'apparent' time, which took into account the altered position of the Moon as seen from a particular locality. The resulting change in the time of conjunction in turn entailed a new zenith distance for the Moon. Consequently, a second parallax correction, what Ptolemy termed *epiparallax*, became necessary.

A logical step that came before this laborious series of calculations was to decide whether a given conjunction was in fact going to be accompanied by a solar eclipse. For a partial solar eclipse with the lowest possible magnitude to occur, the angular distance between the centres of the solar and lunar disks had to be just below the sum of their respective radii (see Figure 1.1 in Chapter 1). According to Ptolemy's own measurements and calculations, this sum did not exceed $0;33,20°$, which value was based on a constant apparent solar radius of $0;15,40°$ and a maximum apparent lunar radius of $0;17,40°$ for the Moon, as was reached when the Moon was at the perigee of its epicycle. The business of converting this angular distance into limits for the Moon's distance from the nearest node was again complicated by parallax, which intervened in altering the apparent positions of the luminaries in both longitude and latitude. Its precise contribution to an eclipse was sensitive to the position of the observer, since the zenith distance (and therefore the apparent positions) of both the Sun and the Moon changed with geographic latitude. In his efforts to address this difficult problem, Ptolemy calculated the maximum effect of the combined lunar and solar parallax at the southern and northern ends of the seven climates—ancient bands of northern latitude covering most of the known inhabited lands—which he took to be located at 16;27° and 48;32°. His overall finding was that, for this entire geographic range, a solar eclipse could theoretically still occur when the *mean* conjunction of the luminaries was at +20;41° or −11;22° in ecliptic longitude from the ascending node, or at −20;41° or +11;22° from the descending node.

The (parallax-corrected) longitudinal distance of the Moon from the respective node, together with its distance from the Earth, served Ptolemy as the key variable in computing the precise circumstances of an eclipse, namely its magnitude and duration. He constructed for this purpose a set of tables in which the magnitude of a solar eclipse was given in terms of the number of digits of the apparent solar diameter, from 0 to 12. The obscuration or eclipsed surface could be obtained from an additional table, which converted these digits into the area of the solar disk. Ptolemy would later significantly revise many of his eclipse-related tables for inclusion in his *Procheiroi Kanones* or *Handy Tables*, which became a key source for computational astronomy in the Graeco-Arabic tradition. Among other things, the *Handy Tables* included a completely new set of tables for parallax, which covered all seven climates and showed the combined parallax of Sun and Moon separated into longitudinal and latitudinal components—a clear sign that these tables were geared towards eclipse calculations.

The assumptions and tools of Ptolemy's eclipse theory would remain a foundational aspect of mathematical astronomy for more than a millennium, as was true both in Greek-speaking Byzantium and in the Islamic world, where the *Almagest* was read in Arabic since the ninth century. A rather different situation arose in the Latin West after the dissolution of the Roman Empire, where scholars had lost access to the more advanced products of Hellenistic astronomy. Among the minor exceptions to this rule is the survival of the *Preceptum canonis Ptolomei*, a sixth-century Latin text and table-set derived from Ptolemy's *Handy Tables*, as well as Theon of Alexandria's *Little Commentary* on the same work, which dates from the mid-fourth century. The extant manuscripts of the *Preceptum* transmit tables for the Sun and Moon as well as some eclipse tables, yet the work's rather obscure language, characterized by excessive amounts of transliterated Greek, must have rendered it practically useless to those few who had access to it during the early Middle Ages.[17]

The lack of access to adequate tools did not prevent Latin thinkers from observing eclipses and speculating about the laws that might underpin their appearance. An anonymous treatise of AD 754, apparently written by an Irish monk, documents an early attempt in this direction. It linked the appearance of solar eclipses to a 30-year cycle dubbed the *annus naturalis*. As a consequence, the author's knowledge that a total solar eclipse had been seen in the British Isles in the year 664 made him expect a further such eclipse in 754 (i.e. 3 × 30 years later). Although this eclipse failed to present itself on the expected date, he was quick to note that this was not necessarily a

---

[17] See ed. David Pingree: *Preceptum Canonis Ptolomei*. Louvain-la-Neuve, 1997. See also Pingree, 'The Preceptum Canonis Ptolomei.' In *Rencontres de cultures dans la philosophie médiévale: traductions et traducteurs de l'antiquité tardive au XIVe siècle*, ed. Jacqueline Hamesse and Marta Fattori, Louvain-la-Neuve, 1990, pp. 355–75.

refutation of his 30-year eclipse cycle. After all, the eclipse in question could have been visible in another hemisphere; or the Sun may have been only partially obscured, rendering the eclipse invisible to certain observers. There was, moreover, the possibility that the event was simply hidden from view by cloudy weather—not an unreasonable hypothesis in an insular climate.[18]

Carolingian scholars of the ninth century may have used similar rationalizations to account for the non-appearance of solar eclipses postulated by the *Liber Nemroth*, a mysterious early medieval collection of cosmological and astronomical lore from Near Eastern sources. In the *Liber Nemroth*, the reoccurrence of solar eclipses is linked to a 24-year cycle that began in the tenth year of the creation of the world. One such period is claimed to have started in the year of the world 6299, which, based on the chronological system used elsewhere in this work, would have placed the previously observed eclipse in AD 807. The obvious candidate would be the annular solar eclipse of 11 February 807, which was best visible over the northern parts of the British Isles.[19]

Attempts to harness the existing astronomical knowledge for a more rationally grounded eclipse theory were made in the mid-eleventh century by the monk Hermann of Reichenau, who performed calculations based on a synodic month of 29 d $12\frac{174}{235}$ h and a sidereal month of 27 d $7\frac{92}{127}$ h.[20] His efforts were still hampered by the absence of an essential piece of knowledge, namely the motion of the lunar nodes, as represented by the draconic month. The first Latin text to discuss this motion in any detail dates from *c.*1120 and is due to Walcher of Malvern, who owed his rudimentary understanding of the matter to Petrus Alfonsi, a converted Jew who had received his education in Muslim Spain. Walcher knew that a solar eclipse occurred if the Sun and Moon were joined together in sufficient proximity to one of the nodes, which he, following established Arabic parlance, dubbed the 'head and tail of the Dragon' (*caput et cauda Draconis*). Understanding this condition allowed for a first approach to identifying ecliptic conjunctions, yet Walcher's ability to predict actual eclipses was still severely limited, not least by his lack of familiarity with the equations that would have made it possible to convert the mean positions of the Sun and Moon into true ones.

The full technical wherewithal to predict solar eclipses from conjunction times and the lunar parallax arrived within a few years from Walcher's treatise with the first sets of astronomical tables translated from Arabic into Latin, which were themselves part of a larger wave of scientific knowledge transferred from the Islamic world to Latin Christendom during the twelfth century. Owing to this momentous influx of knowledge, Latin scholars became acquainted with the intricacies of Ptolemaic eclipse theory and at the same time received notice of some of the advances that had been made in this area by Ptolemy's Arabic successors. One of the more

[18] Immo Warntjes, 'An Irish Eclipse Prediction of AD 754: The Earliest in the Latin West.' *Peritia* 24/25, 2013–14, pp. 108–15.

[19] C. Philipp E. Nothaft, 'Dating the *Liber Nemroth*: The Computistical Evidence.' In *Computus in the Carolingian Age*, ed. Immo Warntjes, forthcoming.

[20] Immo Warntjes, 'Hermann der Lahme und die Zeitrechnung: Bedeutung seiner Computistica und Forschungsperspektiven.' In *Hermann der Lahme: Reichenauer Mönch und Universalgelehrter des 11. Jahrhunderts*, ed. Felix Heinzer and Thomas Zotz, Stuttgart, 2016, pp. 285–321.

obvious flaws detectable in Ptolemy's work concerned his measurements of the apparent diameters of the Sun and Moon. According to the *Almagest*, the Sun—despite its eccentric orbit—had a constant angular diameter at conjunction of 0;31,20°, whereas the Moon's diameter could vary between 0;31,20° and 0;35,20°. This left no room for the occurrence of annular solar eclipses, as the Moon was never small enough to leave the Sun's so-called 'ring of fire' unobscured.

Rather than perpetuating this error, Islamic astronomers operated with alternative diameter values, which they initially received from Indian sources (see Chapter 5). One of the most important Arabic astronomical works to reflect this Indian influence was Muḥammad ibn Mūsā al-Khwārizmī's astronomical tables known as the *Zīj al-Sindhind*—a title derived from the Sanskrit term *Siddhānta* ('doctrine')—which were compiled in Baghdad during the first half of the ninth century. In these 'Indian tables,' the angular diameter of the Sun was not constant, but could change between 0;31,20° and 0;33,48°, while the lunar diameter ranged between 0;29,16° and 0;34,34°, which allowed for the occurrence of annular eclipses. Subsequent Islamic astronomers such as Abū 'Abd Allāh Muḥammad ibn Jābir ibn Sinān al-Battānī (d. 929) incorporated similar diameter-ranges into the framework of Ptolemaic astronomy, thereby ridding Ptolemy's eclipse theory of a significant shortcoming.[21] Al-Battānī's great work of computational astronomy, the *Ṣābi' Zīj*, became available in Latin in the first half of the twelfth century owing to the translation efforts of scholars on the Iberian Peninsula. The same is true of the Khwārizmīan *Zīj al-Sindhind*, which is extant in three different twelfth-century Latin versions.[22]

When it came to the actual practice of eclipse prediction, the most influential scientific Arabic-to-Latin translation in use during the Middle Ages was the so-called *Toledan Tables*. First assembled in Toledo in the second half of the eleventh century, the *Toledan Tables* drew on both the Ptolemaic tradition represented by al-Battānī and al-Khwārizmī's 'Indian' tables. They came with a broad range of tables that had been created specially for eclipse calculations as well as with detailed instructions concerning their use.[23] For the underlying theory, Latin readers could refer to Ptolemy's *Almagest*, which was translated at least three times during the twelfth century (once from Greek, twice from Arabic), as well as other works derived from it. The content of the first six books of the *Almagest* was in fact soon refashioned into an original Latin treatise, known as the *Almagesti minor* (before 1220), which presented Ptolemy's eclipse theory in a rigorous 'Euclidian' format composed of

[21] Hamid-Reza Giahi Yazdi, 'Al-Khwārizmī and Annular Solar Eclipse.' *Archive for History of Exact Sciences* 65, 2011, pp. 499–517; S. Mohammad Mozaffari, 'A Case Study of How Natural Phenomena Were Justified in Medieval Science: The Situation of Annular Eclipses in Medieval Astronomy.' *Science in Context* 27, 2014, pp. 33–47; Mozaffari, 'Annular Eclipses and Considerations About Solar and Lunar Angular Diameters in Medieval Astronomy.' In *New Insights from Recent Studies in Historical Astronomy: Following in the Footsteps of F. Richard Stephenson*, ed. Wayne Orchiston, David A. Green, and Richard Strom, Cham, Switzerland, 2015, pp. 119–42.

[22] Raymond Mercier, 'Astronomical Tables in the Twelfth Century.' In *Adelard of Bath: An English Scientist and Arabist of the Early Twelfth Century*, ed. Charles Burnett, London, 1987, pp. 87–118; repr. as ch. 7 in Mercier, *Studies on the Transmission of Medieval Mathematical Astronomy*, Aldershot, 2004.

[23] Fritz S. Pedersen, *The Toledan Tables: A Review of the Manuscripts and the Textual Versions with an Edition*, 4 vols. Copenhagen, 2002.

propositions and proofs.[24] Inspired by this major infusion of technical knowledge on eclipse prediction, authors in the Latin West went on to produce a wide variety of new astronomical texts in which the complicated procedures involved in calculating the time, duration, or magnitude of an eclipse were conveyed to a wider audience.[25]

While the fundamentals of Ptolemaic eclipse theory were not dramatically altered during this period, some medieval astronomers went to great lengths to render the computational process more user-friendly, for instance by devising new procedures for finding the interval between mean and true conjunction. Works such as the *Tabulae permanentes* created by the French astronomers Jean des Murs and Firmin de Beauval in the 1340s attest to the high degree of sophistication medieval Latin astronomy attained in this area.[26] Given the cumbersome nature of the calculations involved in eclipse predictions, it is also understandable that some sought assistance from mechanical aids. Among the documented examples from medieval Europe is the disc-shaped *eclipsorium* described around 1300 by the Danish astronomer Peter Nightingale, to which rotatable volvelles were mounted on both sides. With its movable cursor and numerous graduated scales, the instrument allowed users to find values for the lunar parallax and other intermediary steps in an eclipse calculation without division or multiplication.[27]

For those who were wary of such short cuts and instead aimed for maximum precision, calculation by hand remained the only viable option. An account of how to compute the details of the annular solar eclipse expected for 3 March 1337, written by the Paris-based astronomer John of Genoa, gives us a sense of the immensity of the labour involved, by specifying over 200 separate computational steps.[28] Once all steps were fully realized, astronomers were in a position not only to predict the time of a solar eclipse, but to depict its expected appearance in the form of a standardized diagram, which revealed which parts of the solar disk were going to be obscured by the Moon. Magnificent examples of such diagrams are transmitted in several manuscripts as part of the astronomical calendar of Walter of Elveden, a member of Gonville Hall, Cambridge. His *kalendarium* came with carefully tabulated numerical data as well

---

[24] Henry Zepeda, *The First Latin Treatise on Ptolemy's Astronomy: The* Almagesti minor *(c. 1200)*. Turnhout, 2018.

[25] A typical example is the thirteenth-century treatise *Ut annos Arabum . . .*, which is edited by Pedersen, *The Toledan Tables*, vol. 2, pp. 552–68 (CbB). Many further texts of this kind survive in unpublished manuscripts.

[26] Richard L. Kremer, 'Cracking the Tabulae permanentes of John of Murs and Firmin of Beauval with Exploratory Data Analysis.' In *Editing and Analysing Numerical Tables: Towards a Digital Information System for the History of Astral Sciences*, ed. Matthieu Husson, Clemency Montelle, and Benno van Dalen, Turnhout, 2021, pp. 363–422.

[27] Fritz S. Pedersen, 'Petrus de Dacia: Tractatus Instrumenti Eclipsium.' *Cahiers de l'Institut du Moyen Âge Grec et Latin* 25, 1978, pp. 1–102; Pedersen (ed.), *Petri Philomenae de Dacia et Petri de S. Audomaro Opera Quadrivialia, vol. 1, Opera Petri Philomenae*, Copenhagen, 1983, pp. 457–512. See also Johannes Thomann, 'Astrolabes as Eclipse Computers: Four Early Arabic Texts on Construction and Use of the Ṣafīḥa Kusūfiyya.' *Medieval Encounters* 23, 2017, pp. 8–44.

[28] This unpublished text appears in MSS Cambridge, University Library, Ee.III.61, fols. 75r–81r; Douai, Bibliothèque Marceline Desbordes-Valmore, 715, fols. 36v–44r; Paris, Bibliothèque nationale de France, lat. 7281, fols. 208v–210v. See Laure Miolo, 'Retracing the Tradition of John of Genoa's *Opus astronomicum* Through Extant Manuscripts.' In *Alfonsine Astronomy: The Written Record*, ed. Richard L. Kremer, Matthieu Husson, and José Chabás, Turnhout, 2022, pp. 343–80, at 371–4.

as illuminated drawings for all solar and lunar eclipses that were expected to occur between 1327 and 1386 (Figure 3.4).

Attempts to assess the accuracy of such predictions were rendered hazardous by the effects of staring directly into the Sun. William of Saint-Cloud, an astronomer active in Paris in the late thirteenth century, reports that the solar eclipse of 4 June 1285, despite its relatively small magnitude, caused a temporary dimming of the eyesight in several local observers.[29] In their efforts to protect themselves against the Sun's glare, medieval observers sometimes looked at solar eclipses through a thin cloth or contented themselves with seeing its reflection in water or in a mirror placed inside a dark vessel.[30] Another available safeguard was to use a pinhole camera or camera obscura, which made it possible to look at the image of a solar eclipse projected through a

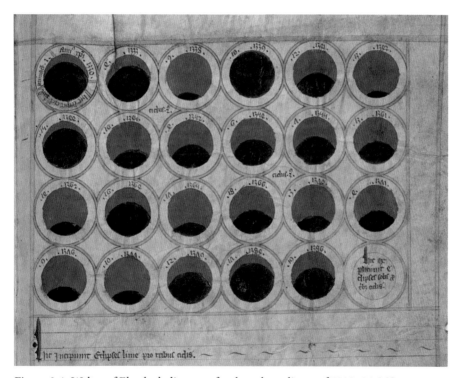

Figure 3.4 Walter of Elveden's diagrams for the solar eclipses of 1330–86, MS Cambridge, Corpus Christi College, 37, fol. 45r.

Source: Reproduced with permission of The Parker Library, Corpus Christi College, Cambridge.

[29] William of Saint-Cloud, *Almanach planetarum* §§35–6, edited in Fritz S. Pedersen, 'William of Saint-Cloud: *Almanach Planetarum*: An Edition of the Canons, a Few Samples from the Tables, and a Foray into the Numbers.' *Cahiers de l'Institut du Moyen Âge Grec et Latin* 83, 2014, pp. 1–133, at 24.

[30] In addition to the testimony of William of Saint-Cloud (previous footnote), see *Investigantibus astronomiam primo sciendum*, Jn301, edited in Fritz S. Pedersen, 'A Twelfth-Century Planetary Theorica in the Manner of the London Tables.' *Cahiers de l'Institut du Moyen Âge Grec et Latin* 60, 1990, pp. 199–318, at 283; Albertus Magnus, *Summa theologiae* 1.3.15.1, in *Alberti Magni Opera omnia*, vol. 34.1, ed. Dionysius Siedler et al., Münster, 1978, 58, ll. 58–65; Guglielmo Ventura of Asti, *Memoriale de gestis*

small aperture. According to William of Saint-Cloud, proper use of the pinhole camera allowed one to observe solar eclipses with a magnitude as small as one half-digit, all without risking danger to the human eye. He recognized the potential utility of this set-up in investigating the magnitude of an eclipse as well as its precise beginning and end points and even suggested that the diameter of the pinhole image could be used to gauge the eccentricity of the Sun's orbit. The first attested application of the latter method can be credited to the fourteenth-century Provençal Jewish scholar Levi ben Gerson (1288–1344), who also used his observations of eclipse magnitudes to propose some highly innovative changes to Ptolemy's lunar theory.[31]

Even the traditional models inherited from Ptolemy and the Arabs put medieval Latin astronomers in a position to forecast the time of an eclipse with remarkable success. An analysis of the eclipse data included in a popular astronomical calendar drawn up in 1386 by the Carmelite friar Nicholas of Lynn suggests that predictions of solar eclipses made in late medieval Europe on the basis of the so-called Alfonsine Tables normally stayed within 30 minutes of their observable mid-point and within 0.10 of the true value for their magnitude.[32] This level of accuracy is also reflected in the work of the celebrated fifteenth-century astronomer Johannes Regiomontanus, who nevertheless remained sensitive to the occasionally serious discrepancies between observation and prediction. To Regiomontanus, they were a sign that the theoretical foundations of mathematical astronomy were in need of a potentially drastic overhaul.[33] This reformist impulse would culminate later on in the publication of Nicolaus Copernicus's *De revolutionibus orbium coelestium* (1543), which proposed a sun-centred planetary system. Despite its revolutionary thrust, the work yielded no immediate consequences for the theory of solar eclipses or the accuracy of their prediction. Things began to improve only with the meticulous observations carried out by Tycho Brahe (1546–1601) and Johannes Kepler's formulation of the familiar three laws of planetary motion (1609–19), which ushered in an entirely new phase in the history of Western astronomy.

civium Astensium et plurium aliorum, c. 3, ed. Coelestinus Combetti, in *Monumenta Historiae Patriae edita iussu Regis Caroli Alberti: Scriptores*, vol. 3, Turin, 1848, col. 705.

[31] William of Saint-Cloud, *Almanach planetarum* §§37–9 (Pedersen, 'William of Saint-Cloud,' 24–6); José Luis Mancha, 'Astronomical Use of Pinhole Images in William of St-Cloud's *Almanach Planetarum* (1929)'. *Archive for History of Exact Sciences* 43, 1992, pp. 275–98; Bernard R. Goldstein, *The Astronomy of Levi ben Gerson (1288–1344)*. New York/Berlin, 1985, pp. 48–50, 140–3; Jarosław Włodarczyk, 'Solar Eclipse Observations in the Time of Copernicus: Tradition or Novelty?' *Journal for the History of Astronomy* 38, 2007, pp. 351–64, at 354–7.

[32] Darren Beard, 'Eclipse Predictions in the Kalendarium of Nicholas of Lynn.' *Journal of the British Astronomical Association* 144, 2004, pp. 343–8. See also Lynn Thorndike, 'Prediction of Eclipses in the Fourteenth Century.' *Isis* 42, 1951, pp. 301–2; Thorndike, 'A Record of Eclipses for the Years 1478 to 1506.' *Isis* 43, 1952, pp. 252–6; Thorndike, 'Eclipses in the Fourteenth and Fifteenth Centuries.' *Isis* 48, 1957, pp. 51–7.

[33] Noel M. Swerdlow, 'Regiomontanus on the Critical Problems of Astronomy.' In *Nature, Experiment, and the Sciences: Essays on Galileo and the History of Science in Honour of Stillman Drake*, ed. Trevor H. Levere and William R. Shea, Dordrecht, 1990, pp. 165–95; John M. Steele and F. Richard Stephenson, 'Eclipse Predictions Made by Regiomontanus and Walther.' *Journal for the History of Astronomy* 29, 1998, pp. 331–44.

4

From Science to Story

Testimony of an Eclipse-Chaser

Mike Frost

Amateur Astronomy, History of Astronomy, Eclipse-Chasing, Umbraphilia

The previous chapters have shown that total solar eclipses feature dramatic and unique phenomena, resulting in a spectacle worth seeing at least once in one's lifetime. Indeed, there are some astronomers—actually, a whole community of eclipse–chasers—who travel the world to see eclipse after eclipse. Why is one eclipse not enough, and why do some people devote so much time, effort, and money to eclipse–chasing?

Let's meet one of them. The author is an astronomer and astronomical historian, who relishes traveling the world to see solar eclipses, and enjoys being part of the community of umbraphiles—those who love being in the moon's shadow.

umbraphile Someone who loves chasing eclipses.

My name is Mike, and I am an eclipse-chaser.[1]

Actually, the statement is not entirely accurate, although the allusion to a personal addiction is appropriate. The Moon's shadow flies across the Earth's surface at speeds in excess of Mach 2, faster than most of us can travel, so chasing an eclipse is a futile endeavour. In 1973, Concorde flew across Africa in an attempt to stay within the Earth's shadow for as long as possible, and even that fastest of commercial aircraft was not able to match the shadow's pace, though it did manage a jaw-dropping 74 minutes of totality, nearly ten times as much as anyone staying on the Earth's surface will ever witness at a single eclipse.

No, the best I can manage is to be an eclipse *interceptor*, travelling the world to place myself in the right place, at the right time, and hopefully with the right weather. At the time of writing, I have been able to do this thirteen times, though not always with the right weather. Eclipse interceptors can be competitive about their total number of eclipses; or even, as a tiebreaker, minutes and seconds of totality witnessed.

In the eclipse fraternity (loosely based around the google group SEML, which stands for Solar Eclipse Mailing List) the term *umbraphile* is preferred. We *love* being beneath the shadow of the Moon. For most of us it is a passion, sometimes all-consuming. It dictates our holiday plans and our birthday wish-lists.

Non-umbraphiles, of course, tend to express their incomprehension with bafflement and exasperation: 'Why travel all that way for two or three minutes of total eclipse?' And sometimes (I am thinking North America in 2017): 'Why go there? We

---

[1] I am referencing Kate Russo's *Total Addiction* (London, 2012), which analyses the psychology of our curious tribe.

saw it here!' Well, no they did not—they probably experienced a partial eclipse, and, as this book has emphasized in every chapter so far, a partial eclipse is not the real thing. The eerie darkness of totality is a qualitatively different experience, and like none other. The night before the English eclipse of 11 August 1999, Ken Livingstone, London politician, declared on BBC TV, 'Well, London will see a 99% eclipse, and 99% is good enough for me, so I'll be staying in London tomorrow' (he may of course have been sweet-talking his electorate). We eclipse-chasers find ourselves searching perpetually for ways to explain that a 99% partial eclipse is not 99% as good as a total eclipse—that it is not even close.

Sometimes objections are framed as thought experiments. A question sent into *The Guardian* newspaper's estimable 'Notes and Queries' section just before the 1999 eclipse: 'Why do I need to go a long way to see a total eclipse of the Sun when a dinner plate held at arm's length would do the job just as well?' Even the question betrays a lack of intuition from the questioner. A dinner plate? Way too large! A closed fist at arm's length easily covers the Sun; indeed, a suitably large thumb will suffice. And, as often happens in physics, actually carrying out the thought experiment is instructive. On a sunny day, try blocking out the Sun with your thumb or fist. What do you see? Nothing that you would not see anyway—and, most importantly, there is no sign of the beautiful phenomena only visible during totality—the gorgeous, subtle, pearly-white corona, the blood-red prominences. These wonders are invisible because they are overwhelmed by the sunlight from the Sun's disk that is scattered by the intervening Earth's atmosphere. To see these beautiful sights, the sun-shield must be above the atmosphere, where it will cut off the Sun's light before any of it reaches the scattering atmosphere. Instead of a dinner plate at arm's length, what is needed is a satellite a quarter of a million miles away.

Another common exasperation amongst umbraphiles arises when we're asked: 'Why travel all that way to see it go dark? It goes dark every night!' (at this point the questioner invariably shakes their head in disbelief and pity). I have two responses here. First, it doesn't go dark during a total eclipse—light levels drop to about that of twilight. My second strategy is to try to turn the question back onto the questioner. So, they think eclipses are like sunsets? Well, aren't sunsets beautiful? Don't they ever spend time watching sunsets? Imagine if they only ever had one chance in their entire lifetime to see a sunset—wouldn't they at least consider travelling to see it?

That is the crux of my argument. Sunsets are overwhelmingly beautiful. So are rainbows; so are aurorae. So are total eclipses; but the difference is that eclipses are rare, and inconveniently located (as opposed to aurorae, which are commonplace and inconveniently located. I go to see aurorae too). A total eclipse offers the briefest of opportunities to see something breathtakingly beautiful, which cannot be seen anywhere else or at any other time. And typically, those places are a long way from home—an eclipse-chaser's life is one dominated by travel. I offer my accounts of some of these journeys as tastes of the raw human encounter that underlies the other chapters of this book.

* * *

My first experience of totality was a short-duration eclipse in India, on 24 October 1995. I had never been to the Subcontinent, and so the destination was very appealing, offering a tour of Jaipur, Agra, and Delhi, viewing the eclipse from the ruined city of Fatehpur Sikri. The tour was memorable. Eclipses, being new Moon phenomena, often coincide with lunar festivals, and the night of the 23rd–24th was Diwali: Jaipur was a riot of colour and excitement. I would have spent longer at the party, but I had a coach heading east just after midnight.

Eclipse day is like no other. We arrived in Fatehpur Sikri to find it full of umbraphiles from across the world, along with a smattering of TV crews. My group had planned to observe the eclipse from the caravanserai—the old marketplace—but our preferred location was already taken by Japanese astronomers, so we rearranged ourselves along the caravanserai walls.

It is hard to explain the thrill of the lead-up to a total eclipse. The first highlight was, as always, 'first contact'—the beginning of the eclipse, when the first tiny bite in the Sun becomes apparent. The excitement grew slowly and inexorably over the next hour. Initially I was content to follow progress through my eclipse glasses, but then began experimenting with projecting the Sun through a pinhole onto a makeshift screen. Gradually, imperceptibly, things became unearthly. Shadows began to behave strangely, sharp in one direction, fuzzy at right angles.[2] A typically scorching Indian day began to cool off, the sky thinning and silvering, as though winter was coming on all at once.

Things sped up in the last few minutes before second contact, the moment at which totality starts. Light levels dropped first steadily, and then dramatically. The edges of the crescent Sun began to break up into Baily's beads—the last views of the bright Sun's disk, glimpsed through valleys between the mountains on the Moon's edge! The corona started to come into view. And then there was no crescent at all, just the last few beads—three, two, one ... the light level plunged, the Moon's shadow shot overhead, and we had totality!

The next 45 seconds were a blur. I picked up my binoculars to study the detail of the corona and the prominences. I had set up my camera to take just one souvenir shot of the eclipsed Sun. I even found time to glance away briefly and spot Mercury for the first time in my life. And then ... the searing brightness of the diamond ring! And totality was over.

Wow! I thought. And then—a common reaction—*when's the next one?*

Shortly after the end of totality there was a loud, low, sudden *bang* as if one of the TV crews had dropped a lighting rig. We found out later it was a sonic boom. Eighty thousand feet above us, Group Captain S. Mukerji of the Indian Air force was flying his MiG-25 at Mach 2.5 in an attempt to chase the eclipse. Even he could only extend totality to ninety seconds.

> Later that day I saw the Taj Mahal for the first time. The eclipse was the bigger deal.

It did not take long to realize that the next total eclipse would be eighteen months later, 9 March 1997, in Mongolia. I had always wanted to go to Mongolia. It struck me then that eclipse-chasing was something that could take me to obscure and wonderful places.

---

[2] This strange shadow anisotropy is a consequence of the 'crescent' form of the nearly-eclipsed Sun. Shadows are sharp in the direction across the thin middle of the crescent, blurred as usual along its length.

Mongolia lived up to my expectations—a huge chunk of nothingness in the middle of Asia. I grew up in the north of England and the steppes reminded me of the bleak, treeless Pennine Hills, except that, instead of containment by the conurbations of Manchester and Sheffield, the Mongolian steppes went on for ever. We were expecting a (very) cold-weather eclipse, but it was warmer than expected—close to freezing. That might sound positive, but it meant we were on a warm front and being snowed on. A convoy of coaches, accompanied by police escort, headed north from our base in Darhan towards the Russian border, in a desperate attempt to outpace the weather. We failed. Totality occurred beneath heavy cloud. The track of totality was wide, and the shadow elliptical, and so in the far distance, to the north-east and south-west we could see dull red glows beneath the clouds, from the territory which wasn't seeing totality.

We returned to Darhan to find that the weather front had passed right over, and everyone there had seen totality! The mood in the town was euphoric; this was the most eventful day they had had for years. I have a picture of youngsters posing for photographs, loving the attention. Sometimes, eclipse-chasers have to make last-minute decisions, and it is all too depressingly easy to jump the wrong way.

The next total eclipse crossed the Caribbean on 26 February 1998. I flew to the Netherlands Antilles. We held a Mardi Gras beach party at Knip Bay, at the top of Curaçao, with an eclipse as the highlight. Heineken, who had a brewery on the island, provided eclipse beers (I still have a souvenir can). This was my first eclipse offering nearby tree cover to take shade in, and I took the opportunity to view the progress of the partial eclipse through the natural pinholes provided by gaps in the leaves—the same method that Aristotle had used to observe eclipses over two thousand years previously.[3] Golden orioles flitted excitedly around. We had a beautiful eclipse in a near-cloudless sky. Afterwards, as everyone celebrated, there was a cheer from one of the yachts in the bay, as a wedding reached its conclusion—one fortunate couple exchanged glances over two diamond rings that day.

I mused about the plot of a suitably dreadful romantic novel set in the eclipse-chasing community. Romance kindled in the Diwali celebrations in Jaipur and visits to the Taj Mahal; Act 2 and break-up in icy Mongolia; then a triumphant conclusion on the beaches of Curaçao. The title? *The Shadow Chasers* of course—no need to spell out the metaphor.[4]

My first three eclipses were but preparation for a long-anticipated event—the only total eclipse on English soil during my lifetime (unless I manage to live until the age of 128). Along with the return of Halley's comet in 1985–6, the Leonid meteor storm of 17 November 1999, and the transits of Venus across the Sun on 8 June 2004 and 6 June 2012, the total eclipse of 11 August 1999 was a predicted celestial event I had been anticipating since childhood. The last total eclipse visible from England had crossed

---

[3] John Hammond, *The Camera Obscura: A Chronicle*. Bristol, 1981, p. 3, describes Aristotle's observations. Hammond references W. S. Hett (trans) *Aristotle—Problems*. London, 1936.
[4] Kate Russo's *Total Addiction* suggests a more explicit metaphor (p. 60): 'Totality also shares an interesting parallel with the sexual response curve, with the euphoria, excitement and pleasure. There is a similar build up of tension that reaches a peak and then rapidly falls off. [...] Totality and orgasm both result in feelings of euphoria as endorphins [...] are released in the brain. [It is] an "intellectual orgasm," for want of a better description.'

the north of the country shortly after dawn on the morning of 29 May 1927. Huge crowds went to see it—some think that day to be the busiest in the entire history of British Railways. Southport, a seaside resort north of Liverpool, was a popular destination, and railway posters advertised overnight or early morning trips. It was even possible to see the eclipse and be back at work in Manchester by 9 a.m. Alas, the famed British weather was not cooperative, and most people saw only cloud during totality.

We hoped for better in 1999. I stayed with friends—among them a solar physicist, who had never seen an eclipse—in the Exeter area, and on eclipse day we travelled south on the train to the town of Totnes. Unfortunately, the weather was not on our side; clouds covered the Sun from seven minutes before totality until two hours afterwards.

* * *

I am the director of the British Astronomical Association's historical section. The Association was founded in 1890, and from the beginning was closely associated with solar observation and eclipses. Two of the prime movers of the Association were solar astronomers. Edward Walter Maunder was the head of the solar observation section at the Royal Greenwich Observatory, and Elizabeth Brown, the first director of our solar section, a noted eclipse-chaser. Her account of her trip to Russia to see the solar eclipse of 1887, 'In pursuit of a Shadow,'[5] is a fascinating one of independent travel; she wrote a similar account of a later trip to Trinidad.

The Association has organized many expeditions to see total eclipses; the first, in 1896, was to the north coast of Norway. As with the 1999 eclipse, the weather was cloudy at all the BAA's observing locations, yet like most of my eclipses, it was a very sociable trip. My predecessor Mary Orr, first director of the BAA's historical section, met her husband John Evershed on the expedition; they spent many years observing the Sun from the Kodaikanal Observatory, Tamil Nadu, where John was director.[6] The Eversheds attended several eclipses under the auspices of the Royal Astronomical Society, including Wallal, West Australia in 1922, where they attempted to replicate Eddington's observation of 1919, which famously validated Einstein's relativity.

Walter Maunder travelled to Norway with his second wife, Annie Scott Dill Russell. Annie Maunder had already proved herself as a solar photographer at Greenwich and became renowned as an eclipse photographer (you can see several examples of her work in Henrike Lange's Chapter 11). The Maunders plotted decades of Greenwich observations and discovered the famous 'butterfly diagram'[7] showing how sunspot locations migrated from higher latitudes towards the solar equator during the 11-year solar cycle. In the expeditions to Norway, India in 1898, and Algeria in 1900, the BAA proved itself to be at the forefront of eclipse science. So, I am following in the footsteps of some of the most forward and accomplished eclipse-chasers. To this

[5] 'A Lady Astronomer,' *In Pursuit of a Shadow*, Gloucester, 1888. Modern-day reprints are available, for example by Kessinger Legacy Reprints or Cambridge University Press.

[6] Whilst at Kodaikanal, Mary Evershed, under her maiden name of M. A.Orr, wrote *Dante and the Early Astronomers* (London and Edinburgh, 1914), explaining the astronomical allusions in Dante's *Divine Comedy* and other works. See Alison Cornish's Chapter 9.

[7] See e.g. Mary Bruck, *Women in Early British and Irish Astronomy*. London, 2009, p. 227.

day, the BAA continues to supply guest astronomers for eclipse expeditions, and I was privileged to fulfil just such a role for the Astro-Trails company at the eclipse of December 2020.

For the first total eclipse of the twenty-first century, 21 June 2001, I joined a tour to northern Zimbabwe, where we saw totality in clear but hazy skies from the grounds of Maname School near Mount Darwin. The school pupils and their teachers were fascinated by our preparations for totality, but they disappeared shortly before second contact—perhaps to leave us alone, but, we suspected, to pray in the school chapel.

Just before totality, we were troubled by swarming insects, who instinctively interpreted the dropping sunlight levels as dusk. People are always interested in reports of animal behaviour during eclipses, but I am sceptical—my contention is that the larger the animal, the less likely I am to trust reports of its behaviour (but see Steven Portugal's Chapter 13 for an authoritative account). Two hundred miles to our west, a BBC film crew was filming hippopotami on the Zambezi River. They reported that the animals were agitated during totality—but would you not be agitated if you had a TV crew filming you with spotlights? Most animals do not usually have a crowd of excitable eclipse-chasers in their vicinity, so their behaviour must not be thought of as entirely due to the eclipse itself. But bugs and maybe even birds probably don't care about humans, so one can trust their more instinctive reactions to totality. (In any case, I suspect the BBC crew simply wanted an excuse to film in Zimbabwe, not always accessible to the foreign press.)

I returned to southern Africa eighteen months later, to view totality on 4 December 2002 from the coast of Mozambique, north of Xai-Xai. Both Zimbabwe and Mozambique are beautiful countries, but there were fascinating social differences between the two nations. A clear example was the official attitude to HIV, endemic to perhaps 25% of the populations of both countries. The government of Mozambique seemed to want to get to grips with the problem; along the road, every tree bore a red heart, advancing AIDS awareness. In Zimbabwe there was silence, the only hint of a national tragedy the many advertisements for coffin-makers in the Harare newspapers.

As totality approached the Mozambique coast, so did a large black cloud; second contact won the race, but only by a fraction of a second, and we missed almost all the rest of the eclipse.

The Mozambique trip was memorable for many reasons, but for totality I should have gone to the other end of the track, where observers saw a short eclipse from South Australia. Many of my friends were on a near featureless plain close to Woomera, where they saw a 23-second total eclipse. Twenty-three seconds is not very long to travel to the other side of the world for, but a short-duration eclipse has its own special features. One of my friends took a wide-angle shot of the scene in which you can see the dark of the Moon's umbra, with the eclipsed Sun in the centre, and brighter skies to either side. The cameraman is literally standing at the tip of the Moon's shadow, the cone of darkness imagined for centuries (see Introduction, Figure I.5), and extending a quarter of a million miles across space to brush the Earth's surface at a desert in South Australia. And he photographed the tip of that shadow.

The next two total eclipses were inconveniently located for me. The only landmass to see the total eclipse of 23 November 2003 was Antarctica, and all options to view this were out of my price range. Trips to Antarctica cost tens of thousands of pounds.

Eclipse flights from Australia were a little cheaper, as there is already a market for tourist flights over Antarctica. Of course, window seats on the correct side of the plane were at a premium! But I have never been very tempted by eclipse flights, which seem to me to lack the atmosphere of eclipse day on the ground. There was also a short-duration eclipse in 2005 over South America, which I passed on.

Instead, I opted for two annular eclipses. To reprise previous explanations, in an annular eclipse the Moon occupies the more distant part of its orbit around the Earth, and as a result its angular size is too small to cover the Sun completely; the best that can be achieved is to see a ring (annulus) of light around the Moon. This sounds aesthetically appealing—and is—but it is nothing compared to totality. The reason is the same as met with several times already in this book in the context of partial eclipses that are nearly total: the Sun's disk is so overwhelmingly bright that, even when the Sun is 99% covered, what remains is still too bright to look at with the naked eye, and therefore dangerous to observe with magnification. Nor are the pearly-white corona or pink solar prominences visible, as their light is overwhelmed by sunlight scattered by the atmosphere, hardly different from normal daylight. Only with a filter is it possible even to appreciate the form of the bright ring, so an annular eclipse lacks the visceral, overwhelming effect of a total eclipse.

There are compensations. Close to and during annularity, shadows take on bizarre forms. Have you ever observed a scene illuminated by an annular light source? It is other-worldly. And if the eclipse is observed just after dawn, or just before sunset, then the attenuation of light means that the eclipse is visible, with care. This is the famous 'ring of fire' which is how annular eclipses are usually described and marketed. Unfortunately, it is difficult to convey that annulars are 'good, but not that good,' especially to a press that confuses viewers by referring to 'total annular eclipses.'

My algorithm is this: if a partial eclipse is occurring where I am, I will observe it. If an annular eclipse is taking place on the same continent as me, I will try to travel to see it. And I will consider going anywhere to see a total eclipse.

My first annular eclipse, 31 May 2003, crossed northern latitudes, starting at dawn in northern Scotland. I opted to observe from the Shetlands, the northernmost islands of Britain, and was rewarded with a view of the partially eclipsed Sun emerging from the sea; unfortunately, it then continued to rise into a bank of cloud which covered it for the remainder of the eclipse. This eclipse, incidentally, crossed the North Pole, and so went backwards through the time zones—finishing the day before it started. My second annular was more successful, seen from a square in Madrid in 2005, in the excited company of many *Madrileños*. Once again, the people made the eclipse experience.

My next total eclipse was 29 March 2006. I chose to observe from the Turkish coast near Side—a short trip from the UK. It was almost too easy. We flew out the previous day, stayed in an all-inclusive hotel. Eclipse day started with a leisurely breakfast, followed by a buffet lunch and then an afternoon eclipse in clear, cloudless skies. After that, things went downhill a little. We had to leave for the airport even before the end of the partial phase of the eclipse; and by the time we landed in England, it was pouring with rain. Had I really seen a total eclipse that afternoon? I felt that such an overwhelming event deserved more commitment from its observers, and that such an extraordinary phenomenon ought to involve extraordinary personal circumstances. My friends who travelled to Libya had to take a 12-hour bus journey to a viewing site in the middle of the Sahara Desert, where they were joined by Colonel Gaddafi (or possibly his body double). That's what an eclipse day should be like!

For work and family reasons, I was unable to see the total eclipses of 2008 (visible in central Asia), 2009 (southern China), and 2010 (the Pacific and South America). I was working in Brazil at the time of the 2010 eclipse and made enquiries as to whether I could travel from suburban Brazil to Chilean Patagonia, where the eclipse ended at sunset—and be back in time for a critical start-up for my employer the next day. Not possible. So, I personally recorded this eclipse as a 0% partial eclipse, occurring after sunset, and cloudy.

The eclipse of 13 November 2012 marked my return to eclipse interception. The totality track started in Northern Australia and crossed Queensland before heading off across the Pacific. I had not been back to Australia after working there for a year in the 1980s and was keen to revisit one of my favourite countries and meet up again with friends.

From Port Douglas in Queensland the eclipse was visible shortly after dawn. At first light, we made our way to the beach, as the view eastwards to the Barrier Reef offered an appropriately photogenic backdrop. Worryingly, the sky was cloudy. Our group set up their equipment on a narrow strip of beach—because the Sun and Moon were pulling in exactly the same direction, the high tide was particularly high. The sea looked appealing, but the combination of sharks, box jellyfish, and saltwater crocodiles put us off going for a swim.

As totality approached, there was still substantial cloud cover, and in the final minutes before second contact I was beginning to judge which direction to sprint to find the gap in the clouds. But then a miracle happened ... well, not really a miracle. Eclipses generate their own local weather, as Giles Harrison explains in Chapter 14; the drop in solar flux due to the eclipse can radically change atmospheric conditions in minutes—sometimes for the worse, but more often than not, for the better. Just before second contact, the clouds began to dissolve, and we were able to witness totality through a conveniently placed gap.

The eclipse of November 3 2013 was a short-duration total eclipse across central Africa (technically a 'hybrid' eclipse as the eclipse began as an annular). I decided not to travel as, at every location along the track, the eclipse was either too short, weather prospects too cloudy, or local conditions too dangerous. But in one sense I had already seen this eclipse. The 2013 eclipse was 18 years, 11 days and 8 hours after my very first eclipse, in India. This is the saros period discussed in Chapter 1, corresponding to 223 synodic months (new Moon to new Moon), 242 Draconic months (Moon crossing the plane of the Earth's orbit round the Sun), and for that matter 239 anomalistic months (Moon perigee to perigee). So, the circumstances of the 2013 eclipse were similar to those for my Indian eclipse of 1995 (also short duration, of course), except that the track was shifted 120 degrees to the west, and slightly south. The saros 'series' continues to the eclipses of 2031, 2049, and so on.

There is a similar but lesser-known series called the inex; 358 synodic months is approximately the same as 388.5 Draconic months, and so eclipses also occur in sequences of 28 years 345 days. As 223 and 358 are mutually prime,[8] every eclipse is in a unique pair of saros and inex series (see Chapter 1). In a classic work on eclipse theory from 1955,[9] George Van den Bergh produced a chart plotting every eclipse for 3,400 years in a grid of saros (vertical) and inex (horizontal)—over time, eclipses occur like a wildfire making its way across the plot. We have better computers now, so John Tilley and Luca Quaglia produced an epic plot of every solar eclipse for the previous and next 13,000 years, classified by saros and inex series, and colour-coded as partial, annular, hybrid, or total. On 3 November 2013, I was attending a recreational maths conference, and at the time the eclipse was in progress, I gave a talk on saros and inex series, illustrated by a 5-metre-long plot of the Tilley/Quaglia chart.

On 20 March 2015, one saros on from my Mongolian eclipse, we had another eclipse in high northern latitudes. 120 degrees west from Mongolia takes us to the North Atlantic, but this time we had an eclipse which managed to avoid almost all land, the track of totality passing between Iceland and the British Isles, both countries enjoying a deep partial eclipse. The only *terra firma* to see totality were the Faroe Islands and the far-north archipelago of Svalbard. Neither option had good weather prospects (the Faroes are notorious for cloudy weather), but I travelled to Svalbard with a small group of aurora-watching friends who were used to cold-weather observation.

The Svalbard eclipse was magnificent, my personal favourite. The event was graced with perfect cloudless skies; just a little windswept snow in the air. A low Sun hung in the south, snow-covered mountains in the foreground. The temperature drop to −22 degrees Celsius during totality (remember, the temperature plunges as the solar flux diminishes) gave me frostnip in my fingers, aggravated by watching totality through binoculars, the metal conducting heat away from my gloved fingers.

Of all the wonders of this particular eclipse, the one that stands out is the shadow bands. These are consequences of the Sun 'twinkling,' as stars do at night, during the seconds before and the seconds after totality. The well-known stellar 'twinkle, twinkle' is not because stars' light output really varies on short timescales, but because that light is refracted on passing through the air cells, of differing temperatures and densities, which make up Earth's atmosphere. The light from Sirius, the brightest star in the night sky, is steady before it hits the atmosphere, but if you observe Sirius low in the sky in a northern-hemisphere winter, it appears to vary rapidly in brightness and even colour, as different amounts of light are directed towards the observer by the ever-churning atmosphere. The Moon shows similar behaviour, appearing to 'boil' in conditions that amateur astronomers call poor 'seeing.' The Sun does exactly the same, but we don't usually notice it because the Sun is far too bright to observe directly. However, in the seconds after a total eclipse finishes, the tiny, uncovered fraction of the Sun is essentially a point source and can be seen to twinkle just as Sirius does.

This manifests itself in a couple of ways. Video of the Svalbard eclipse appears to show strobing of the sunlight as totality approaches and recedes. More noticeable to

---

[8] Two numbers are mutually prime if they have no common factors (apart from, trivially, 1).
[9] Van den Bergh *Periodicity and Variations of Solar (and Lunar) Eclipses*. Tjeenk Willink and Haarlem, 1955.

those present were the effects around them, where waves of light and darkness appear to sweep across the ground. It is a very disconcerting experience—as though the Earth is spinning off its axis—and has been known to disorient observers.

Shadow bands don't happen at every eclipse; turbulent air cells are required, and winds blowing in the right direction. I had only seen hints of them at two previous eclipses, Curaçao in 1998 and Zimbabwe in 2001. They can also be difficult to observe as they are low-contrast effects. Before eclipse observations I usually set up a 'shadow band detection system' consisting of a white bath towel or bedsheet borrowed from my hotel. In Svalbard we were observing from the middle of a large snow field, so no towels were needed. The shadow bands could be seen in their entirety, sweeping along the ground in long parallel lines. I have seen shadow bands on three subsequent eclipses (very strongly in 2017 and less so in 2019 and 2020), but the setting in Svalbard made them especially memorable.

My next total eclipse, 9 March 2016, was 55°C warmer. Minus 22°C in Svalbard, plus 33°C in Indonesia, virtually on the equator. We watched the eclipse from the Spice Island of Tidore, with the largest island of the Moluccas, Halmahera, in the foreground. We were on the edge of a cloud system, but most of totality was cloud-free, bar a few scudding clouds at the end. This eclipse offered the remarkable sight of a huge, detached prominence on the 'trailing edge' of the Sun. But to me the most extraordinary visual feature was provided by the gorgeous skies to the south; a warm orangey-yellow glow reminiscent of a beautiful sunset (see Figure 4.1). The beauty brought with it a puzzle—at sunsets the warmer colours accompany a setting sun as it nears or crosses the horizon, and are produced by its light, from which the atmosphere

Figure 4.1 Phillipe Lopez's eclipse and false sunset from Ternate, Indonesia in 2016.

has removed the colder blue end of the spectrum. So, what produced these colours for our eclipse? And why were they not visible to the north?

It is always fun to socialize with the locals at eclipse time, but the people of Tidore were exceptionally friendly. They laid on shade, refreshments, and entertainments in the form of traditional dancing, and were fascinated to view the Sun through filters, or by our demonstrations of pinholes and shadows. Indonesians, it seems, love selfies, and we were treated like film stars—everyone wanted a picture!

The Indonesians referred to the eclipse as a 'Gerhana Mata Hari,' though it was initially difficult to see what Mata Hari might have to do with it. The famous exotic dancer, courtesan, and spy, executed by firing squad in 1916, was born Margarethe Zolle in the Netherlands, allowing her to travel freely to both sides of the lines during the First World War. During her unhappy marriage she lived in Java in the Dutch East Indies, and she took her stage name from the Malay for the Sun, literally 'eye of the day.' So, a Gerhana Mata Hari is a solar eclipse.

Many readers of this book will have experienced their first solar eclipse in 2017, when the great American eclipse crossed the United States from coast to coast. For that widely broadcast event I flew to Seattle and then drove to Pendleton, Oregon, to assess conditions along the track of totality. Weather looked favourable for eclipse day right across Oregon, but there was a little smoke from forest fires in the centre of the state. More importantly, roads south from Pendleton were single carriageway, and traffic jams were possible. I opted instead to drive down I-84 to Huntington and observe the eclipse from the confluence of the Snake and Burnt Rivers, on the Oregon–Idaho border. The location had been additionally recommended to me by Dr Donald Kennerly, who had reconnoitred the entire totality track, scouting for suitably scenic locations.

Dr Kennerly's recommendation was a good one. The Snake River, which has made its way across a fertile plain, turns north at Farewell Bend, named by the pioneers on the wagon trains which, 150 years ago, opened up the American West. At Farewell Bend, the wagons left the protection of the riverside and pushed upwards into the austere and unforgiving hills. The Burnt River threads its way through those hills.

The site was not crowded but was occupied by an eclectic mix of fellow observers to keep me company. Day trippers from Oregon and Idaho, campers from Seattle and further afield, two experienced imagers from Germany, an enthusiastic group of pensioners from Canada. Once again, we had cloudless skies and a beautiful eclipse, although the altitude of the Sun in the sky deprived us of a backdrop. Once again, shadow bands appeared, now streaming across the hillsides. The solar corona was magnificent, with three separate streamers (see Figure I.4), and the trailing limb of the Moon uncovered a series of prominences along the rim of the Sun as the eclipse progressed. My favourite moment, captured on the soundtrack of my video, was just after totality, when a little girl a few feet away from me announced, 'Gee mommy, that was pretty.' Out of the mouths of babes . . ..

The great American eclipse, of course, was the next in a saros cycle from my not-so-great British eclipse. The two southern African eclipses at the start of the millennium were followed, one saros later, by two total eclipses in South America. In July 2019 I travelled to Bella Vista in San Juan province, northern Argentina, where our group found a viewing site on a llama farm, boasting a view to die for across to the Andes.

The llamas, determined to disprove my theories about animals, were completely unfazed by the appearance of hundreds of eclipse-chasers and seemed completely unaffected by the eclipse. We were more concerned about the hot air balloon being inflated in the neighbouring field, which threatened to block our view of totality, but fortunately the wind got too high to launch. We were also buzzed by CNN's drone, until our tour organizers had a word with them.

For this eclipse, a highlight was the motion of the Moon's shadow. This travels more-or-less west to east at temperate latitudes, but as the observers are looking at the Sun, its apparent motion comes from different directions depending on the time of day. In Australia and Indonesia, watched in the morning, the shadow came over our heads; in Svalbard and Oregon, midday eclipses, it moved from side to side across our field of view. Here in Argentina, for a late afternoon eclipse, we could watch the shadow coming towards us. Photographs often show the shadow as a sharp dividing line between totality and partiality, but my experience is of a more diffuse view, a sense of impending doom sweeping towards us across the Andes. We spoke to other groups who were in the hills above Bella Vista, and they reported seeing the shadow sweep across the valley floor—something for me to watch out for at a future eclipse.

Thousands of people travelled to see the 2019 eclipse, and thousands more hoped to travel back to South America to see the next eclipse, 14 December 2020, which crossed Patagonia. Unfortunately, the COVID-19 pandemic foiled most people's plans, as both Chile and Argentina closed their borders. However, umbraphiles are resourceful people. Several tour groups negotiated with the Argentinian government and secured permits for 94 people to enter specifically to see the eclipse; the first foreign tourists to enter the country in nine months. I was one of the ninety-four! My tour travelled as a 'bubble,' like a sports team, interacting as little as possible with anyone else.

We were rewarded with a great view of the eclipse from an *estancia* (ranch) at Fortin Nogueira, Neuquén Province. In Chile, the weather on eclipse day was terrible, but the Argentinian side of the Andes enjoyed a rain shadow, so the skies were largely cloudless. Interestingly, the temperature-drop as totality approached did form some light cloud, and the Sun was slightly obscured for the first thirty seconds of totality. A side-effect was the formation of a meteorological corona, rings of light in rainbow colours, around the eclipsed Sun (see Figure 4.2). A corona formed from corona light seemed apposite in the year of the coronavirus.

Our expedition to Patagonia, achieved against overwhelming bureaucratic obstacles, and perhaps an object lesson in foolhardiness, nevertheless sums up the spirit of eclipse-chasers. We do not give up easily.

* * *

What conclusions can be drawn about the behaviour of this curious tribe of umbraphiles? We are adventurous, loyal to our cause. We are a very sociable bunch— I have good friends I only ever meet at eclipses, in remote locations around the world. We never tire of totality.[10]

---

[10] I attended my fourteenth total solar eclipse on 20 April 2023, in Exmouth, Western Australia. This was a short-duration total eclipse; I saw 56 seconds of totality from my location in central Exmouth. The

Figure 4.2 Andreas Moeller's photograph of the 'corona corona', a meteorological corona formed by light from the solar corona, at the Patagonia eclipse of 2020.

We live in the middle of a procession of eclipses, starting not long after the formation of the Moon billions of years ago, and due to continue for a billion years to come (total eclipses will dwindle as the Moon recedes gradually from the Earth, becoming extinct in something like a billion years). When I see an eclipse, I feel myself to be standing shoulder to shoulder with eclipse-chasers from years gone by—Elizabeth Brown, Walter and Annie Maunder, Mary and John Evershed. And on into the future—if you haven't yet seen a total eclipse, I hope you'll be shoulder to shoulder with me at an eclipse to come.

See you there!

short duration of totality meant that the sky did not get especially dark, but the progress of the Moon's shadow across my field of view was noticeable, especially on video. Two unexpected aural experiences stood out for me—the roar of the crowd, watching from the public viewing area at Exmouth beach, which reached us a second or two after totality began; and the deafening shrieks of a flock of galahs (pink and grey coloured cockatoos) who decided to fly over us to roost immediately after totality began. For more information on the behaviour of animals, including birds, see also Steven Portugal's chapter on animal behaviour (Chapter 13).

II
HISTORY AND RELIGION

5

Solar Eclipses across Early Asia

John Steele

History of the Exact Sciences in Antiquity, Assyriology

Between the ancient astronomer and the modern eclipse-chaser extend centuries and millennia of human engagement with total solar eclipses all around the world—many of them heavily investing their religions, beliefs, and political import into both prediction and experience. In this chapter, John Steele presents a brief tour of the observation, prediction, and interpretation of solar eclipses across ancient and medieval Asia up to about the year AD 1000. This is no easy task: Asia covers an enormous geographic area, home to a wide range of cultures, languages, and ancient and modern countries. Covering almost three thousand years of human history, Steele focuses his discussion on four case studies: ancient Assyria and Babylonia, India, and China. Apotropaic enterprises such as the practice of the substitute king ritual are excellent examples of human attempts to make sense of a cosmic event and to control its effects on earth.

anomalistic month The period between successive arrivals at an apsis (perigee or apogee for the Moon).

apotropaic rituals Rituals for the aversion of evil.

conjunction When the Moon has the same celestial longitude as the Sun; this is when a solar eclipse can occur.

double dawn When a solar eclipse occurs just before dusk or just after dawn, so that light levels drop because of the eclipse and then recover once the eclipse has finished.

lunar nodes The two points in the Moon's orbit where it crosses the ecliptic (the plane of the Sun's orbital motion). Eclipses can only happen when the Moon is close to the nodes.

magnitude The fraction of the Sun's angular diameter covered by the Moon in an eclipse.

opposition When the Moon has the opposite celestial longitude to the Sun (i.e. it is on the opposite side of the celestial sphere). This is when a lunar eclipse can occur.

portent, omen An indication of something momentous about to happen.

substitute king ritual An example of an apotropaic ritual, where a substitute king is imposed for the duration of an eclipse.

synodic month The average period between successive new Moons.

On 27 May 669 BC, a solar eclipse was seen across Asia. In China, the eclipse reached a maximum magnitude of about 90% at a local time of about a quarter past ten in the morning in the city of Qufu, the capital of the Lu state. The eclipse was observed, rituals were performed in response, and a brief account of the eclipse was written down and later found its way into the fifth century BC *Spring and Autumn Annals*.[1] More than six thousand miles to the west, the Sun rose already eclipsed at Nineveh, the capital of the Assyrian empire. Less than a quarter of the Sun's surface was still covered when it rose. Although only of small extent, the eclipse occurred shortly after a period of political unrest and from preserved correspondence between the Assyrian king and his scholarly advisors we can see that the event caused considerable consternation, resulting in dramatic measures being taken to protect the king. The eclipse was also visible from other parts of Asia. In central India, for example, it would have appeared as a fairly extensive partial eclipse shortly after sunrise and, although we do not have any records preserved of any eclipses from this early period in India, we can presume based upon later parallels that, if it was seen, then it was similarly viewed as a significant event that portended ominous things and may have resulted in the performance of rituals.

This chapter presents a brief tour of the observation, prediction, and interpretation of solar eclipses across ancient and medieval Asia up to about the year AD 1000. This is no easy task: it is important to remember that Asia covers an enormous geographic area, home to a wide range of cultures, languages, and ancient and modern countries. I am also covering almost three thousand years of human history. I therefore focus my discussion on four case studies: ancient Assyria and Babylonia, India, and China. These cultures are those from which we have significant textual evidence concerning solar eclipses from before about AD 1000. Eclipses were of course seen and interpreted in other parts of Asia, but most of our evidence comes from later times.[2] From these four case studies we will see both similarities and differences in how people observed, interpreted, and responded to solar eclipses in these different cultures.

Assyria

A rich body of documentation is preserved from the court archives of kings Esarhaddon and his son Ashurbanipal who ruled the Assyrian empire from 681–669 BC and 669–639 BC respectively. Centred on its capital of Nineveh in what is now northern Iraq, at its height the Assyrian empire controlled territory stretching from the Persian Gulf to the Mediterranean coast, reaching as far north as parts of modern Turkey and at times conquering Egypt. Along with military and political advisors and bureaucrats, the kings employed a group of around twenty scholars as advisors.

---

[1] F. Richard Stephenson and Kevin K. C. Yau, 'Astronomical Records in the *Ch'un-Ch'iu* Chronicle.' *Journal for the History of Astronomy* 23, 1992, pp. 31–51.

[2] In particular, I note that we have many records of solar eclipses from Korea, Japan, and Vietnam from the early second millennium AD onwards. For these records, see F. Richard Stephenson, *Historical Eclipses and Earth's Rotation*. Cambridge, 1997; John M. Steele, *Observations and Predictions of Eclipse Times by Early Astronomers*. Dordrecht, 2000; and Akira Okazaki, 'Solar and Lunar Eclipse Records in Vietnam from Ancient Times Through to the Nineteenth Century.' In *Exploring the History of Southeast Asian Astronomy*, eds Wayne Orchiston and Maynank N. Vahia, Cham, 2021, pp. 163–225.

Among other activities, these scholars kept watch for astronomical phenomena which held ominous meaning for the king. They sent written reports of their observations to the palace in which they quoted the relevant astrological omen. In answer to questions sent by the king, the scholars would often follow up by writing letters explaining how the omen should be interpreted and recommending a course of action that the king might wish to follow such as the performance of a ritual to prevent the omen's ill effects.[3] Any observed celestial event can usually be associated with several different omens depending upon which aspects of the observation were considered the most important by the interpreter. Furthermore, many omen predictions are ambiguous in meaning, forcing the scholar to tailor the interpretation to the current circumstances. By employing several scholars, therefore, the king ensured that he had a range of interpretations to choose from. The omens were just one source of information that could be used by the king as he made decisions, which, in the end, were based more upon the existing political circumstances and his own desires. Indeed, it is possible that, in some cases, the omens were used to provide a justification for actions that the king had already decided to take.[4]

Reports of three or four solar eclipses are found in the preserved correspondence between the scholars and the king.[5] Of these, the eclipse of 27 May 669 BC was probably the most significant. The eclipse took place during the final year of the reign of Esarhaddon. Esarhaddon came to the throne in 681 BC in the aftermath of the murder of his father Sennacherib by two of Esarhaddon's brothers who were angered by Sennacherib's decision to name Esarhaddon as his heir and successor. The brothers tried to seize the throne for themselves but Esarhaddon raised an army and defeated them in a short but brutal civil war. A little over a month later Esarhaddon was recognized as king and established himself in the capital, Nineveh. But this experience left Esarhaddon mentally scarred and throughout his reign he suffered from ill health and a tendency to view others with suspicion. His suspicion was not without good reason: the year before the 669 BC eclipse Esarhaddon uncovered a conspiracy to kill him and seize the throne, in response to which he executed the conspirators and their allies and undertook a radical reshuffle within his government.[6]

At sunrise on 27 May 669 BC (the 29th of the month Iyyar in the Assyrian calendar), Assyrian scholars saw that a small portion of the Sun was eclipsed when it rose over the eastern horizon. One of these scholars, Akkullanu, wrote to the king to report the

---

[3] Such rituals, which are believed to have the power to avert evil, are known as 'apotropaic rituals.'

[4] For an overview of the role and activity of these scholarly advisors, see Karen Radner, 'Royal Decision-Making: Kings, Magnates, and Scholars.' In *The Oxford Handbook of Cuneiform Culture*, eds Karen Radner and Eleanor Robson, Oxford, 2011, pp. 358–79. For editions and translations of the correspondence between the scholars and the king, see Hermann Hunger, *Astrological Reports to Assyrian Kings*, Helsinki, 1992; and Simo Parpola, *Letters from Assyrian and Babylonian Scholars*, Helsinki, 1993. Open-access digital editions and translations of these texts, based upon the books by Hunger and Parpola, are available at http://oracc.museum.upenn.edu/saao/saa08/ and http://oracc.museum.upenn.edu/saao/saa10/.

[5] The exact number is uncertain as some of the reports are damaged and hard to date and so it is unclear whether they refer to the same or different eclipses.

[6] Karen Radner, 'The Trials of Esarhaddon: The Conspiracy of 670 BC.' In *Assur und sein Umland: Im Andenken an die ersten Ausgräber von Assur*, eds Peter A. Miglus and Joaquín Ma Córdoba, Madrid, 2003, pp. 165–84.

eclipse. His report is not preserved, but another scholar quoted it when he wrote to the king:

> May [Nabû and Marduk] bless the k[ing]! A certain Akkullanu has written: 'The sun made an eclipse of two fingers at sunrise. There is no apotropaic ritual against it, it is not like a lunar eclipse. If you say, I'll write down the relevant interpretation and send it to you.'[7]

Akkullanu simply noted that the eclipse had been seen and told the king that no action was necessary. As he pointed out, whereas apotropaic rituals should be performed in response to certain lunar eclipses, which could, among other things, foretell the death of the king, no such ritual needed to be performed for a solar eclipse because these were generally considered to be less serious omens. The king, however, did not agree. Another scholar, Rašil, sent a dreadful warning to the king:

> If there is a solar eclipse in Iyyar [Month II] on the 29th day, it begins in the north and becomes stable in the south, its left horn is pointed, its right horn long: the gods of all four quarters will become confused; great . . . will be spoken by the gods; rise of a rebel king; the throne will change within five years; there will be rebellion in Akkad.[8]

For a king like Esarhaddon, who had just survived an attempt to usurp him, the threat of the rise of a rebel king and a change on the throne within five years must have been very worrying. Rašil went on further to say that the eclipse predicted famine, enemy attack, floods, war, and the theft of cultic objects from the temples—in other words, just about everything bad that could be predicted. Indeed, among the whole corpus of letters and reports sent to the Assyrian kings, Rašil's report is one of the most extensive and bleak in its predictions. Coming on the back of the previous year's conspiracy, Esarhaddon's ongoing health problems, and concerns over the succession, it is perhaps not surprising that Esarhaddon decided to ignore Akkullanu's advice that no action was needed. A letter sent jointly by three of Esarhaddon's scholars reveals how Esarhaddon decided to act:

> [To] our [lord: your servants] Issar-šumu-ereš, [Urad]-Ea, and [Marduk]-šakin-šumi. The best [of he]alth [to] our lord! [May Nabû and] Marduk bless our lord!
> [In accordance with what] our lord [wr]ote to us: '[on the 29]th day, a solar [eclipse took place,'—we shall p]erform the pertinent [apotropaic ritual; somebody should s]it (on the throne) [and remo]ve [your evil].[9]

---

[7] SAA 10 148; translation by Parpola, *Letters from Assyrian and Babylonian Scholars*. In this and subsequent translations, text in square brackets is lost on the original due to damage and has been restored by the translator. Ellipses indicate words or passages that cannot be understood and translated (usually due to damage). Text in parenthesis has been added by the translator to aid in reading.

[8] SAA 8 384 Obv. 9–13; translation by Hunger, *Astrological Reports to Assyrian Kings*.

[9] SAA 10 25; translation by Parpola, *Letters from Assyrian and Babylonian Scholars*.

The ritual that Esarhaddon decided to perform was the so-called 'substitute king' ritual. During this ritual, a substitute (usually a convicted criminal, a political enemy, or a commoner) would be placed on the throne and given the trappings of office so that the evil associated with the eclipse would be taken on by the substitute. Meanwhile, the real king, Esarhaddon, would be cloistered away in a special house where he would be addressed as 'the farmer' and undergo a range of cleansing rituals. After a set time period, in theory lasting one hundred days but in practice often somewhat shorter, the substitute king would be executed ('sent to his fate'), allowing the real king to return to the throne unharmed by the evil of the eclipse. The substitute king ritual involved intricate preparation and considerable resources to perform, and could interfere with the smooth running of the government (although it is clear that the real king was still in change even when playing the role of the farmer). As a consequence, it was normally only performed on the occasion of select lunar eclipses which predicted the death of the king. Late in his reign, Esarhaddon, however, performed it several times as he grew increasingly concerned for his personal safety. Ultimately, despite his efforts, it was not sufficient to secure Esarhaddon's long-term future as he died only a few months later.

In addition to providing an insight into the political importance of eclipses in the Assyrian empire during the seventh century BC, Akkullanu's initial report of this eclipse also tells us something about the astronomical understanding of eclipses by the scholars of this time. Akkullanu reported that when the Sun rose 'two fingers' of the Sun was eclipsed. The Babylonians and Assyrians measured the magnitude of eclipses using a unit of linear measure called fingers, with a total eclipse corresponding to 12 fingers. Thus, when the Sun rose, only one-sixth of the Sun was eclipsed (i.e. it had a magnitude of less than 20%). It is very unlikely that an observer would notice a solar eclipse of this small size by chance. Akkullanu's report, therefore, indicates that he and other scholars were watching for the eclipse. In other words, they expected that an eclipse might happen because they had predicted it in advance.

Various hints about how the Assyrian scholars identified when to look for an eclipse are found in their letters and reports.[10] It is important to note that the Assyrians attempted merely to predict eclipse possibilities—that is, occasions when an eclipse might be seen as opposed to those when an eclipse definitely will not be seen. Predicting the local circumstances of a solar eclipse was beyond the ability of the Assyrians (and, as we will see, the Babylonians). The letters and reports include statements that eclipses can only occur a certain period after another eclipse. This period is usually a multiple of six months, but can occasionally be five months, 11 months, or 17 months. In a letter written by the scholar Nabû-aḫḫe-eriba we find reference to these periods:

[Eclipses] cannot occur [dur]ing certain periods. [After] 5 months, there was a watch in Marchesvan (Month VIII), and now, in the month Kislev (Month IX) we will (again) keep watch.[11]

---

[10] John M. Steele, 'Eclipse Prediction in Mesopotamia.' *Archive for History of Exact Science* 54, 2000, pp. 421–54.
[11] SAA 10 71 Rev. 4–10; translation adapted from Parpola, *Letters from Assyrian and Babylonian Scholars*. Parpola read the number before the word months as 4, but noted that the sign was damaged. The

Other letters indicate that the scholars knew that lunar and solar eclipses could take place within half a month of one another. In other words, a solar eclipse is possible at the conjunction either immediately before or immediately after the opposition at which a lunar eclipse is possible, but not at other conjunctions.

Babylonia

South of the Assyrian heartland lies Babylonia. Although under Assyrian domination during the seventh century BC, Babylonia, whose history as a nation already stretched back more than a thousand years, remained the cultural capital of the area. It was from Babylonia that Assyria inherited much of its tradition of astronomy and astrology, and several of the scholars who advised the Assyrian king were based in Babylonian cities and were themselves Babylonian. Late in the seventh century BC, Babylonia and its eastern allies succeeded in overthrowing Assyrian rule, and Babylonia again became the dominant political power in the region. This power lasted for only a hundred years or so, however, before Babylonia fell to the Persians and became part of the Achaemenid empire in the late sixth century BC, and was subsequently captured by Alexander who brought the region under Greek control in the late fourth century BC. Greek rule lasted until the second century BC when the region was captured by the Parthians. Throughout all of this political turmoil, however, Babylonian culture, and in particular Babylonian traditions of astronomy and astrology, remained strong, with texts continuing to be written in Akkadian cuneiform.

The beginnings of astronomy and astrology in Babylonia can be traced back to at least the early second millennium BC, when we have the earliest preserved texts to contain astrological omens—most of the preserved examples contain lunar eclipse omens—lists of stars and constellations, and simple numerical schemes for modelling the change in the length of daylight over the course of the year. It is not until the first millennium BC, however, that we find evidence of extensive astronomical practice in Babylonia. This practice included the regular and systematic observation of lunar and planetary phenomena, the development of methods to predict these same phenomena in advance, and new methods of astrology which sought to make predictions both for the civil realm (e.g. business, weather, and agriculture) and for personal matters (e.g. the fate of an individual, medical treatment, and the birth of sons and daughters).

Regular astronomical observation seems to have started in Babylonia in the middle of the eighth century BC and by the end of the seventh century BC had become a standardized practice not only in what was observed and how the observations were made but also in recording those observations in texts known to modern scholars as 'astronomical diaries' (their ancient name was 'regular watching').[12] These observations were made in the city of Babylon, which is located about 90 kilometres south

number 5 makes more sense here and is confirmed as possible on inspection of the original tablet (collation courtesy C. B. F. Walker).

[12] Abraham Sachs, 'Babylonian Observational Astronomy.' *Philosophical Transactions of the Royal Society of London* A 276, 1974, pp. 43–50; John Steele, 'The Early History of the Astronomical Diaries.' In *Keeping Watch in Babylon: The Astronomical Diaries in Context*, eds Johannes Haubold, John Steele, and Kathryn Stevens, Leiden, 2019, pp. 19–52.

of present-day Baghdad. Several hundred astronomical diaries have been preserved, the vast majority of which date from between about 400 BC and 50 BC, with a very small number of earlier examples.[13] They typically cover half of a Babylonian year and are arranged into sections for each month. Most of each monthly section is taken up with a day-by-day (or night-by-night) record of what was observed. This chronological presentation is followed by four short summary sections which present data on the value of certain staple commodities in the market, the river level as it flows through Babylon, certain planetary data, and selected historical events. Contrary to the letters and reports sent to the Assyrian kings discussed in the previous section, the Babylonian astronomical diaries are very dry, anonymous documents which present in a systematic and formulaic fashion the astronomical data with no comment on its astrological meaning or anything personal about the scholar who either made the observations or compiled the record.

More than a hundred records of solar eclipses are preserved in the astronomical diaries or in texts which were compiled from them.[14] Of particular interest are two detailed accounts of an observation of the total solar eclipse on 15 April 136 BC. The eclipse took place on the morning of the 29th day of the second (intercalary) Month XII of year 175 of the Seleucid Era in the Babylonian calendar. The first account reads as follows:

> The 29th, at 24 (uš) after sunrise, solar eclipse; when it began on the south and west side, [. . . Ve]nus, Mercury and the Normal Stars were visible; Jupiter and Mars, which were in the period of invisibility, were visible in its eclipse [. . .] it threw off (the shadow) from west and south to north and east; 35 (uš) onset, maximum phase, and clearing; in its eclipse, the north wind which had set [to the west side blew . . .][15]

The account states that the eclipse began at 24 uš (= 1 hour 36 minutes) after sunrise,[16] on the south-west of the Sun's disc. During the maximum phase of the eclipse, the four planets Venus, Mercury, Jupiter, and Mars were visible, these last two being at that time in their invisible phase when they cannot be seen during the night. The 'Normal Stars', a group of reference stars distributed unevenly around the zodiacal band, were also seen during the eclipse. The visibility of stars and planets implies that the eclipse was almost certainly total. Unfortunately, the text breaks away at this point so we are not told how big the eclipse was. When it resumes, we learn that the shadow cleared towards the north-east; that the whole eclipse lasted for 35 uš (= 2 hours 20 minutes), and that the north wind blew during the eclipse.

The second account of this eclipse, which is found in a Goal-Year Text which was compiled from data in the diaries, seems to contain an abbreviated account of the

---

[13] All of the known astronomical diaries have been edited and translated by Abraham J. Sachs and Hermann Hunger, *Astronomical Diaries and Related Texts from Babylonia*, vols. 1–4. Vienna, 1988–2022. Open-access digital editions and translations of these texts, based upon the books by Sachs and Hunger, are available at http://oracc.museum.upenn.edu/adsd/.

[14] For detailed studies of these records, see Steele, *Observations and Predictions*, and Peter J. Huber and Salvo de Meis, *Babylonian Eclipse Observations from 750 BC to 1 BC*. Milan, 2004.

[15] ADART III −136B Rev, 13′–16′; translation adapted from Sachs and Hunger, *Astronomical Diaries and Related Texts from Babylonia*, vol. 3.

[16] The uš is a Babylonian unit of time equivalent to 4 minutes.

eclipse. Crucially, however, this second account preserves a clear statement that the eclipse was total:

> The 29th, solar eclipse, [when it began] on the south and we[st] side, in 18 (uš) of daylight to the inside [of the Sun], it was complete, and made a total (eclipse). At 24 (uš) after sunrise.[17]

This account provides the additional information that it took 18 uš (= 1 hour 12 minutes) for the eclipse to become total.

The two reports of the total solar eclipse of 136 BC are remarkable for the precision with which they record the circumstances of the eclipse—the time at which it began, the duration of the various phases, and the direction of the obscuration—and also for the lack of either dramatic language used in its description or any attempt at astrological interpretation. For want of a better way of putting it, they are scientific reports of a scientific observation made by someone with a scientific mind. And the precision and detail of the observation reports have made them invaluable to the observer's modern scientific counterparts. This eclipse observation has played a crucial role in modern scientific studies of the long-term changes of the Earth's rate of rotation and its geophysical causes.[18]

The only other possible report of a total solar eclipse from Babylon is preserved on a tablet so badly damaged that the identification of the eclipse is not fully certain. The account reads as follows:

> The 28th day, solar eclipse; from [. . .] it began; 23 (uš) of day to the inside of the sun [. . .] its [. . .] were clear; 2 (uš) [maximal phase;] Venus, Mercury, Mars [. . .] the remainder [. . .] Sirius, which has set in its non-visibility [. . .] In its eclipse [. . .] stood there [. . .] people broke pots [. . .] they broke. In 23 (uš) of day it cleared from north [and west] to south and east. 48 (uš) onset, [maximal phase,] and clearing. In its eclipse, the north and west winds blew. At 1,30 (uš) of day before sunset.[19]

The extensive damage to the text makes it hard to follow all of the details of what was observed. But comparing the account with those of the total solar eclipse of 136 BC we find several similarities: the eclipse reached the 'inside of the sun', planets and stars were visible during the eclipse, and the eclipse has a maximum phase which lasted for 2 uš (= 8 minutes), which is slightly too long for a total solar eclipse but this probably just reflects the fact that time measurements were rounded to the nearest whole uš (4 minutes). It seems very likely, therefore, that this report refers to a total solar eclipse, and if that is the case then the only total solar eclipse which was visible in Babylon at the right time of day between 750 BC and AD 100 is the eclipse of 30 June 10 BC, which had a maximum length of totality of 6 minutes

---

[17] ADART VI 68 Rev. I 21–5; translation adapted from Sachs and Hunger, *Astronomical Diaries and Related Texts from Babylonia*, vol. 6.

[18] F. Richard Stephenson and Leslie V. Morrison, 'Long-term Fluctuations in the Earth's Rotation: 700 BC to AD 1990'. *Philosophical Transactions of the Royal Society of London* A 351, 1995, pp. 165–202.

[19] ADART V 32 2′–12′; translation adapted from Sachs and Hunger, *Astronomical Diaries and Related Texts from Babylonia*, vol. 5.

14 seconds; computation puts Babylon right on the edge of the path of totality of this eclipse, which would reduce the local duration.[20] An interesting aspect of the report is the mention of people breaking pots, which seems to be a reference to a spontaneous ritual response to the eclipse by the general population. There is no suggestion in the report that the author participated in any such ritual or was alarmed or surprised by the occurrence of the eclipse. Total solar eclipses are, of course, impossible not to notice unless the sky is covered by clouds. The Babylonian observers, however, also recorded several very small partial eclipses. Such observations imply that the possibility of an eclipse was predicted in advance and a watch kept for it. The Babylonians developed two main methods for predicting eclipses. In the first method, they cleverly exploited their identification of the saros eclipse cycle of 223 synodic months, which closely equals 242 draconitic months and 239 anomalistic months, not only to predict eclipses which occur one saros after a visible eclipse, but, by combining this knowledge with the fact that eclipse possibilities occur at either six-month or (less often) five-month intervals, also to predict all of the 38 eclipses within a saros period of 223 months in a pattern that then repeats for several cycles.[21] They were also able to use their knowledge of the saros cycle to predict the time and likely visibility of the eclipse. The second method for predicting eclipses developed by the Babylonians used arithmetical functions which take into account the variable motion of the Sun and Moon and the motion of the node of intersection of the orbits of the Sun and Moon as seen from the Earth to calculate the Moon's position in celestial longitude and latitude at each syzygy.[22] From this information, they could select the syzygies at which eclipses were possible, the time at which they would occur, and their maximum magnitude. Both of these methods—the simpler saros cycle computation and the more complicated mathematical functions used to compute the Moon's position—gave excellent results and the two approaches were used side by side by the Babylonian astronomers during the last few centuries BC.

India

India has a long and complex history of astronomy which stretches back to the Vedic period (late second and first millennia BC) and whose traditional astronomy continued to be practised up to the nineteenth century AD.[23] Indian astronomy is an amalgam of practices coming out of early Indian religious and cosmological traditions; astronomical knowledge that came from Babylonia, Greece, Persia, and, later, the Islamic world and was adapted to an Indian context; and later Indian developments in astronomy and, especially, mathematics. The majority of the sources for

[20] Steele, *Observations and Predictions*, 37–41.

[21] Steele, 'Eclipse Prediction in Mesopotamia.' For the saros, see also the discussion in Chapter 2.

[22] Babylonian mathematical astronomy operated with arithmetical functions rather than geometrical models of the kind found in ancient Greek astronomy and discussed by Nothaft in Chapter 3. For the Babylonian methods, see Mathieu Ossendrijver, *Babylonian Mathematical Astronomy: Procedure Texts*. New York, 2012.

[23] For a detailed study of Indian material on eclipses, see chapter 5 of Clemency Montelle, *Chasing Shadows: Mathematics, Astronomy, and the Early History of Eclipse Reckoning*. Baltimore, MD, 2011.

Indian astronomy date from the middle of the first millennium AD onwards (and almost all of these exist only in much later copies), but some include references to earlier material.

According to early tradition, eclipses of the Sun and Moon were caused by the demon Rāhu devouring the luminary, the eclipse ending when the luminary subsequently fell out of a gaping hole in the neck of the demon. Over time, the head of Rāhu and Ketu, the lower part of his body which had been cut off by Viṣṇu's sword, came to be associated with the lunar nodes, the two points of the intersection of the Moon's orbit with the ecliptic, the path of the Sun. The idea that the demon Rāhu caused eclipses was clearly considered problematical by many astronomers, but directly challenging this tradition was difficult for cultural reasons. One person that did challenge it, however, was Varāhamihira, who lived in the sixth century AD. Describing Rāhu, he wrote:

> His disc is similar in form to those of the Sun and Moon, but he is not visible in the heavens except on Parva days (New and Full Moon days) on account of his blackness in colour. Owing to a boon conferred by Brahma, he is visible only at the time of eclipses and not on other days. One school of learned men says that Rahu—son of Sumhika—is of a serpentine form with only the face and the tail; while another class maintains that he is formless and of the nature of pure darkness.[24]

Varāhamihira lists several problems with the idea that Rāhu is responsible for eclipses:

> For, if Rahu has a form, travels in the zodiac, possesses a head and has a circular orb, how is it that he whose movement is fixed and uniform seizes the two luminaries who are situated 180° away from him? If his gait has not been fixed, how is his exact position determined by calculation? If he is not to be distinguished by his tail and his face, why should he not seize them at other intervals (instead of only when 180° apart)? For, if this Rahu who is of the form of a serpent is able to seize the Sun or the Moon through his tail or his mouth, why should he not conceal or hide half of the zodiac which is the interval between his head and his tail? If there should be two Rahus, when the Moon has set or risen and is eclipsed by one Rahu, the Sun (who is 180° from the Moon) should also be eclipsed by the other Rahu whose rate of motion is also similar.[25]

These are all sound arguments, arguments that all of Varāhamihira's contemporary astronomers, who could calculate the occurrence of eclipses using geometrical methods, would have agreed with. In place of Rāhu, Varāhamihira explains that the real causes of lunar and solar eclipses are shadows:

> In her own eclipse, the Moon enters the shadow of the Earth, and in that of the Sun, the solar disc. Therefore it is that the lunar eclipse does not commence at the western

[24] Varāhamihira, *Bṛhatsaṃhitā* V 2–3; trans. Panditabhushana V. Subrahmanya Sastri and Vidwan M. Ramakrishna Bhat, *Varahamihura's Brihat Samhita, with an English Translation and Notes*, Bangalore, 1946.

[25] Varāhamihira, *Bṛhatsaṃhitā* V 4–7; trans. Sasti and Bhat, *Varahamihira's Brihat Samhita*.

limb, nor the solar at the eastern limb. Just as the shadow of a tree goes on increasing on one side, so is the case with the Earth's shadow every night by its hiding the Sun during its rotation. If the Moon, in her course towards the East and placed in the 7th house from the Sun, does not swerve much either to the north or the south, she enters the shadow of the earth. The Moon moving from the west conceals the solar disc from below just like a cloud. The solar eclipse is therefore different in various countries according to the visibility of the eclipsed disc. In the case of a lunar eclipse, the concealing agency is very big, while in that of the solar, it is small. Hence in semi-lunar and semi-solar eclipses, the luminous horns are respectively blunt and sharp. Thus the cause for the eclipses has been given by our ancient masters possessed of divine sight. Hence, the scientific truth is that Rahu is not at all the cause of it.[26]

In another work, the *Pañcasiddhāntika* which summarizes the astronomy of five earlier *siddhāntas* (astronomical treatises), Varāhamihira provides detailed geometrical methods for calculating the shadow and the resulting eclipse.

Similar arguments against the notion that the demon Rāhu caused eclipses were given by other astronomers including the seventh-century astronomer Brahmagupta and his ninth-century commentator Pṛthūdakasvāmin, and the eight-century astronomer Lalla.[27] These authors are somewhat more circumspect, however, explaining why Rāhu the demon cannot be the cause of an eclipse and what the true cause is, before the explaining that earlier people connected Rāhu to the eclipse because it is a name for the node, a rather unconvincing after-the-fact justification made in order not to contradict the authority of the Vedas and related texts.

Several Sanskrit astronomical treatises from the first millennium AD include detailed explanations of how to calculate lunar and solar eclipses. The methods are based upon the geometrical astronomy of the Greek tradition.[28] They use trigonometry to take into account the effect of parallax and the geographical location of the observer, as well as lunar and solar models which account for the various anomalies in the motion of these bodies and a model for the motion of the lunar nodes. They also recast the Greek methods into a form using the trigonometric sine function rather than the Greek trigonometric chord function, and transform the calculations in order use the traditional Indian astronomical epoch of 17/18 February 3102 BC, the start of the current *yuga* period, the *Kaliyuga* in which we are currently living.

The first millennium AD astronomical treatises are all theoretical and/or computational in nature, providing the methods by which eclipses can be predicted. They do not preserve any records of observed lunar or solar eclipse, or indeed reports of observations of any astronomical phenomena.[29] For reports of observed eclipses, we must look elsewhere. Many inscriptions carved into stone, copper plates, and other media attached to temples etc. are preserved which record grants and gifts by kings and other

---

[26] Varāhamihira, *Bṛhatsaṃhitā* V 8–13; trans. Sasti and Bhat, *Varahamihira's Brihat Samhita*.

[27] Montelle, *Chasing Shadows*.

[28] For the circulation of astronomical knowledge between the Greek world and India, which seems to have taken place in the early centuries of the first millennium BC, see David Pingree, 'The Recovery of Early Greek Astronomy from India.' *Journal for the History of Astronomy* 7, 1976, pp. 109–23.

[29] A handful of reports of observations are found in later, second millennium AD astronomical treatises, but even these are rare.

high members of society.[30] Often these gifts were given in response to the observation of celestial events, including solar eclipses. For example, on the equivalent of 20 August 993 AD, a solar eclipse is recorded in the Janjirā Plates of Aparājita:

> On Sunday, the fifteenth tithi in the dark fortnight of Śrāvaṇa in the cyclic year Vijaya in the expired year nine hundred and fifteen of the era of the Śaka king on the very holy occasion of a solar eclipse when the disc of the hot-rayed one (the Sun) was swallowed by the planet Rāhu, and when the sun was in the rāśi Siṃha.[31]

Modern computation shows that the eclipse was total near Jinjirā. It began around noon and ended shortly before 3 pm and the Sun was indeed located in the zodiacal sign (rāśi) Siṃha (= Leo) at that time.

An inscription located in the Saligrama Hassan District in south-west India refers to the annular solar eclipse of 26 June 819 AD. During this eclipse, about 95% of the Sun's disc was covered. The inscription seems to refer to the annular nature of the eclipse using a term drawn from a Sanskrit word meaning 'defect,' presumably referring to the Moon being too small to completely cover the Sun.[32]

China

Like India, China has a long history of astronomy which can be traced back to the second millennium BC. However, with the exception of the late second millennium BC Shang dynasty oracle bones—animal bones and turtle plastron which were used in divinatory practice and which were sometimes inscribed with texts which in small proportion of cases seem to refer to astronomical events—and a handful of texts written during the early and mid first millennium BC, written evidence for Chinese astronomy comes more or less exclusively from texts written after the unification of China by the first Qin emperor, Shi Huangdi, in the late third century BC. Unfortunately, the interpretation of the oracle bone inscriptions is a difficult and evolving process, especially with regard to any astronomical references they contain. A few oracle bones have been claimed to contain references to solar eclipses, but both the identification of these as eclipses and their date remain uncertain.[33]

The earliest references to solar eclipses are found in the *Bamboo Annals*. This text has a complicated history. The original text was buried with king Xiang of Wei in the early third century BC and then rediscovered in AD 281 when a tomb robber broke into Xiang's tomb. The text was lost again during the Song dynasty (AD 960–1279). Two versions of the text circulated after the sixteenth century: a 'current' version

---

[30] Montelle, *Chasing Shadows*; B. S. Shylaja and Geetha Kydala Ganesha, *History of the Sky—On Stones*. Bangalore, 2016.

[31] *CII* VI 5, p. 17, lines 57–9; translation by Montelle, *Chasing Shadows*.

[32] Shylaja and Ganesha, *History of the Sky*.

[33] For these records, see Zhentao Xu, Kevin K. C. Yau, and F. Richard Stephenson, 'Astronomical Records on the Shang Dynasty Oracles Bones.' *Archaeoastronomy* 14, 1989, pp. S61–S72. On the problems of identifying eclipses in these records, see F. Richard Stephenson, 'How Reliable are Archaic Records of Large Solar Eclipses?' *Journal for the History of Astronomy* 39, 2008, pp. 229–50.

(which some modern scholars think was a forgery) and a partial 'ancient' version reconstructed in the late nineteenth and early twentieth centuries by scholars who combed pre-Song works for quotations of the *Bamboo Annals*. The *Annals* present a history of China from very ancient, mythical times (the age of the Yellow Emperor), down to 299 BC. Two possible references to eclipses are found in this text. The first concerns a solar eclipse supposedly seem during the reign of Zhong Kang of the Xia dynasty. The record reads as follows:

> 5th year of Emperor Zhong Kang of Xia, autumn, 9th month, day *gengxu* (47), the first day of the month; there was an eclipse of the Sun.[34]

The Xia dynasty was considered the first Chinese dynasty in traditional Chinese historiography, preceding the Shang dynasty, traditionally assumed to have ruled between 2205 and 1766 BC. No contemporaneous documents are known from the Xia dynasty and some modern scholars consider it mythical. The Xia eclipse report seems very precise, giving the exact date of the eclipse using both the lunisolar calendar and the 60-count cycle of days. As a consequence, scholars in both China and Europe have long tried to identify this eclipse in order to provide an absolute date for the reign of Zhong Kang. For example, the late seventh and early eighth century AD Chinese scholar Yixing used astronomical calculation to try to identify the eclipse, deciding that it fitted a solar eclipse in 2128 BC.[35] Almost a thousand years later, the Jesuit Antoine Gaubil in Beijing wrote to the French astronomer Joseph Nicholas Delisle for help with identifying the eclipse, and his letter was passed on first to Leonard Euler and then to the astronomer Tobias Mayer, who had made a significant leap forward in producing accurate lunar tables. Mayer wrote:

> I have taken the effort to calculate this eclipse according to my tables, though I was almost assured well in advance that my efforts would be of no particular utility. According to this calculation I find for two reasons that this eclipse was invisible throughout China. In the first place the latitude of the moon was overly to the north for the shadow or the penumbra to have touched even the smallest part of China; and secondly the time of conjunction of the moon with the sun fell just before the rising of the sun. I have full reason to believe that this calculation is more correct than those which indicate that the eclipse was visible in China, because in my tables, the motion of the moon was slower in former times than now, which others do not have. And so I have grounds that the entire report of the seen and yet nonetheless invisible solar eclipse is a highly suspect and completely false account. Perhaps a Chinese astronomer in a much later time has brought out a solar eclipse from a very imperfect backwards calculation of the abovementioned year, and with it given that it appeared real at that time. Ignoring that, it appears as well from the above that even this year of the eclipse is still highly doubtful. [...] I conclude from all this that a sharp judgement is necessary of the report that is brought to us from China. A half-scholarly

---

[34] *Bamboo Annals*; trans. Zhentao Xu, David W. Pankenier, and Yaitiao Jiang, *East Asian Archaeoastronomy: Historical Records of Astronomical Observations of China, Japan, and Korea*. Amsterdam, 2000.
[35] Weixing Niu, 'Liu Xin and Ancient Astronomical Chronology.' In *The Studies of Heaven and Earth in Ancient China*, ed. Xiaoyuan Jiang, New York, 2021, pp. 55–93.

and in astronomical matters moderately experienced people like the Chinese were before the Europeans came to them, could easily forge such false eclipses, and to make their origins respectable and their erudition old supply inaccurate calculations for the observations; because they believed that he was not smart enough to discover the fraud.[36]

Mayer's judgement here was sound. Despite various modern attempts to rehabilitate this eclipse record and try to date it, it seems certain that this is nothing more than a literary story that is embedded within the mythical early history of the *Bamboo Annals*. The *Annals* also refer to an unusual event during the reign of king Yi of the Western Zhou (early first millennium BC):

> 1st year of King Yi (of W. Zhou), the 1st month of Spring [. . .] the day dawned twice at Zheng.[37]

Many authors have suggested that this 'double dawn' event refers to a solar eclipse which became total (or at least very large) just after sunrise, thus producing a second darkening of the sky just after dawn, and have identified the eclipse as the annular solar eclipse of 23 April 899 BC.[38] This proposal is attractive in that two similar 'double dawn' events are described in medieval European chronicles. Ultimately, however, we cannot be certain in this identification, or indeed whether this report even refers to a solar eclipse. As pointed out by Stephenson, there are Chinese and Korean reports of aurorae which also refer to dawn-like effects, even, in one case, referring specifically to dawn.[39]

The earliest reliable reports of solar eclipses date from about 700 BC onwards and are found in the *Spring and Autumn Annals*. These reports include the solar eclipse of 27 May 669 BC with which I started this chapter and which we saw had significant repercussions in Assyria. We know less about how this specific eclipse was interpreted in China: the record itself merely states that the eclipse took place and that 'drums were beaten and oxen were sacrificed at the temple.'[40] It is likely, however, that the eclipse was interpreted politically—this is certainly the case with later eclipses. Indeed, by the end of the first millennium BC, a political philosophy had developed that saw unexpected celestial and terrestrial events as criticisms of the rule of the emperor sent by heaven—a direct sign that the emperor was losing his mandate to rule. Reports of observed solar eclipses recorded in the treatises of the official dynastic histories often allude to this fact; indeed, we can read the compilations of

---

[36] MS Mayer 9 §54–5; ed. Eric Forbes, *The Unpublished Writings of Tobias Mayer. Vol. 1: Astronomy and Geography*, Göttingen, 1972. See the detailed discussion in John M. Steele, *Ancient Astronomical Observations and the Study of the Moon's Motion (1691–1757)*, New York, 2012.

[37] *Bamboo Annals*; trans. Xu, Pankenier, and Jiang, *East Asian Archaeoastronomy*.

[38] See, for example, Chaoyang Liu, 'Examination of Solar and Lunar Eclipses During the Last Years of Yin and the Early Years of Zhou.' *Proceedings of Chinese Culture* 4, 1944, pp. 85–119 and Ciyuan Liu, Zueshun Liu and Liping Ma, 'Examination of Early Chinese Records of Solar Eclipses.' *Journal of Astronomical History and Heritage* 6, 2003, pp. 53–63.

[39] Stephenson, 'How Reliable Are Archaic Records of Large Solar Eclipses?'

[40] *Spring and Autumn Annals*; trans. Stephenson and Yau, 'Astronomical Records in the *Ch'un-Ch'iu* Chronicle.'

astronomical observations given in the 'Heavenly Patterns' treatises of the dynastic histories as political commentaries as much as—perhaps more than—astronomical reports. For example, the report of the solar eclipse on 5 August 221 AD recorded in the Heavenly Patterns treatise of the *Jinshu* (Book of Jin) only briefly describes the eclipse before discussing the political situation:

> Emperor Wen of Wei, 2nd year of the Huangchu reign period, 6th month, day *wuchen* (5), last day of the month, the sun was eclipsed. The officials petitioned the emperor to dismiss the Prime Minister (putting the blame for the eclipse on him), but the imperial rescript said, 'Since portents are warning for the Head of State, to put the blame on my Minister would be against the spirit of self-penance exercised by Emperors Yu and Tang. I (hereby) order all my officials to devote themselves diligently to their respective duties and refrain from impeaching my senior minister in the event of future celestial and terrestrial portents.'[41]

Wen was the first emperor of the Wei state during the Three Kingdoms period which followed the collapse of the Han. It seems that his officials were keen to pin the blame for the eclipse on his Prime Minister rather than the emperor himself, but the emperor refused.

Political commentary continues with other eclipse records given in this chapter of the *Jinshu*. The record for the solar eclipse of 24 April 247 reads as follows:

> 8th year (of the Zhengshi reign period), 2nd month, day *gengwi*, first day of the month, the sun was eclipsed. In those days the government was under the despotic control of Cao Shuang and the laws were changed by Ding Mi and Deng Yang. Upon the occurrence of the eclipse (the Emperor) Cao Fang summoned together his officials and asked them to point out his faults. Whereupon Jiang Ji submitted a memorial saying, 'Formerly when the great Shun assisted (Emperor Yao) in his government he avoided all contacts with the wicked, and when Zhou Gong helped (Emperor Wu Wang) he was cautious in selecting his associates. The Marquis of Qi once enquired what calamities (would follow a solar eclipse) and was advised by Yan Zi to practise more charity, and when the ruler of Lu asked about the same portent Zangsun advised him to ease the labour of his people. The only response to the portent in the heavens can thus be: nothing more can be done within human capabilities.' The advice and illustrations given by Ji were correct and of vital importance. However, from the Emperor to the Ministers none was enlightened (by his advice), until ultimately ruin and disaster fall upon (the Kingdom of Wei).[42]

The author of this passage makes a number of points: he explains the basic philosophy of omens and the interpretation of eclipses in particular, he presents examples of past responses to an eclipse by wiser emperors, and he notes that these examples were

---

[41] *Jinshu* 12; translation adapted from Peng Yoke Ho, *The Astronomical Chapters of the Chin Shu*. Paris, 1966.
[42] *Jinshu* 12; translation adapted from Ho, *The Astronomical Chapters of the Chin Shu*.

not followed by Cao Fang and his despotic government, which ultimately led to the downfall of the kingdom.

This practice of astronomical record-keeping as political commentary in the Heavenly Patterns chapters of the official dynastic histories was not the only way that Chinese astronomers worked with eclipses, however. Alongside the interpretation of portents as signals from heaven about the state of the emperor's rule, the sky needed to be regulated through the advance prediction of celestial phenomena in order to maintain the heavenly mandate. Beginning in the Han dynasty, Chinese astronomers developed systems of mathematical astronomy capable of predicting the conjunction of the Sun and Moon (which marked the beginning of the month), the motion of the Sun, Moon, and planets through the constellations, and eclipses.[43]

The earliest systems relied upon basic astronomical periods, but later systems incorporated the effects of lunar and solar anomaly, the Moon's latitudinal motion, and an approximation to parallax into their calculations. It was well known that eclipses were the hardest to calculate. As Guo Shoujing wrote towards the end of the thirteenth century:

> The test of an astronomical system's exactitude is its treatment of eclipses. In this art of pacing the celestial motions, exactitude is hard to come by. There is always (uncertainty about) whether the (predicted) time of day is early or late, and whether the immersion is too shallow or too deep. If exact agreement (with the phenomena) be the goal, there can be no room for happenstance.[44]

In other words, if there is any inaccuracy in a system of astronomical calculation, such as an inaccurate parameter whose effects will build up over time, this will first become evident in the prediction of eclipses. Eclipses were observed, therefore, both to determine the parameters of astronomical theories and to test the accuracy of those theories. These tests often involved comparing predictions of upcoming eclipses with their subsequent observation as well as retrocalculating the circumstances of eclipses that were recorded in earlier texts and could, at least in principle, lead to the replacement of the official state system of astronomy with a new one that had performed better in these tests.[45]

Detailed observations of several solar eclipses are preserved in the dynastic histories. It seems that the Chinese astronomers observed solar eclipses by looking at the reflection of the Sun in water or oil, which allowed them to observe eclipses where only a small part of the Sun's disc was obscured. A reference to this practice is found

---

[43] For a translation and analysis of the three earliest systems of mathematical astronomy, see Christopher Cullen, *The Foundations of Celestial Reckoning: Three Ancient Chinese Astronomical Systems*. Abingdon, 2017. For contextual studies of these systems and more broadly of astronomy in early China, see Christopher Cullen, *Heavenly Numbers: Astronomy and Authority in Early Imperial China*. Oxford, 2017 and Daniel P. Morgan, *Astral Sciences in Early Imperial China: Observation, Sagehood and the Individual*. Cambridge, 2017.

[44] *Yuanshi* 53; translation by Nathan Sivin, *Granting the Seasons: The Chinese Astronomical Reform of 1280, With a Study of Its Many Dimensions and an Annotated Translation of Its Records*. New York, 2009.

[45] Yuzhen Guan, 'Calendrical Systems in Early Imperial China: Reform, Evaluation and Tradition.' In *The Circulation of Astronomical Knowledge in the Ancient World*, ed. John M. Steele, Leiden, 2016, pp. 451–577; Morgan, *Astral Sciences in Early Imperial China*.

in a report of an observation of a solar eclipse on 11 April 1176 by Cheng Dachang. He explains that it has been predicted that the solar eclipse will have a magnitude of 1.5/10ths. In order to check the accuracy of the prediction, he looked at the reflection of the Sun in a dish filled by oil. Of course, when it came time to make the observation, it was cloudy. When the clouds cleared, less than one-tenth of the disc was covered.[46]

The report of the solar eclipse of 12 December 429 AD notes that the eclipse was total in some places but only partial elsewhere:

> Emperor Wen of Jin, 6th year of the Yuanjia reign period, 11th month, day *jichou* (26), first day of the month. There was an eclipse of the Sun; it was not complete but like a hook. During the eclipse the stars appeared. By the hour of *fu* it ended. In Hebei province the earth was in darkness.[47]

A second record reports that 'stars were seen in daytime.'[48] Modern computation shows that the track of totality passed just to the north of the capital Jiankang, resulting in the 'hook'-like eclipse where the sky darkened enough to make out stars, and was total in Hebei.[49]

* * *

Along this brief tour of solar eclipses across early Asia we have encountered both similarities and differences in the way that people have observed, calculated, theorized, interpreted, and responded to eclipses of the Sun, both between the four cultures studied here and between each one of them and modern western society. For example, while all four of the cultures studied here developed mathematical methods to predict eclipses, the methods themselves differed considerably. The Babylonian astronomers identified and exploited the 223-synodic month saros cycle in order to predict eclipses using arithmetical methods combined with previous observations. The earliest Chinese methods to predict eclipses also relied upon the identification of an eclipse cycle. But instead of the saros cycle, they used a 135-synodic month cycle. Later Chinese methods instead used arithmetical models which took into account the variation in lunar and solar velocity. A very different approach was adopted by Indian astronomers who generally predicted eclipses using geometrical models for the motion of the Sun and Moon and a consideration of the geometry of the eclipse shadow.

One common feature most of the cultures studied here, and more broadly across the ancient and medieval world, is that eclipses were frequently seen as omens. They way that these interpretations were used, however, differed. In Assyria and China, for example, eclipses (and celestial omens more generally) were interpreted through a political lens and could be used as a way of supporting, criticizing, and/or changing a ruler's behaviour. Although eclipses were also interpreted as portents in Babylonia

---

[46] I thank Christopher Cullen for pointing me to this report.
[47] *Songshu* 34; trans. Xu, Pankenier, and Jiang, *East Asian Archaeoastronomy*.
[48] *Nanshi* 2; trans. Xu, Pankenier, and Jiang, *East Asian Archaeoastronomy*.
[49] Stephenson, *Historical Eclipses*.

and India, and could have important political impacts, their interpretations seem to have been less directly connected to the royal court.

Contrary, perhaps, to our modern preconceptions, across most of the ancient world, the same scholars who interpreted eclipses as omens also understood them as an astronomical phenomenon which could be studied, explained, and predicted using mathematical methods. It is a modern fallacy to assume that there is an inherent tension between these different ways of interacting with astronomical phenomena.[50] In particular, the ability to understand eclipses mathematically did not always remove their ominous significance. Rather, it provided a way to be prepared ahead of time, enabling rituals to be preformed efficiently and effectively and, in some cases, giving the scholars some measure of power in state matters through the guidance they could provide to the ruler. Just like today, in most ancient cultures science and politics went hand in hand.[51]

---

[50] As will be clear from some of the other contributions in this book, for example those by Bentley Hart (Chapter 8) and Pasachoff (Chapter 2), our modern understanding of the physics of solar eclipses does not stand in the way of emotional and other (non-scientific) types of responses to seeing an eclipse, even amongst scientists.

[51] John M. Steele, 'Astronomy and Politics'. In *Handbook of Archaeoastronomy and Ethnoastronomy*, ed. Clive L. N. Ruggles, New York, 2015, pp. 93–101.

6

'The Face of the World was Wretched, Horrifying, Black, Remarkable'

Solar Eclipses in the Middle Ages

Giles E. M. Gasper

Medieval History

We are now moving on from early Asia to medieval Scandinavia and Europe — from China, India, and Babylonia to Norway, Ireland, and England. The cultural focus will be placed on sources from the Christian tradition moving over a wide geographical range from the lands of the medieval Rus' to the Atlantic archipelago in the west, crossing linguistic, political, and religious boundaries. Christian scholars in the European Middle Ages developed their own responses to the science, experience, and interpretation of total solar eclipses. Most medieval thinkers of the period had some connection to the church. Knowledge of the material world was held as part of broader understanding of a divinely created cosmos and Christian teaching. The description of the 1133 total solar eclipse as simultaneously 'Wretched, Horrifying, Black, Remarkable' reveals the variety of responses to these celestial events.

Ptolemaic system The geocentric (Earth-centred) model of the cosmos.

Facies mundi miserabilis, horribilis, nigra, mirabilis.[1]

These words, written in 1133, recorded a solar eclipse. They appear in a set of annals compiled probably by Magnus of Reichersberg, one of a community of priests following the Augustinian rule in the diocese of twelfth-century Salzburg. The note of trepidation does not obscure the expression of wonder at the event. The annals go on to state that some took the eclipse as a sign and portent, although of exactly what is left unclear. In what follows something of the ways in which medieval authors recorded and interpreted total solar eclipses will be explored using evidence from across medieval Europe. The evidence is varied, reflecting the different

---

[1] 'The face of the world was wretched, horrifying, black, remarkable,' Magnus of Reichersperg, *Annales* (921–1167), s.a. 1133, ed. W. Wattenbach, Monumenta Germaniae Historica, Scriptores 17. Hanover, 1861, p. 454.

responses to solar eclipses.[2] Most of it appears in the context of chronicles and annals, for reasons which will be outlined below, but histories of human society and of the natural world include description of and deliberation on solar eclipses, as do, more rarely, hagiographies (descriptions of the lives of saints). The cultural focus will be placed on sources from the Christian traditions moving over a wide geographical range from the lands of the medieval Rus' to the Atlantic archipelago in the west, crossing linguistic, political, and religious boundaries. By comparing the responses of the Catholic (and in some cases only recently converted) communities of Latin Christendom and those of Eastern Orthodoxy, as well as those of authors from different institutional backgrounds, monastic and non-monastic, differences and similarities in the interpretation of solar eclipses can be indicated.

The chronological span is centred on the twelfth century, a period in which source material survives in sufficient measure to allow more detailed comparisons to be made across Christian cultures. The level of detail also allows consideration of mode of record, the sources of information, and the connection between written text and observation. Whose observation is being recorded is a question easy to ask but more difficult to establish. The relationship between event and record is crucial to the assessment of why and how medieval accounts were made, but it is not always easy to distinguish a medieval chronicler's own experiences from those of their sources. How medieval authors understood the physical effects of the solar eclipse, how they related it to other similar phenomena, and how they attributed metaphorical and prophetic meaning to these events are central to the discussion. The combination of awe, curiosity, dread, and astonishment provoked by solar eclipses is common across human cultures and finds a particular expression in the works and writers of the period and places below.

Solar eclipses in the period were understood as events rare and marvellous, whatever else was attributed to them. This cultural inflection is captured well by the Anglo-Welsh scholar Gerald of Wales who noted, in around 1185–7 in his *The History and Topography of Ireland*, that:

> human nature is so made that only what is unusual and infrequent excites wonder or is regarded of value. We make no wonder of the rising and the setting of the sun which we see every day; and yet there is nothing in the universe more beautiful or more worthy of wonder. When, however, an eclipse of the sun takes place, everyone is amazed—because it happens rarely.[3]

Gerald's remarks form part of a distinctive reflection about the notion of prodigy and the ordinary across his writings on the conquest of Ireland including the *Expugnatio*

---

[2] See Roos (Chapter 7) and Steele (Chapter 5).

[3] Gerald of Wales, *Topographica Hibernica*, Distinctio I, cap. XV, ed. J. S. Brewer, *Giraldi Cambrensis Opera*, vol. 5 of 8. London, 1861, p. 49: 'Sic enim composita est humana natura, ut nihil praeter inusitatum, et raro contingens, vel pretiosum ducat vel admirandum. Solis ortum et occasum, quo nihil in mundo pulchrius, nihil stupore dignius, quia quotidie videmus, sine omni admiratione praeterimus. Eclipsim vero solis, quia raro accidit, totus orbis obstupescit.' English translation from Gerald of Wales, *The History and Topography of Ireland*, trans. John J. O'Meara. London, 1982, §11, p. 42.

Hibernica. Discussion of natural marvels finds place, though not extensively, in his *Lives* of St Remigius of Lincoln, St Hugh of Lincoln, and St Ethelbert of Hereford, some aspects of which will be discussed in what follows.[4]

The rarity of the solar eclipse stands alongside the significance of the Sun in medieval understanding and practice of astronomy and astrology. These two disciplines were closely related. Hugh of St Victor defines astronomy as the law of the stars and astrology as the discourse concerning the stars respectively in his *Didascalicon* (*c.*1129). Astronomy was 'the discipline which examines the spaces, movements, and circuit of the heavenly bodies at determined intervals'; astrology was partly natural, as it concerns the temper of complexion of physical things, health, illness, storm, calm, productiveness, unproductiveness, and mostly superstitious, as pertaining to freedom of choice. A significant influx of newly translated texts from Greek and Arabic over the twelfth century increased substantially the store of learning available to Latin Christian scholars in both areas.[5] These range from the translations of Ptolemy of Alexandria's compendia of astronomy and astrology, the *Almagest* and *Tetrabiblos* Latinized as the *Quadripartitus*, by Gerard of Cremona, Plato of Tivoli, and others to the Persian scholar Abu Ma'shar's *Kitāb al-madkhal al-kabīr ilá 'ilm aḥkām al-nujūm* (Latinized as the *Introductorium in astronomiam* or *Introduction to Astronomy*), written in the late ninth century in Baghdad, which was translated into Latin twice in full versions in the 1130s and 1140s, by John of Seville and Hermann of Carinthia, with an epitome by Adelard of Bath from the same period.[6]

Both astronomy and astrology flourished and evolved in the Latin West over the course of the twelfth century. Despite the reservations of many churchmen such as Hugh of St Victor, on the dangers to a faithful Christian life posed by astrology in its contradiction of human free will, it is clear that others in society, including some clerics, held fewer concerns.[7] The monastic chronicler William of Malmesbury noted Gerard, Archbishop of York's (1100–8), fascination with the writings of Julius Firmicus, a source for ancient astrology well known to scholars of the twelfth century

---

[4] Whether frequently experienced or not, solar eclipses are in some decades relatively common, in others not. See the catalogue maintained by NASA: https://eclipse.gsfc.nasa.gov/eclipse.html (viewed 1 June 2022).

[5] Charles Burnett and David Juste, 'A New Catalogue of Medieval Translations into Latin of Texts on Astronomy and Astrology.' In Faith Wallis and Robert Wisnovsky (eds.), *Medieval Textual Cultures: Agents of Transmission, Translation and Transformation.* Berlin, 2016, pp. 63–76.

[6] David Juste, 'Ptolemy, Almagesti (tr. Sicily *c.*1150)' (update: 4 March 2021), *Ptolemaeus Arabus et Latinus. Works*, URL: http://ptolemaeus.badw.de/work/21, and his 'Ptolemy, Almagesti (tr. Gerard of Cremona)' (update: 7 May 2021), *Ptolemaeus Arabus et Latinus. Works*: http://ptolemaeus.badw.de/work/3 (accessed 13 July 2021). On Abu Ma'shar see *Kitāb al-madkhal al-kabīr ilá 'ilm aḥkām al-nuju—Liber introductorii maioris ad scientiam judiciorum astrorum*, ed. Richard Lemay, 9 vols. Naples, 1995–6, and his *Abu Ma'shar and Latin Aristotelianism in the Twelfth Century: The Recovery of Aristotle's Natural Philosophy through Arabic Astrology*. Beirut, 1962; Abū Ma'shar, *On Historical Astrology (On the Great Conjunctions)*, ed. and trans. Keiji Yamamoto and Charles Burnett, 2 vols. Leiden, 2000; Charles Burnett, 'John of Seville and John of Spain: A *Mise au Point*,' in his *Arabic into Latin in the Middle Ages, Variorum Collected Studies Series*. Farnham, 2009, pp. 59–78. An abbreviated version containing only the technical information, leaving out the philosophical justification, had already been translated by Adelard of Bath in the 1120s: Abu Ma'shar, *The Abbreviation of the Introduction to Astrology: Together with the Medieval Latin Translation of Adelard of Bath*, ed. and trans. Charles Burnett, Keiji Yamamoto, and Michio Yano. Leiden, 1994.

[7] David Runciman, 'Bishop Bartholomew of Exeter (d. 1184) and the Heresy of Astrology.' *The Journal of Ecclesiastical History* 70, 2019, pp. 265–82: DOI: https://doi.org/10.1017/S0022046918001306.

and earlier.[8] Personal horoscopes were commonly produced across high medieval society in this period.[9] Astrological prediction in connection to natural events also underwent considerable development, not least in the field of medical prognostication.[10] Robert Grosseteste in his treatise *On the Liberal Arts* (*c*.1195) devoted the greater part of his discussion of astronomy and its service to natural philosophy to three exemplary areas: when plants should be planted, when alchemy should be attempted, and when medicine should be prepared.[11] The influence of that which is above, in this case the planets and stars, on that which is below, including the human body and the spheres of the four elements which compose this sensible world, was a central tenet of the way in which Grosseteste and his contemporaries understood their universe.

The shifts in astronomical and astrological knowledge are essential background for understanding how solar eclipses were interpreted. The mechanism of the solar eclipse, as the conical shadow of the Moon falling on the Earth, was perfectly well understood by the twelfth century. What other meaning it was deemed to hold is a different question. Linking celestial to earthly events was a natural instinct within a system of knowledge rationally conceived and long studied. How the linkage between events was made by individual authors is more open to individual interpretation.

Recording Celestial Phenomena

Solar eclipses are mentioned in a wide variety of sources. Who recorded them, and in what form, are key questions to consider. Within the twelfth century the majority of these sources emerge from a religious context, by dint of education and by the needs and priorities of monastic communities. Changes across modes and institutions of higher learning are part of the wider developments of intellectual life in the period.[12] These are the decades in which the Cathedral Schools, and the first universities, emerge in response to the needs of the church, clergy and congregations, in its provision of pastoral care.[13] To support itself, the Christian community needed leaders with education to explain the teachings of the church; to navigate the system of

---

[8] William of Malmesbury, *Gesta pontificum anglorum*, ed. and trans. R. M. Thomson and M. Winterbottom. Oxford, 2007, 118.2.

[9] John North, *Horoscopes and History*. London, 1986, and his 'Some Anglo-Norman Horoscopes.' In Charles Burnett (ed.), *Adelard of Bath: An English Scientist and Arabist of the Early Twelfth Century*. London, 1987, pp. 147–61.

[10] Roger French, 'Foretelling the Future: Arabic Astrology and English Medicine in the Late Twelfth Century.' *Isis* 87, 1996, pp. 453–80.

[11] Robert Grosseteste, De artibus liberalibus, §11–13, ed. and trans. Sigbjørn O. Sønnesyn, in Gasper et al., *Knowing and Speaking: Robert Grosseteste's De artibus liberalibus 'On the Liberal Arts' and De generatione sonorum 'On the Generation of Sounds.'* Oxford, 2019, pp. 89–95, see also pp. 166–95.

[12] Amongst a significant literature see R. W. Southern, *Scholastic Humanism and the Unification of Europe*, Vol. 1, Foundations. Oxford, 1995; Stephen Jaeger, *Envy of Angels: Cathedral Schools and Social Ideals in Medieval Europe, 950–1200*. Philadelphia, PA, 1994.

[13] Another significant area of modern research, a classic account remains that of Leonard Boyle, *Pastoral Care, Clerical Education, and Canon Law, 1200–1400*. London, 1981; see also Ronald Stansbury (ed.), *A Companion to Pastoral Care in the Late Middle Ages (1200–1500)*. Leiden, 2010, and A. Firey (ed.), *A New History of Penance*. Leiden, 2008.

penance and forgiveness for sin; to forge, question, and apply the law of the church; and to construct its annual calendar. Calendrical science was a subject of consistent interest and importance across the medieval period, not without its controversies, and essential for the calculation of the primary Christian feast, namely Easter, which required a familiarity with both lunar and solar calendars.[14] Knowledge of astronomy was therefore a practical requirement for medieval clergy. Grosseteste in his guide for priests, *On the Temple of God*, written *c.*1220, insists that they should have 'a book of compotus [calendrical science] so that they may know the moveable and immovable feasts.'[15]

The need for educated clergy to serve the church led to the development of educational institutions, the graduates of which served equally the needs of secular government. These well-trained men populated the courts and other locations of royal and comital power. Clerical interest in, and domination of, the levers of social order, meant that clerical, Latin learning, including in astronomy and astrology, was far from absent within secular circles. Nevertheless, monastic houses retained, certainly for the twelfth century and, in different modes for the rest of the medieval period, an important role in higher learning. The library collections of older and generally richer Benedictine houses in western Christendom were complemented by those of the new orders which emerged from the late eleventh century onwards: the Cistercians (reformed Benedictines) and the Carthusians above all. With the same liturgical needs as the rest of the church, male and female monastic communities were active users of computistical and astronomical learning. Monasteries were far from passive organizations in the pursuit of new knowledge, whether culled from authorities or from contemporary experience, and many orders gradually built up presence at universities.[16] In the regions of Orthodox Christianity monastic communities continued to play an important role in the provision of education, including in Byzantium where higher education came more directly under imperial administrators.[17] Amongst the Rus', clerical learning was largely the preserve of monasteries or episcopal courts.[18] Astronomy, as for western scholars, remained a popular subject.[19]

Chronicles provide a fundamental source for the record of solar eclipses in both eastern and western Christianity. One of the most common genres of historical writing for the Middle Ages across cultures, chronicles are organized in an annalistic style (i.e. around the events of a single year).[20] The length and level of detail range from very short to highly specific entries. Chronicles were often associated with a specific community but equally often were composite documents incorporating records from many different authors from earlier periods. A useful contemporary twelfth-century

[14] C. Philipp E. Nothaft, *Scandalous Error*. Oxford, 2019.

[15] Robert Grosseteste, *Templum Dei*, X.3, ed. J. W. Goering. Toronto, 1984, p. 50.

[16] J. G. Clark, 'Monks and the Universities, *c.*1200–1500.' In Alison I. Beach and Isabelle Cochelin (eds), *The Cambridge History of Medieval Monasticism in the Latin West*. Cambridge, 2020, pp. 1074–92.

[17] Athanasios Markopoulos, 'Education.' In Robin Cormack, John F. Haldon, and Elizabeth Jeffreys (eds), *The Oxford Handbook of Byzantine Studies*. Oxford, 2008, pp. 785–95.

[18] T. Guimon, *Historical Writing of Early Rus (c.1000–c.1400) in a Comparative Perspective*. Leiden, 2021.

[19] Anne Tihon, 'Numeracy and Science,' *Oxford Handbook to Byzantine Studies*, pp. 803–19, esp. 805–10.

[20] D. N. Dumville, 'What is a Chronicle?' In E. Kooper (ed.), *The Medieval Chronicle II: Proceedings of the Second International Conference on the Medieval Chronicle Drieberger/Utrecht*, 16–21 July 1999, Costerus n.s. 144. Amsterdam, 2002, pp. 1–27.

assessment of the work of the chronicler is given by Gervase, monk of Christ Church Cathedral Priory, Canterbury, in the opening of his own chronicle. This was framed around the years 1135, the year in which Henry I of England died, and 1199, the death of Richard I. Up to the late 1170s Gervase is dependent on earlier sources, the writings of William of Malmesbury, for instance, and Henry of Huntingdon. Of his distinctive task as chronicler Gervase states that:

> To some extent the historian and the chronicles have the same aim and subject matter, but their way of dealing with it is different, as is the form. They share a common aim, since both strive for truth. The form of their work is different, because the historian proceeds at length and in an elegant way, while the chronicler in a simple and brief manner.[21]

A chronicle, he goes on to underline, deals with the computation of time as well as with the deeds of kings and princes and the record of miracles and portents, and it does so succinctly.[22] As the passage by Magnus of Reichersberg quoted above suggests, associated with the final category of the miraculous and auspicious are the records of celestial phenomena, including eclipses. Chroniclers did not always set out to provide a comprehensive record of such events; as compilations, chronicles are limited to the information in the sources they include. And chroniclers made mistakes; in the case of eclipses on date and location. For all of that, and in some ways because of it, chronicle record offers longitudinal evidence for the continuities and subtleties of interest in the solar eclipse.

Understanding the Solar Eclipse

While solar eclipses were wondrous and could be taken as signs or portents, they were also understood as natural phenomena and as rationally explicable, at least for those who recorded them. By the twelfth century, this was the inheritance not only of ancient authorities like Lucretius and Seneca, but also Christian writers like Isidore of Seville and Bede. The task of 'demystifying natural phenomena' through reason was undertaken by both classical and Christian authors to counter superstitious interpretation of nature and to emphasize the order and beauty of the universe and its operations, and in the case of the Christians to underline that this was ordained by the Creator.[23] Although Bede was wrong in the details of how the solar eclipse occurs (his contention, following Pliny the Elder, was that the Moon must be larger than the Earth, based on the mistaken assumption that solar eclipses are visible everywhere),

---

[21] Gervase of Canterbury, *Chronica*, in *Opera Historica*, ed. W. Stubbs. Longman, 1876, p. 87: 'Historici autem et cronici secundum aliquid una est intentio et materia sed diversus tractandi modus est et forma varia. Utriusque una est intentio, quia uterque veritati intendit. Forma tractandi varia, quia historicus diffuse et eleganter incedit, cronicus vero simpliciter, graditur et breviter.'

[22] For a summary of Gervase's works see, Michael Staunton, *The Historians of Angevin England*. Oxford, 2017, pp. 51, 53, 108.

[23] The phrase is that of Faith Wallis and Calvin B. Kendall in their translation and commentary: *Bede, On the Nature of Things and On Times*. Liverpool, 2010, p. 2.

the instinct to explain the phenomenon is paramount, as well as the trust placed in authority.[24] Later glosses to Bede's work from the tenth century, drawing on the methods of Carolingian commentators from the previous century, excise the mistake, probably, as Wallis and Kendall point out, because of access to the different, and in this case more accurate, traditions of Martianus Capella and Chalcidius.[25]

An episode from three centuries later illustrates well the mixture of reason and wonder provoked by solar eclipses. It appears in an unusual crime report, namely the account left by Galbert of Bruges of the murder of Count Charles of Flanders in 1127. As part of the background for his description of the startlingly violent end of Count Charles and its social and political ramifications, Galbert includes a portentous note on events in 1124.

> In the year 1124 from the incarnation of our Lord, in the month of August, an eclipse in the body of the sun appeared to all the inhabitants of the lands around the ninth hour of the day, and an unnatural lack of light so that the eastern part of the circle of the sun was obscured and sent little by little into the other parts strange clouds, which did not, however, obscure the entire sun all at once but in part, and this same cloud wandered similarly over the whole circle of the sun, traveling all the way from the east to the west but only in the circle of the solar essence. As a result, those who kept an eye on the state of the peace and the wrongs in law courts threatened everyone with the danger of coming famine and death.

> When men were not corrected in this way, neither lords nor serfs, the starvation of unexpected famine arrived, and the lashes of death followed hard upon its heels. Whence the Psalm: *He summoned a famine on the land, and broke every staff of bread.*[26]

Whether Galbert's reference to the visibility of the eclipse to all the inhabitants 'of the lands' might imply that the ideas of Seneca and Bede were not so quite so easily abandoned is an intriguing question. However, it might simply indicate the extent of Galbert's network of correspondents, as in this instance a partial or total eclipse was indeed visible throughout the whole of the lands in Europe, Asia, and Africa

---

[24] Bede, *De natura rerum*, c. 22, ed. Charles W. Jones, Corpus Christianorum series latina, 123A, Turnhout, 1975, pp. 189–234; see also *On the Nature of Things*, Wallis and Kendall, p. 154.

[25] Bede, *On the Nature of Things*, Wallis and Kendall, pp. 40 and 153.

[26] Galbert of Bruges, *De multro, traditione et occisione Karoli comitis Flandriarum*, c. 2, ed. J. Rider. Turnhout, 1994: 'Inmisit ergo flagella famis et postmodum mortalitatis omnibus qui in regno degebant nostro, sed prius terrore signorum revocare dignabatur ad penitendum quos pronos praeviderat ad malum. Anno ab incarnatione Domini milleno centeno vicesimo quarto, in Augusto mense, universis terrarum habitatoribus in corpore solari circa nonam diei horam apparuit eclipsis, et luminis non naturalis defectus ita ut solis orbis orientalis obfuscatus paulatim reliquis partibus ingereret nebulas alienas, non simul tamen totum solem obfuscantes, sed in parte, et tamen eadem nebula totum pererravit solis circulum, pertransiens ab oriente usque ad occidentem tantummodo in circulo solaris essentiae. Unde qui statum pacis et placitorum injurias notabant, futurae famis et mortis periculum minabantur universis. Cum que neque sic correcti sunt homines, tam domini quam servi, venit repentinae famis inedia et subsequenter mortalitatis irruerunt flagella. Unde in psalmo: Et vocavit famem super terram et omne firmamentum panis contrivit.' English translation from *The Murder, Betrayal, and Slaughter of the Glorious Charles Count of Flanders*, trans. Jeff Rider. New Haven, CT, 2013, pp. 6–7.

then known to Europeans.[27] There are oddities nevertheless in the account. While Galbert includes, as is common in an annalistic style, the year, month, and hour of the eclipse, he does not include the day, which was 11 August. Nor does the Moon, the cloud crossing the circle of the Sun in Galbert's narrative, move east to west in a solar eclipse but in the opposite direction. All of which suggests that Galbert, a notary and not necessarily a priest, although possibly a cleric in minor orders, was not especially knowledgeable of celestial matters. Moreover, this is the only mention of an eclipse, solar or lunar, in the whole narrative. Other portents such as fires, famine, floods, and bloody water are repeated at intervals, punctuating the record of events playing out amongst the protagonists in the uprisings associated with the killing of the count. The account of the eclipse nevertheless has a clear function in the structure of Galbert's work, unaffected by his omissions and inaccuracies.

By the time Galbert was writing, different sources for astronomical learning were becoming available to scholars of Latin Christendom in translations from Greek and Arabic. A notable figure in this process was Petrus Alfonsi, a convert from Judaism to Christianity from south-western al-Andalus (Muslim Iberia) trained in the Judaeo-Arabic intellectual traditions with a particular focus on astronomy.[28] After his conversion in 1106 he moved to northern Europe to Henry I's England, and at a later point to northern France. Petrus translated the astronomical tables, the *Zîj al-Sindhind*, of al-Khwārizmī (d. 850) into Latin, an important component in the transformation of Latin astronomy.[29] Petrus's influence over one of his known pupils, Walcher, prior of Malvern, near Worcester, provides an illustration in small of the larger effect of the new knowledge, albeit with respect to lunar rather than solar eclipses.[30] Walcher produced a remarkable early work on lunar computation developing a coordinate system to track the Sun and Moon across the zodiac and predict their eclipses. The system was not successful empirically, and the observation of the Moon's variable motion lay at odds with the contemporary orthodox theory which stressed the non-irregular motion of Sun and Moon. After encountering Petrus in c.1120 Walcher was able to deploy knowledge of the Moon's orbital nodes in his treatise *On the Dragon* to give a fuller, if still rudimentary, account of the Moon's motion, in turn allowing a higher degree of accuracy in the prediction of its courses.

Petrus was also responsible for an assertion of the inherent rationality of the universe and the benefits of astronomical knowledge in his *Dialogue Against the Jews*. This was a work of unprecedented polemic for Latin Christendom, and which would go on to be a much-copied work as Christian attitudes towards and relations with

---

[27] X. M. Jubier, F. Espenak, and J. Meeus, 'Five Millenium Canon of Solar Eclipses: −1999 to +3000.' http://xjubier.free.fr/en/site_pages/solar_eclipses/5MCSE/xSE_Five_Millennium_Canon.html (accessed 3 August 2022).

[28] On Petrus see John Tolan, *Petrus Alfonsi and his Medieval Readers*. Gainsville, FL, 1993 and the essays collected in Carmen Cardelle de Hartmann and Philip Roelli (eds), *Petrus Alfonsi and his Dialogus: Background, Context, Reception*. Florence, 2014. For general context, see Brian Catlos, *Kingdoms of Faith: A New History of Islamic Spain*. London, 2018.

[29] Tolan, *Petrus Alfonsi*, 55–61 and 66–8; Otto Neugebauer, *The Astronomical Tables of al-Khwarizmi*. Copenhagen, 1962.

[30] C. Philipp E. Nothaft, *Scandalous Error: Calendar Reform and Calendrical Astronomy in Medieval Europe*. Oxford, 2018, pp. 80–5, for a useful summary. For Walcher's treatises see Walcher of Malvern, *De lunationibus and De Dracone*, ed. C. Philipp E. Nothaft. Turnhout, 2017.

Jewish communities worsened over the later Middle Ages.[31] The first part of the dialogue involves showing the irrationality of his Jewish opponent in a discussion of the sphericity of the world, centred on arguments about the relative location of east and west to latitude. As an example Petrus uses the Indian city of Arim, frequently advanced in Islamicate calendrical science as located at the centre of the world.[32] The difference in time between the observation of a solar eclipse at Arim and other cities to its east and west is deployed in this connection.[33] Knowledge of the observation of a solar eclipse in different places underpins Petrus's arguments here.

The extent of the transformation of Latin astronomical knowledge and the motions of the Sun and Moon is shown in the appearance at the beginning of the thirteenth century of *On the Sphere* by John Sacrobosco. Even less is known about Sacrobosco than Petrus Alfonsi, which stands in contrast to the huge success of *On the Sphere*, an introduction to astronomy, in the later medieval and into the Early Modern periods.[34] Sacrobosco's work was informed by the complex inheritance of Ptolemy which inspired a considerable body of Arabic commentary and critique from the eighth century onwards, and this was, mutatis mutandis, mirrored in Latin Christendom with numerous epitomes and attendant texts, translated from Arabic and created in Latin.[35] Sacrobosco, for example, made heavy use of the *Compilation on the Science of the Stars* by al-Farghānī (d. 861) (Latinized as Alfraganus), a summary and revision of Ptolemy.[36] *On the Sphere* includes an account of the causes of eclipses in its final sections (see Figure 6.1). These combine understanding of the Sun and Moon's motions known to Petrus and then Walcher:

> When the moon is in the head or tail of the dragon or nearly within the limits and in conjunction with the sun, then the body of the moon is interposed between our sight and the body of the sun. Hence it will obscure the brightness of the sun for us and so the sun will suffer eclipse—not that it ceases to shine but that it fails us because of the interposition of the moon between our sight and the sun.[37]

Sacrobosco notes that a solar eclipse should occur therefore at the conjunction of the two celestial bodies or at the new Moon. And that, unlike an eclipse of the Moon, which is visible everywhere, that of the Sun is not. This flows into the final reflection

---

[31] Petrus Alfonsi, *Dialogus*, ed. and trans (German), Peter Stotz. Florence, 2018. English translation: *Dialogue Against the Jews*, trans. Irven Resnick. Washington, DC, 2006.

[32] Petrus Alfonsi, *Dialogue Against the Jews*, 55, n. 22.

[33] Petrus Alfonsi, *Dialo]gus*, I. 69–70, pp. 38–9. And see discussion in Giles E. M. Gasper et al., *Mapping the Universe: Robert Grosseteste's De sphera—On the Sphere*. Oxford, 2023.

[34] See summary in *Mapping the Universe*, ch. 1.

[35] George Saliba, *Islamic Science and the Making of the European Renaissance*. Cambridge, MA, 2007.

[36] Alfraganus, *Compilatio Astronomica*. Ferrara, 1493 and also al-Farghani, *Differentie*, trans. John of Seville, ed. Francis J. Carmody. Berkeley, CA, 1943; Gerard of Cremona's translation is found in Alfraganus, Il 'libro dell'aggregazione delle stelle,' ed. Romeo Campani. Città di Castello, 1910.

[37] John Sacrobosco, *On the Sphere*, Bk. 4, ed. and trans. Lynn Thorndike, *The Sphere of Sacrobosco and Its Commentators*. Chicago, 1949, p. 116: 'Cum autem luna fuerit in capite vel in cauda draconis vel prope infra metas et in coniunctione cum sole, tunc corpus lunare interponetur inter aspectum nostrum et corpus solare,unde obumbrabit nobis claritatem solis. Et ita sol patietur eclipsim, non quia deficitsia a lumine sed quia deficit nobis propter interpositionem lune inter aspectum nostrum et solem.' English translation at p. 149.

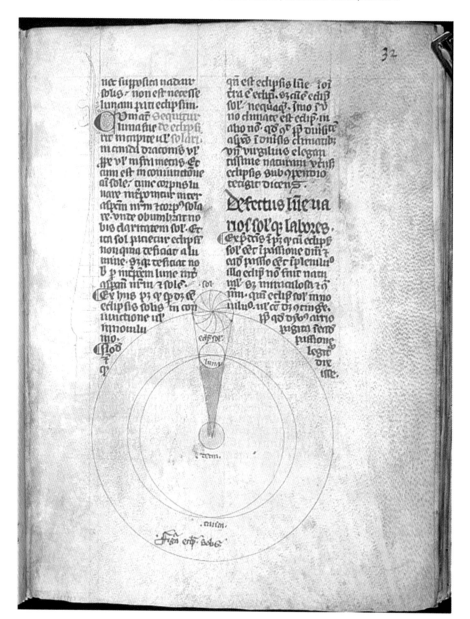

Figure 6.1 A diagram of a solar eclipse from the first quarter of the fourteenth century, in the treatise *De sphaera—On the Sphere* by John of Sacrobosco.

Source: The British Library MS Royal 12.C.XVII, f. 32 (released to public domain): https://www.bl.uk/catalogues/illuminatedmanuscripts/ILLUMIN.ASP? Size=mid&IllID=33456.

of the treatise, on the nature of the solar eclipse that took place during the final stages of the crucifixion of Jesus Christ. The Gospel of Luke (23.44–45) records darkness over all the Earth from the sixth to the ninth hour and that the Sun was darkened.

Sacrobosco points out that, combined with the fact that the passion took place at full Moon, this would mean that what he took to be a solar eclipse was miraculous, in this instance, rather than natural.

That *On the Sphere* should end with this discussion is a reminder of the fundamental purpose of knowledge of natural phenomena for Christian scholars of the period. Not only was the created world an expression of divine rationality, allowing the work of the creator to be understood, the sinful state of humanity notwithstanding, but the Bible, since it too mentions natural phenomena, provoked their fuller explanation. And, as Sacrobosco lays out in detail, the solar eclipse at the crucifixion, which took place at full Moon, serves to underline the importance of astronomy to calendrical study and the cardinal place of the calculation of Easter, dependent on both solar and lunar cycles, within medieval Christendom.

The Eclipse of Kings and Colourful Observation

The miraculous solar eclipse associated with the death of Christ in the medieval period offered a framework for Christian writers for a similar association of the cosmic phenomenon with earthly rulership. As the example from Galbert of Bruges's account of the murder of the Count of Flanders illustrates, solar eclipses and what they portended could play an important function in the author's narrative. How the eclipse is described is also significant. The wider knowledge of the phenomenon acquired by individual chroniclers across the twelfth century, as well as care for the accuracy of their record, emerges in sometimes very detailed description of a celestial event.[38] None of this is to suggest that solar eclipses were associated exclusively with rulers or that this was a cultural phenomenon tied to twelfth-century Christendom. The association was cross-cultural, as examples from the lands of the Rus' explored below indicate, and cross-chronological. The ninth-century biographer of Charlemagne, Einhard, included a catalogue of celestial and temporal signs preceding the death of the emperor which started with solar and lunar eclipses in the last three years of his life and a black spot on the Sun that remained for a week.[39] Nevertheless, the combination of narrative association, technical knowledge, and historical context makes twelfth-century chroniclers' descriptions particularly interesting as cultural responses.

---

[38] A good example of the careful record of an unusual phenomenon is Gervase of Canterbury on the Moon apparently splitting in two—see Giles E. M. Gasper and Brian K. Tanner, '"The Moon Quivered Like a Snake." A Medieval Chronicler, Lunar Explosions, and a Puzzle for Modern Interpretation.' *Endeavour* 44, 2021, 100750 (and, with editorial correction, 46, 2022, 100813).

[39] Einhard, *Vita Karoli Magni*, Bk. IV, c.32, ed. G. Waitz, Monumenta Germaniae Historica, Scriptores rerum Germanicarum, 25. Hannover, 1911, p. 36. Together with reports from 807 (B. Hetherington, *A Chronicle of Pre-telescopic Astronomy*. Chichester, 1996, p. 90) the description of a sunspot is one of the earliest convincing European reports of the phenomenon; there is some debate concerning Classical Greek sightings (J. M. Vaquero, 'Sunspot observations by Theophrastus revisited.' *Journal of the British Astronomical Association* 117, 2007, p. 346). Visibility for a week is consistent with the 807 reports and represents a quarter of the rotation period of the Sun about its axis. A quarter of the sphere is the area within which a sunspot might be clearly visible from Earth. The years 811–14 corresponds to the early part of a maximum in solar activity (R. Arlt and J. M. Vaquero, 'Historical Sunspot Records.' *Living Reviews in Solar Physics* 17, 2020, article 1, https://doi.org/10.1007/s41116-020-0023-y) that peaked about 850.

A good example is the narrative response to the death of Henry I of England, especially in the aftermath of the succession crisis that followed. Henry, who died without a male heir, was succeeded as King of England and Duke of Normandy by his nephew Stephen (whose tenure of the throne was contested by Henry I's daughter Matilda and her supporters).[40] What followed was a reign of nineteen years of political instability which induced high levels of anxiety amongst contemporary chroniclers, especially those in areas most affected by the disruption of royal governance or on the borderlands between the rival parties. The *Anglo-Saxon Chronicle*, maintained at the abbey of Peterborough up to the end of Stephen's reign, memorably described the period as one in which people 'said openly that Christ and His Saints slept.'[41] A cluster of monastic chronicle entries from England and Wales, in Latin, Old English, and Welsh, use solar eclipses to frame their concerns on the change of regime. Although annalistic in spirit, the chronicles were carefully curated by their composers and compilers, making their narrative choices all the more intriguing.

John, a monk of Worcester (d. *c*.1140), made a clear connection between a solar eclipse on 2 August 1133 and Henry I's death two years later and the subsequent political turmoil. At the outset he was keen to emphasize that the eclipse took place on the same day in the annual cycle that his brother and predecessor, William Rufus, had been killed in the New Forest, leaving nothing of the implication to be misunderstood or missed by the reader. John went on to describe the celestial event itself, which took place as Henry, surrounded by his guards, was about to cross the English Channel to Normandy, probably from Dover, at about noon.

> Suddenly a cloud appeared in the sky, which was visible throughout England, though of varying size. In some places indeed the day only appeared darkened, but in others it was so dark that men needed the guidance of candlelight to do anything. The king and his followers and many others walked about, marvelling greatly, raised their eyes to heaven, and saw the sun shining as though it were a new moon, though it did not keep the same appearance for long. One moment it was broader, the next it was narrower, now curved, now straight, now steady as usual, now moving, and seemed quivering and liquid like quicksilver. Some claim that an eclipse of the sun had taken place. If this was the case, then the sun was in the head of *Draco* and the moon in the tail or the sun was in the tail and the moon in the head of the fifth sign, *leo*, in the seventeenth degree of that sign. The moon was twenty-seven days old. At the same day and hour many stars appeared.[42]

---

[40] Amongst an extensive literature see Edmund King, *King Stephen*. New Haven, CT, 2010.

[41] *The Anglo-Saxon Chronicle, A Collaborative Edition*, Vol. 7, *MS. E*, ed. Susan Irvine, s.a 1137. Cambridge, 2004, p. 135: 'Hi saeden openlice ðat Crist slep and his halechen.'

[42] John of Worcester, *Chronicle*, s. a. 1133, ed. and trans. P. McGurk, vol. 3 of 3. Oxford, 1998, vol. 3, pp. 208–11: 'subito in aeres nubes apparuit, que tamen unius eiusdem quantitatis per universam Angliam non comparuit. In quibusdam enim locis quasi dies obscurus uidebatur, in quibusdam uero tante obscuritatis erat, ut lumine candele ad quodlibet agendum ipsa protecti homines indigerent. Vnde rex latusque regium ambientes et alii complures mirantes, et in celum oculos leuantes, solem ad instar noue lune lucere conspexerunt, qui tamen non diu se uno modo habebat. Nam aliquando latior, aliquandiu subtilior, quandoque incurior, quandoque erectior, nunc solito modo firmus, modo mouens, et ad instar uiui argenti motus et liquidus uidebatur. Asserunt quidam eclypsim solis factam fuisse. Quod si uerum est, tunc *sol*

Several features might be noted here, which suggest that John had read Walcher of Malvern's *On the Dragon* but had not fully understood its content. First, John is equivocal about the nature of the event, reporting only a claim, by some, that an eclipse had taken place. Modern computation indicates that the eclipse would have been of magnitude 0.95 at Worcester, so even with cloud cover and assuming he was in Worcester, John would have been able to experience personally visible dimming of light level, albeit the reason for it not being fully understood. Second, and importantly, the mechanics are wrong: what John presented as a solar eclipse was confused with the conditions for a lunar eclipse.[43] Although he reports that the Moon was in the twenty-seventh day of the lunar cycle, he does not appear to have realized what this implies for the relative positions of Sun and Moon. A reasonable conclusion would be that John was not present at the events he describes in detail, and that he was not totally expert in the subject.

That said, the vividness of the language that he employs is striking. His date and timing are correct; modern computation showing maximum obscuration at Dover, at 11:41 UT on 2 August 1133, is entirely consistent with John's description of the events occurring about noon.[44] His source, who was presumably with the king at Dover, accurately describes the appearance of the Sun as a new Moon; computation of the appearance at maximum obscuration at Dover (Figure 6.2) shows the visible part of the Sun's disc in that orientation. The appearance of stars is recorded, an observation made frequently by medieval writers on (usually total) solar eclipses. In general, for an eclipse of maximum magnitude of 0.96 and obscuration of 96.4%, Venus, Jupiter, Sirius, Canopus, Mercury, and possibly alpha Centauri, Arcturus, and Vega could be visible. For this eclipse at Dover, Venus, Mercury, and Mars were all close to the Sun, Arcturus well above the horizon, and Sirius and Vega just above the horizon (so probably not bright enough).[45]

The description of the changes of shape of the observable part of the Sun's disc reflects not only the changing shape but also that these do not proceed uniformly over time. Dramatic changes occur close to maximum obscuration in very short time periods compared with changes near first contact. The likening of the appearance to that of quicksilver may indicate that John had an additional source who observed the total eclipse, perhaps in the north of England or Scotland (the same source might also have reported the visibility of stars). The effect, often referred to as Baily's Beads, arises from the non-uniform shape of the Moon when very close to totality and a good recent example of the light appearing to ripple downwards like mercury is from the 2012 total eclipse in Australia. All of this would appear to suggest that John reported accurately what he was told about the eclipse but was using Walcher of Malvern's *On the Dragon* incorrectly as a framework for his understanding. As such, it represents

erat in capite draconis, et luna in cauda, uel sol in caude at luna in capite in .v. signo leonis in .xvii. gradu ipsius signi. Erat autem tunc luna .xxvii. Eodem etiam die et eadem hora, stele quamplurime apparuere.' See also McLeish and Frost (Chapter 1), and Nothaft (Chapter 3), in this volume.

[43] John of Worcester, *Chronicle*, p. 210, n. 2.
[44] Jubier, Espenak, and Meeus, 'Five Millenium Canon of Solar Eclipses −1999 to +3000.'
[45] G. P Können and C. Hinz, 'Visibility of stars, halos and rainbows during solar eclipses.' *Applied Optics* 47, 2008, H14–H24.

Figure 6.2   Reconstruction of the appearance of the Sun at maximum obscuration at Dover in 1133.

Source: Image created by Brian Tanner using X. M. Jubier's interface to data and predictions courtesy of Fred Espenak, NASA/Goddard Space Flight Center, from eclipse.gsfc.nasa.gov.

a verifiable moment in the transmission of ideas on astronomy from east to west.[46] New knowledge led to different ways in which natural phenomena such as eclipses were explained. The description of the eclipse was followed by other unexplained phenomena: the ship throwing its anchor and moving on a calm sea, earthquakes, a sighting of two moons. Henry, John states, 'was never to return to England or to see it alive.'[47]

The same set of circumstances are recorded by another monastic chronicler, William of Malmesbury (*c*.1090–after 1142), though with differences in detail, tone, and style, in his *History of New Events*, a continuation of his *Deeds of the Kings of England*. William also made clear that the solar eclipse of 1133, and subsequent earthquake, presaged the death of the king. Henry's channel-crossing took place, according to William, on Wednesday 5 August (the dates are not quite accurate), on the day of the eclipse, at the sixth hour. The Sun he described with a quotation from

---

[46] See Anne E. Lawrence-Mathers, 'John of Worcester and the science of history.' *Journal of Medieval History* 39, 2013, pp. 255–74; Kathy Bader, 'A Culture of Inquiry: Scientific Thought and its Transmission in the Severn Valley, *c*.1090–*c*.1150.' Unpublished PhD Thesis, Durham University, 2022.

[47] John of Worcester, *Chronicle*, s.a. 1133, p. 211: 'non ulterius uita comite rediturus uel uisurus Angliam.'

Virgil's *Georgics*, 'covered its shining head with gloomy rust.'[48] Where William seems to have accepted, at least poetically, Virgil's phrase as a description of a solar eclipse, this was to follow the fourth-century grammarian Servius, the first to make that identification.[49] No eclipse actually accompanied Caesar's demise. That of 1133, however, put fear into the minds of those who experienced it, as William recounts, and was followed by an earthquake two days later with violent shaking and a dreadful noise. To these events William was an eyewitness: 'In the eclipse I saw myself the stars around the sun, and in the earthquake the wall of the house in which I sat lifted up by two shocks and settling down at a third.'[50] The section ends with William foreshortening the next two years, remarking that Henry never returned to England, and died in 1135, after falling ill while hunting in the forest of Lyons some twenty miles to the west of Rouen, the capital of the Duchy of Normandy.

Although the basic elements of the narrative are similar, William's account includes none of the new astronomical learning shown by John of Worcester. The eyewitness claim by William is of interest particularly in the insistence on the stars appearing around the Sun and the definite statement that the event experienced was an eclipse of the Sun. As the magnitude of the eclipse was 0.94 at Malmesbury, a few stars and planets should have been discernible. The *Anglo-Saxon Chronicle* records many of the same elements, although its compiler seems to have conflated the year of Henry's passing with the report of the solar eclipse. The eclipse is, once more, seen in close connection to the king's death.

In this year, the king Henry went over sea at Lammas [1 August]. And the second day when he lay asleep on [his] ship, then the day darkened over all lands and the sun became such as if it were a three-nights old moon, and stars about it at midday. Men were greatly astonished and afraid, and said that a great matter ought to follow hereafter: so it did, for that same year the king died the second day after Saint Andrew's mass-day in Normandy. Then at once these lands darkened, for every man who could at once ravaged another.[51]

Although this account lacks the depth of description in terms of the colour and movement of the Sun, the comparison of the latter to a three-day old Moon is noteworthy (Figure 6.2). The mention of stars at midday also coheres with the other two chronicles. So does the report of intense darkness. The laconic style of the *Anglo-Saxon*

---

[48] William of Malmesbury, *Historia novella*, Bk. 1, §457, ed. K. R. Potter. Edinburgh, 1955, pp. 11–12: 'tetra ferrugine.' Virgil, *Georgics*, I.467, ed. R. A. B. Mynors. Oxford, 1990, p. 34: 'cum caput obscura nitidum ferrugine texit.'

[49] Virgil, *Georgics*, I.467, ed. Mynors, p. 34, notes to ll. 466–8.

[50] William of Malmesbury, *Historia Novella*, Bk 1, §457, p. 12: 'Vidi ego et in eclipsi stellas circa solem; et in terre motu parietem domus in qua sedebam, bifario impetus eleuatum, tertio resedisse.'

[51] *The Anglo-Saxon Chronicle, A Collaborative Edition*, Vol. 7, MS. E, ed. Susan Irvine, s.a 1135. Cambridge, 2004, p. 133: 'On þis gære for se king Henri ouer sæ æt te Lammase. And Ðat oþer dei þa he lai an slep in scip, þa þestrede þe dæi ouer al landes and uuard þe sunne suilc als it uuare thre niht ald mone, an sterres abuten him at middæi. Wurþen men suiðe ofuundred and ofdred, and sæden ðat micel þing sculde cumen herefter: sua dide, for þat ilc gær warth þe king ded ðat oþer dæi efter Sancte Andreas massedæi on Normandi. Þa þestreden sona þas landes, for æuric man sone ræuede oþer þe mihte.'

Chronicle makes the clear and unambiguous association of the eclipse with the death of Henry and its consequences all the more powerful. A metaphorical darkness preceded by the eclipse of the Sun.

The association of solar eclipses with the deeds of those who would be king, or queen, after Henry's death is continued in the Welsh *Brut y Tywysogyon* (*Chronicle of the Princes*). Covering 682–1282 and compiled in the thirteenth century at the Cistercian house of Strata Florida, its twelfth-century material drew on sources from Llanbardan Fawr, a Benedictine house briefly between 1111–35, and a Welsh parish until the later medieval period. The entrance of Henry's daughter Matilda is mentioned in immediate proximity to an eclipse of the Sun:

> In the following year [1136/7] the empress came to England to subdue the kingdom of England for Henry, her son; for she was a daughter of Henry the First, son of William the Bastard. And then there was an eclipse of the sun on the twelfth day from the Calends of April.[52]

The entry here contains a more typical chronicle listing of month and day in annalistic style, with far less description of the phenomenon itself. It is also inaccurate in terms of date as there was no solar eclipse on 21 March in either 1136 or 1137. There was a partial eclipse visible from London on 1 June 1136 to which the entry may refer, but if so, its timing has been adjusted to suit the narrative. Even within the spare style usually adopted, more detail is sometimes included in other references. For instance, a notice under the years 1184–5 highlights the visit of the Patriarch of Jerusalem to Henry II of England. These were the final years before the ejection of the Latin Christian kings from the Holy City by Salah al-Din, and the occasion of increased diplomatic contact with leading figures and allies in Christendom. Not only is the solar eclipse recorded with a royal connection, it also opens up resonance with the deeper spiritual lordship of the Kingdom of Jerusalem, and with the universal lordship of Christ. Any portent of the eclipse is left implicit, but, given the expected attack on the city and its fall in 1187, recorded in the *Brut* under 1186–8, an association is not hard to make. While no horror on the part of those witnessing the event is recorded, the shifting colour of the Sun is emphasized: 'In that year on the day of the Calends of May the sun changed its colour; and some said that it was under an eclipse.'[53] A very similar description of the same eclipse was offered in annals of the Cistercian abbey of Margam, some eighty miles to the south 'with the sun after the eclipse a blood-like colour reddened in a marvellous manner.'[54] However

---

[52] *Brut y Tywysogyon*, s.a. 1137–8, ed. and trans. Thomas Jones (Cardiff, 1955), s, pp. 116–17: 'Yn y ulwydyn racwyneb y doeth yr amherodres y Loegyr yr darestwg brenhinyaeth Loegyr y Henri, y mab; kanys merch oed hi y Henri Gyntaf vab Gwilim Bastartt. Ac yna y bu diffic ar yr heul y deudecuetyd o Galan Ebrill.'

[53] Ibid. s.a. 1184–5, pp. 168–9: 'Yn y ulwydyn honno dyw Calen Mei y sumudawd yr heul y lliw; ac y dywat rei uot arnei diffyc.'

[54] *Annales de Margan*, in *Annales monastici*, ed. Henry Richards Luard, vol. 1 (London, 1864), p. 17; 'sole post eclipsim colore sanguineo quodam mirabili modo rubente.' On the annals themselves see Robert B. Patterson, 'The author of the "Margam annals": early thirteenth-century Margam abbey's compleat scribe.' *Anglo-Norman Studies* 14, 1991, pp.197–210.

gathered, the material presented by later compilers, anonymous in this case, speaks to the importance of particular details for the record of the solar eclipse.

The solar eclipse of 1185 recorded in the *Brut y Twysogion* was also described in a chronicle from the lands of the Rus', the *Laurentian*, written in Old Church Slavonic, where the eclipse was experienced as total. A similar emphasis on colour is evident, although with the sense of wonder, awe, and dread explicitly included.

> In the year 6694 (1186)[55] on the first day of the month of May, on the feast day of the holy prophet Jeremiah, on Wednesday during vespers there was a sign in the sun. And it was very dark, so that the stars too were visible to people, their eyes saw [things] as if green, and as though a moon was forming in the sun. It was as if hot embers were coming from its horns. It was frightening for people to see a sign of God. In the same year in the same month of May, on the 18th day, on the feast day of the holy martyr Potapius, on Saturday, a son was born to the grand prince Vsevolod.[56]

The episode comes within years of a struggle for power amongst the Rurikids, the ruling dynasty of the Kyivan Rus'. The Rus' emerged as rulers of territories from Novgorod in the north to Kyiv in the south from the ninth century onwards. By the twelfth century, Novgorod was more or less independent of the ruling dynasty in Kyiv; the former survived the incursions of the Mongols in the mid-thirteenth century where the latter did not.[57] Vsevolod at this point was in the process of building his authority, as prince of Chernigov and ultimately grand prince of Kyiv.

Although representing a different linguistic tradition, and although using the Byzantine dating system (*annus mundi*—from the beginning of the world), the approach of the chroniclers and compilers further east was not dissimilar to those of Latin Christendom.[58] While only two chronicles survive from the medieval period (the *Laurentian* and the *Hypatian*), the same processes of compilation can be identified, and the same interest in celestial phenomena and in what they might portend. The *Laurentian Codex* took its final form in 1377, probably at the monastery at Vladimir (near Moscow) with material from Kyiv, Pereyaslavl-on-Dnieper, Rostov, and Tver.[59] The segment from 1185–8 was probably the responsibility of a scribe

---

[55] The year is a mistake for 6693/1185, an issue covered in more detail in Giles E. M. Gasper and Brian K. Tanner, '"In the shape of a cooking pot over the fire": Records of solar prominences in the 1180s.' *Endeavour*, 47, 2023, 100875, https://doi.org/10.1016/j.endeavour.2023.100875.

[56] The Laurentian Chronicle [Lavrent'evskaia letopis], ed. Evfimii E. Karskii, parts 1–3, Polnoe sobranie russkikh letopisei, 1 (Leningrad, Arkheograficheskaia komissia1926–8); Laurentian Codex, 1377, digital edition with transcription and modern Russian translation (http://expositions.nlr.ru/LaurentianCodex/eng/index.php), fol. 134: 'В год 6694 (1186) месяца мая в 1 день, на память святого пророка Иеремии, в среду во время вечерни, было знамение в солнце. И темно было очень, так что и звезды были видны людям, в глазах словно зелено было, и в солнце образовался словно месяц. Из рогов его как уголь жаркий исходил. Страшно было видеть людям знаменье Божье. В том же году того же месяца мая в 18 день, на память святого мученика Потапия, в субботу, родился сын у великого князя Всеволода.'

[57] See, S. Franklin and J. Shepard, *The Emergence of Rus, 750–1300*. London, 1996; J. Martin, *Medieval Russia, 980–1584*. Cambridge, 1995; M. Dimnik, *The Dynasty of Chernigov, 1146–1246*. Cambridge, 2003.

[58] Guimon, *Historical Writing of Early Rus*. Leiden, 2021.

[59] Ibid., p. 38.

attached to Archbishop Luka of Rostov, appointed in 1186.[60] The proximity of reference between the solar eclipse and the affairs of the dynasty is to be noted. So too the visibility of the stars, which is common across chronicle accounts of Latin Christendom as mentioned above, and the description of the event as though a Moon was forming in the Sun, and the hot embers coming from the Sun. This latter description has been attributed to the appearance of solar prominences—an interpretation that is entirely plausible as, in this total eclipse, solar and lunar disc diameters were almost equal, a requirement for observation of the prominences.[61] More unusual are the remarks on people seeing things as if green. During an eclipse, beyond about 80% obscuration, the landscape often takes on a metallic blue-grey hue. It may well be this phenomenon that the chronicler was trying to describe.[62]

The eclipse, total in Rostov, is referred to in the *Hypatian Chronicle*, the other from the region to survive in its medieval form.[63] The eclipse is mentioned in a broader record of a raid by Igor Svyatoslavich, one of the Rurikid princes vying for power in the second half of the twelfth century, and to whom Vsevolod had sent military support.[64] The raid was directed against the Polovsty in the south-east steppe lands. When crossing the river Donets on 1 May, Igor saw the Sun appearing as a crescent of the Moon; the eclipse of 1185 would indeed have been partial over the Donets.[65] The eclipse was recorded in the probably twelfth-century poem, *The Lay of Prince Igor* as well: 'Igor looked up at the bright sun, / and saw that his warriors / became enveloped in darkness.'[66] Igor's men, according to the *Hypatian Chronicle*, took the eclipse as a sign of misfortune to which their leader replied that God had created the sign, just as he created the world, and that no-one could know the divine purpose. The raid met with disaster; Igor was captured, although he later escaped. The description of the solar eclipse juxtaposes the fear of the warriors with the assertion of divine agency. The vicissitudes of rulership linked, in this context, more closely to the mysteries of the Creator.

Light and the Holy: St Óláfr and St Ethelbert

If the solar eclipse at the crucifixion offered a framework for the association of the phenomenon with earthly rulership, then this was even more so in the particular category of royal saints. As pointed out by Martin Chance, the eclipse, solar or lunar, is a rare feature in the lives of the saints.[67] Three examples that he identified date to

---

[60] Alan Timberlake, 'Who Wrote the Laurentian Chronicle (1177–1203)?' *Zeitschrift für Slavische Philologie* 59, 2000, pp. 237–65.

[61] A. N. Vyssotsky, 'Astromical Records in the Russian Chronicles from 1000–1600 AD,' *Meddelande från Lunds Astronomiska Observatorium*, ser. II no. 126, Historical Notes and Papers 22. Lund, 1949, p. 11.

[62] F. Espenak, 'What to look for in a total solar eclipse' *Earth Sky*, 2017, https://earthsky.org/astronomy-essentials/stages-of-a-total-eclipse-what-to-look-for/.

[63] Guimon, *Historical Writing*, p. 40.

[64] Dimnik, *Dynasty of Chernigov*, pp. 163-77.

[65] *The Hypatian Chronicle [Ipat'evskaia letopis']*, s.a. 6693, ed. Aleksei A. Shakhmativ, *Polnoe sobranie russkikh letopisei*, 2. Saint Petersburg, Tip. M. A. Aleksandrova, 1908, cols. 636–51; Vyssotsky, 'Astromical Records,' p. 10. See also Jay Pasachoff in this volume.

[66] *The Lay of Igor's Campaign*, in *Medieval Russia's Epics, Chronicles, and Tales*, trans. Serge A. Zenkovsky. New York, 1974, p. 170.

[67] Martin Chance, *Einarr Skúlason's Geisli: A Critical Edition*. Toronto, 2005, p. 32. What follows explores the two examples that Chance identifies.

the twelfth century: an Old Norse life of St Óláfr (*c*.995–1030) by Einarr Skúlason, probably dating from 1153, and two Latin *Lives* of St Ethelbert, one from an early point in the century and the other by Gerald of Wales from the 1190s. In both lives the saints are presented as martyrs, a solar eclipse marking their moment of death or its prefiguration, completing the imitation of Christ and drawing on the same understanding of the darkness at the crucifixion in Luke's gospel. All three authors were skilled and practised in their craft; both Einarr and Gerald are known to have been working to commission, which may very well have been the case for the earlier text too.

Einarr, an Icelander, probably the most productive poet of the period, found favour at the Norwegian court, and enjoyed the patronage of king Eysteinn (*c*.1125–57) who co-ruled with his two brothers, and who commissioned Einarr's life of Óláfr. 1153 marked a significant change in the status of the Norwegian church with the confirmation of Nidaros (Trondheim) as an independent archbishopric, a context important for the poetic presentation of the recent royal martyr, which Einarr recited in the cathedral.[68] This, according to the earliest Icelandic chronicle of the Norwegian kings, *Morkinskinna*, compiled in the mid-twelfth century, was accompanied by miracles intimating approval from St Óláfr and the king.[69]

Óláfr ruled Norway from 1016 until 1028 when he was ousted and exiled to the Kyivan Rus' by local rivalries exploited by Cnut the Great of Denmark and England. Óláfr returned to reclaim his kingship in 1030, where he was defeated and killed at the battle of Stiklestad. Presented in the twelfth century and later as a defining moment in Norwegian identity—especially with respect to the widespread adoption of Christianity in the context of relatively recent conversion to the faith, emphasized all the more after Óláfr's canonization in 1164—the reality was more prosaic. The conversion of the region was a slower process and the battle more to do with rival claims to rule.[70] Einarr's poem, however, was a key part of the presentation of the king as Christian hero and martyr.

Light is present throughout the poem, which bears the title *Geisli*, meaning a ray or a beam. Though a common Christian metaphor, Einarr uses light in a striking and structured manner. The opening sections use the metaphor of Sun and sunbeam to explain the relationship between God the Father and God the Son.[71] This is revisited in the description of Óláfr's fall.

The bright sun was not permitted to shine then, when the desirer of the ringshield lost his life; the guardian of the hall of earth showed his signs. It was previously that the blessing-rich shining of the sun ceased at the death of the king of earth's roof; speech-tools are of use to me.[72]

---

[68] Chance, *Geisli*, pp. 9–10. On the promotion of Nidaros see Anders Bergquist, 'The papal legate: Nicholas Breakspear's Scandinavian mission.' In Brenda Bolton and Anne Duggan (eds), *Adrian IV the English Pope (1154–1159)*. Aldershot, 2003, pp. 41–8.

[69] *Morkinskinna*, trans. Theodore M. Andersson and Kari Ellen Gade. Ithaca, NY, 2000, p. 393.

[70] Magnus Rindal, *Fra hedendom til kristendom: perspektiver på religionsskiftet i Norge* (Oslo, 1996); Anders Winroth, *The Conversion of Scandinavia: Vikings, Merchants, and Missionaries in the Remaking of Northern Europe*. New Haven, CT.

[71] *Geisli*, 1–3, pp. 51–3.

[72] *Geisli*, 19, p. 69: '<Nædit> biartr þa er beidir / baugskiallda lauk alldri / syndi saluordr grundar / sin takn rodull skina. / Fyrr var hitt at harra / haudrtiallda <bra> dauda / happ nytaz mer mætt[u] / mæltol <skini> solar.'

The solar eclipse took place, in Einarr's narrative, after the death of Óláfr in battle. That martyrdom could be achieved in this manner was a question with sharper interest in the twelfth century in the wake of Christian crusade to the Holy Land from 1095 and in the Baltic against Slavs, Balts, and Finns, identified as pagan, from the mid-twelfth century.[73] The connection to the eclipse at the death of Christ 'previously', as Einarr phrased it, is made clear by the juxtaposition of the two events, though left to the listener to fill out. As Chance remarks, Óláfr 'the beam of the Sun of Righteousness' stood in typological relationship to Christ, the martyrdom a post-figuration of the crucifixion with its attendant celestial marker.[74] An intriguing question remains as to whether there was really an eclipse at the battle of Stiklestad.[75] None can be calculated for the traditional date 29 July 1030, but one did take place on 31 August which was of magnitude of 0.99, that is almost total, at Stiklestad. Other contemporary evidence would support the 29 July date for the battle; one other early source also mentions the eclipse but its attribution is not entirely secure. It is possible that the whole episode is a literary device, or, perhaps, that the two events, death in battle and eclipse, were elided in short order.

The solar eclipse motif used in the *Lives of St Ethelbert* occurs in a more direct formulation, but carries the same typological function. The relationship between the two lives of the saint is curious. Gerald's *Life* was a reworking and abbreviation of a mid-twelfth-century version by Osbert of Clare (d. in or after 1158), prior of Westminster, which had been composed for Gilbert Foliot, bishop of Hereford.[76] This original text no longer survives in a complete form, but is represented in a later compilation by another monk of Westminster, Richard of Cirencester (before 1340–1400). Gerald wrote, he states, at the behest of the chapter of the cathedral, who had asked for something more succinct.[77] Gerald, sometime archdeacon of Brecon, enjoyed good relations with the chapter of Hereford Cathedral, one of whose number tried at about the same time to get him to settle in their city on account of his reputation for learning. Although the attempt failed, the connection was not lost and it is possible that Gerald was in Hereford when he died.[78]

The earlier anonymous life is much shorter and there is little to suggest that it was consulted by Gerald. The author was certainly familiar with Hereford, and a date of composition in the first third of the twelfth century would coincide with the heightened interest in the saint in the later eleventh and twelfth centuries. Ethelbert

[73] H. E. J. Cowdrey, 'Martyrdom and the First Crusade.' In P. W. Edbury (ed.), *Crusade and Settlement*. Cardiff, 1985, pp. 46–56; Eric Christiansen, *The Northern Crusades*. London, 1997; Alan V. Murray, *Crusade and Conversion on the Baltic Frontier 1150–1500*. London, 2017.

[74] Chance, *Geisli*, p. 35.

[75] See discussion in Chance, *Geisli*, p. 36.

[76] M. R. James, 'Two Lives of St Ethelbert, King and Martyr.' *English Historical Review* 32, 1917, pp. 214–41, analysis at 214–21.

[77] James, 'Two Lives,' p. 216; Frank Barlow, 'Clare, Osbert of (d. in or after 1158), prior of Westminster Abbey and ecclesiastical writer.' *Oxford Dictionary of National Biography* 23, September 2004, (accessed 2 August 2022). https://www.oxforddnb.com/view/10.1093/ref:odnb/9780198614128. 001.0001/odnb-9780198614128-e-5442; Matthew Mesley, 'Depicting the Bishop: Hagiography and Religious Communities in England, c.1070–c.1215.' Unpublished PhD Thesis, University of Exeter, 2009, p. 198.

[78] See Gasper et al, *Knowing and Speaking*, pp. 18–20 and 31–5; R. Bartlett, *Gerald of Wales: A Voice of the Middle Ages*. Oxford, 1982, reprinted Stroud, 2006.

Figure 6.3 Hereford Cathedral.
Source: Photograph by Giles E. M. Gasper.

(779/80–794) is an obscure figure, king of the East Angles, whose execution on the orders of King Offa of Mercia is recorded laconically in the *Anglo-Saxon Chronicle* and for whom there is almost no additional evidence before the mid-eleventh century.[79] The promotion of Ethelbert as a royal martyr is really a phenomenon of the twelfth century with shorter accounts of his life in the chronicles of John of Worcester and William of Malmesbury, and the three *Lives* as mentioned. The cult centre was Hereford, (Figure 6.3) whose cathedral was dedicated to Ethelbert and in which, according to the twelfth-century narrative, his relics were deposited (Figure 6.4).

The *Lives* create the context for the martyrdom, with Ethelbert attending Offa's court, sitting near to Hereford, to marry his daughter. On the journey, according to the anonymous life, an earthquake and then a solar eclipse occurred, causing consternation amongst Ethelbert's council:

To the sign of the earth the sign of heaven soon responded. The sun scattering rays throughout the world had flashed brightly, and mark! the whole council is darkened in mid-journey. The density of the clouds arose suddenly, [amongst] the travellers themselves one prevented from seeing the other. Only by the sound of the voice did anyone know each other. King Ethelbert was shocked when the radiant sun thus

[79] Andy Todd, 'Æthelberht [St Æthelberht, Ethelbert] (779/80–794), king of the East Angles.' *Oxford Dictionary of National Biography*, 23 September 2004 (accessed 2 August 2022). https://www.oxforddnb.com/view/10.1093/ref:odnb/9780198614128.001.0001/odnb-9780198614128-e-8903. *The Anglo-Saxon Chronicle, Collaborative Edition*, Vol. 7, *MS. E*, s.a. 792 [*recte* 794].

Figure 6.4 St Ethelbert painted by Peter Murphy (born 1959) on the Shrine to St Ethelbert the King in Hereford Cathedral.

Source: Photograph by Giles E. M. Gasper.

became darkened. To the stunned council he took it upon himself to exclaim. 'We bend our knees' he said 'we ought to urge the heavens with prayer that the almighty God might have mercy upon is.' What is to be supposed, dearest brothers, prefigured by these signs? The truly glorious King, martyred for the name of Christ, removed from the light of the present time and crowned in heavenly glory.[80]

The terror at the phenomena is turned to interrogation of what they might portend, in this case the king's death for Christ. The association with the crucifixion eclipse is not drawn out explicitly; the text rather addresses the movement from the light of the world to the light of heaven.

The events of Christ's death are explicitly connected to Ethelbert's by Gerald, who goes on to explain what the earthquake and solar eclipse signify.

> Nor is it an extraordinary thing if the signs which appeared in the death of Christ and before the death of this member of the body of Christ [Ethelbert], and that they foretold the same for Christ's beloved. Seeing as the earth was moved as if shuddering at a sin that now threatened, while the sun, having been darkened, turned away its face as though not to see. Or the earthquake that could clearly signify the agitation and devastation which endured from the birth of Ethelbert for the round of many years under petty kings and tyrants up to the time of King Edmund: moreover, the sun, withdrawing light itself, made clear with an obvious sign that he was to be withdrawn from that light very soon after.[81]

Gerald's emphasis on shifting the interpretation of what is extraordinary to what is explicable can be compared usefully to his remarks on how solar eclipses were perceived: the amazement of the viewer is amplified because of the rarity of the event. That the eclipse foretold the same for Christ and for Ethelbert underlines the holiness of the latter and the multi-layered understanding of the drama of the celestial world.

Where Magnus of Reichersberg recorded the uncertainty amongst the inhabitants of Salzburg as to what the eclipse of 1133 might portend, other chroniclers found more coherent narratives to tell in which what happened above was associated with what happened below. All of the narratives considered show the same mixture of wonder and fear, and the ubiquity across different Christian cultures of the features recorded by chroniclers is striking. Nevertheless, it is the variation of description

---

[80] *Passio sancti Athelberhti regis et mariris*, ed. James 'Two Lives,' pp. 236–41, at 238: 'Terre signo celi mox respondit signum. Sol per orbem radios spargens fulserat lucide, et ecce obscuratur toti curie medio in itinere. Densitas nebularum subito oborta itinerantes sesc alterutrum uidere negat. Dumtaxat uocis per sonum quislibet alterum nouit. Obstupescit rex Æðelberhtus dum sic radiosus phebus obtenebrescit. Ad stupidam curiam clamare cepit. "Genua," inquit, "flectemus: prece polum pulsemus ut nostri misereatur omnipotens deus." Quid putandum fratres karissimi his signis prefiguratum? Regem sane gloriosum pro Ihesu Christi nomine martyrizandum, presentiarum luci subtrahendum, celestique gloria coronandum.'

[81] Gerald of Wales, *Vita Regis et martiris Æthelberti*, ed. James, 'Two Lives,' pp. 222–36 at p. 225: 'Nec mirum si signa que in morte Christi apparuerunt et ante mortem huius membri Christi eiusque dilecti eandem presagiencia contigerunt. Terra quippe quasi scelus abhorrens quod iam iminebat mota est, sol autem obscuratus tanquam ne uideret faciem auertit. Uel terremotus ille terre regni tocius commocionem et desolacionem que ab ortu Ethelberti multis annorum circulis sub regulis et tirannis usque ad regis Edmundi tempora durauit potuit aperte significare: sol vero lucem subtrahens ipsum ab hac luce in proximo subtrahendum manifesto indicio declarauit.'

between author-compilers that stands out in sharper relief. The twelfth century saw considerable change to the conceptual frameworks for how the heavens worked and the reasons why solar eclipses occurred. For all of that, the understanding of the phenomenon was cast in a Christian vision, which links, ultimately, its distinctive appearance to the central moment of the faith, the crucifixion.[82] The solar eclipse was a metaphor for the Creator to creation as well as a physical sign, and it is this, perhaps, that ensured its faithful record amongst those charged with chronicling the days and years of their communities.

---

[82] See David Bentley Hart's chapter in this volume.

7

Annus Tenebrosus

Black Monday, Faith, and Political Fervour in Early Modern England

Anna Marie Roos

History of Early Modern English Science, History of Medicine

Moving out of the Middle Ages, we reach a turning point in European astronomy. The ideas of Nicolaus Copernicus, Johannes Kepler, and, not much later, Isaac Newton would lastingly revolutionize the subject. The changes in religious history, material and technological history, and media all impacted the production and distribution of knowledge about total solar eclipses, as evident in the discussion of the almanac, an annual publication containing calendrical and astronomical information. The interpretation of eclipses then was caught in the same suspension between an understanding of world, life, and human beings ruled by God, or by the confidence and anxieties of the human mind, or by the increasingly measurable scientific knowledge emerging from a proto-enlightened age of observing, categorizing, and theorizing. On the brink of the clockwork universe, the famous total solar eclipse over England in 1652 was surrounded by the political exploitation of popular fear on 'Black Monday' in the mid-seventeenth century. Its historical record serves as a reminder that Enlightenment and superstition would remain in conflict for a long time to come.

almanac An annual publication containing calendrical and astronomical information.

Black Munday / Mirk Monday Common names in England and Scotland for the total solar eclipse of 29 March 1652 (OS).

humoral medicine Medicine relating to the fluids of the body following the ancient Greek physician Hippocrates (*c*.460 BC–370 BC), involving the theory of the four humours (blood, yellow bile, black bile, and phlegm) and their influence on the body and its emotions.

millenarianism The belief in a future millennium following the second coming of Christ.

> The Sun Eclipsed 12. Deg. The 29. Day at 10 before noon. Not a man now living ever saw the like before in England, nor shall hardly ever see the like again; therefore it will be worth your Observation.
>
> *Vincent Wing's, Ouranizomai or an Almanack and Prognostication for the Year of Our Lord, 1652 (London, 1652), sig. A7 recto*

In 1652, the 'Black Monday' solar eclipse was the seventeenth-century equivalent of 'fake news,' containing enough epistemic authority to be plausible (an eclipse did occur), yet used as a vehicle to promote political or religious beliefs.[1] Many sectarians in the English Civil War believed in millenarianism and signs of nature became extremely significant to their prophesies of the end of the world, as well as to their promotion of their political and religious agenda. As the importance of chiliastic beliefs grew in the 1650s, the emanations of the luminaries, as well as their interruption in eclipses, for some held even more significant powers. A solar eclipse lasting for 169 seconds in totality on 29 March 1652 OS/8 April 1652 NS, called 'Black Monday' in England, or 'Mirk Monaday' in Scotland, was utilized by Fifth Monarchists, as well as others with apocalyptic beliefs, as a sign predicting the fall of government and monarchy, the end of the world, and the second coming of Christ.[2] As David R. Como stated:

> There was a sense here, anxious and raw, yet palpable, that England was moving into a phase of revelation, in which new truths of church and state would be unveiled ... we see here a marked leakage between the religious and political realms, for this sense surely drew heavily, if in an informal and half-conscious way, upon theological ideas about progressive revelation, 'new light,' and indeed the eschatological unfolding of God's divine plan. Just as God's church was now to be purified [by sectarians], so too the state would be cleansed and rebuilt.[3]

Simply, the eclipsed, invisible sun made fears about the stability of the state and the future of the world even more visible.

This chapter, however, complicates and enriches Como's point as well as one of this book's major themes—that an eclipse is more of an unveiling than a hiding, an apocalypse rather than an eclipse.[4] My analysis indicates that in the case of 'Black Monday,' an eclipse could be for some a presage of the world's ending, while at the same time

---

[1] This chapter builds and draws upon some of my earlier work in *Luminaries in the Natural World: Perceptions of the Sun and Moon in England, 1400–1720*. Bern and Oxford, 2001, pp. 135–7, 174–7. Quoted text from primary sources is left in its original transcribed form, and not modernized. For more information about this eclipse and the concept of 'epistemic authority,' see Don Hertzog, *Cunning*. Princeton, NJ, 2006, pp. 105, 117.

[2] The Fifth Monarchy referred to the beliefs of the Fifth Monarchy Men, a sect that believed that the Second Coming of Christ predicted in the Books of Daniel and Revelation was fast approaching, and that King Jesus would intervene in English politics to bring about a democratic and utopian English Commonwealth. See Christopher Hill, *The World Turned Upside Down: Radical Ideas During the English Revolution*. Harmondsworth and New York, 1985, p. 72. See also Bernard Capp, *The Fifth Monarchy Men; A Study in Seventeenth-Century English Millenarianism*. London, 1972. OS is equivalent to 'Old Style' dating or the Julian Calendar, in use in England until 1752; NS is 'New Style' or the Gregorian Calendar, today's international calendar in current usage. The English were late adopters of the Gregorian Calendar (France, for instance, had adopted it in 1582). To align the English calendar with that in use in Europe, it was necessary to correct it by 11 days, which accounts for the date difference between 'Old Style' and 'New Style.'

[3] David R. Como, 'Print, Censorship and Ideological Escalation in the English Civil War.' *Journal of British Studies* 51, October 2012, pp. 820–57, on p. 845.

[4] See the Introduction to this book by Henrike Lange and Tom McLeish. My comments about astrology and astronomy are adapted from my article: 'Astronomy and Space Science: Astronomy Emerges from Astrology.' *Scientific Thought in Context*, 3 vols. Detroit, 2009, vol. 1, pp. 61–70. The article has recently appeared on Encyclopedia.com, https://www.encyclopedia.com/science/science-magazines/astronomy-and-space-science-astronomy-emerges-astrology (accessed 16 October 2021).

for others provided reassurance that the natural cycles of the universe continued in regular, self-sustaining order.

This state of affairs explains why, on one hand, although the totality of 'Black Monday' was only 169 seconds, its apocalyptic effects were feared by some audiences to last far in the future. Astrologer William Lilly noted in 1651 in his almanac about 'Black Monday':

> The Sun performs his course in a year, the Moon in Moneths. So also the effects of a Solar Eclips continues for yeers, the Moone's onely Months. Those of the Sunne may continue three or foure years . . . the effects of the two superiors have their Portents for a long season, by reason of the so many and infinite transits of themselves and other Planets through the severall Regions of Heaven.[5]

English almanacs, pocket-size pamphlets with astrological forecasts for the forthcoming year and sometimes medical advice or political gossip, increased in popularity in the early modern period due to cheaper printing costs and increasing literacy:[6]

> By 1659 London astrologer William Lilly's (1602–1681) political almanacs alone were selling nearly 30,000 a year. The readers of these almanacs could also cast their own charts by using the astrological tables they provided. The seventeenth century, sometimes known as the Age of Iron, was a time of often brutal warfare in which post-Reformation religious and political motivations were intertwined. Almanacs thus became a means of understanding or accepting these calamitous events for the general public.[7]

This apocalyptic stance afforded to 'Black Monday,' however, did not necessarily indicate ignorance of its cause amongst most early modern astronomers, natural philosophers, and literary figures, who considered eclipses merely as planetary interpositions, easily predicted by lunar and solar cycles. Some almanacs even inserted simple explanations and diagrams into their tracts and almanacs to show that eclipses were the result of planetary interposition, not magical occurrences.[8] Chamberlaine's 1649 almanac, for instance, stated:

> the eclipse of the sun is nothing else but the direct interposition of the body of the Moon between the Sun and us: and the cause of the Moon eclipsed is nothing else but the direct putting of the dark body of the Earth between the Sun and the Moone:

---

[5] William Lilly, *Annus Tenebrosus, or the Dark Year*. London, 1651, p. 42.

[6] Ann Geneva, *Astrology and the Seventeenth-Century Mind: William Lilly and the Language of the Stars*. Manchester, 1995, p. 197.

[7] Roos, 'Astronomy and Space Science: Astronomy Emerges from Astrology,' p. 67.

[8] Some of this material is taken from my book Anna Marie Roos, *Luminaries in the Natural World: The Sun and the Moon in England, 1400–1720*. Bern, Berlin, Brussels, New York and Oxford, 2001. William E. Burns has also analysed the impact of Black Munday on astrological beliefs in 'The Terriblest Eclipse That Hath Been Seen in Our Days': Black Monday and the Debate on Astrology during the Interregnum.' In Margaret Osler (ed.) *Rethinking the Scientific Revolution*. Cambridge, 2000, pp. 137–52.

for the moon having no light of herself but that she borroweth the Sun, is hindered of that borrowed light by the shadow of the Earth.[9]

Although Edward Pond's 1652 almanac conceded that eclipses were thought to cause 'many mischiefs in the world,' he not only described their regular cycles but provided accurate instructions for curious readers on how to 'behold an eclipse of the Sunne without hurt to the eyes':

> Take a burning-glass, such as men use to light tobacco within the sunne; or a spectacle-glasse that is thick in the middle, such as is for the eldest sight; and hold this glasse in the sunne as if you would burn through it a pastboard or white paper book, or such like; and draw the glasse from the board or book, twice so farre as you do to burn with it: so by direct holding it nearer or further, as you shall see best, you may behold upon your board, the round body of the sun, and how the moon passeth between the glasse and the sun during the whole time of the eclipse.
>
> This mayest thou practise before the time of an eclipse, wherein thou shalt discern any cloud passing under the sunne: or by putting or holding a bullet or his fingers end betwixt the sun and the glasse at such time (The sunne shining) as though holdest the glasse as before thou are taught.[10]

Dr John Wybard, using such methods, gave a very matter-of-fact observation of the eclipse in Carrickfergus, Ireland:

> The Moon, as if at that moment, and unexpectedly, threw itself so very nimbly between the entire path or circuit of the Sun's disc (in so far as it appeared to our sight); so that it seemed to move in a circle or roll around, like a plate or upper mill-stone; with the Sun, glowing or rather shimmering, all around its rim or edge.[11]

Eclipses were also extensively used by early modern intellectuals such as Joseph Scalinger (1540–1609) whose *Opus novum de emendation temporum* (editions: 1583, 1598, 1629) and *Thesaurus temporum* (1606, 1658) combined 'scientific with philological studies' to 'build chronological arguments around dated eclipses'; Scalinger's aim was to ground historical enquiry in a 'hard core of indisputable fact,' namely using eclipses as date markers, transforming historical method.[12] More prosaically, John Taylor, a miner of lead ore who began work in the pit at nine years 'dressing lead

---

[9] Joseph Chamberlaine, *Chamberlaine, 1649. A new almanacke*. London, 1649, Wing #A1392, fol. C4 v.

[10] Edward Pond, *Pond an Almanack for the year of our Lord God 1652*. Cambridge, 1652, sig. C1v. See also 'Black Monday: 1652,' The Pulter Project, Northwestern University, https://pulterproject. northwestern.edu/curations/c1-black-monday-1652.html (Accessed 22 July 2021).

[11] Vincent Wing. *Astronomia instaurata, or, A new compendious restauration of astronomie in four parts*. London, 1656 [Luna momento quasi, et eximproviso, totam se intra Disci Solis orbitam seu ambitum (quatenus conspectui nostro appareret) tam agiliter injiciebat; ut circumagere aut circumvolutare videretur, sicut catillus, seu lapis molaris superior; Sole tunc circum-circa, ejus limbum seu marginem splendidulo vel corusco, apparente]. See Elizabeth DeBold, 'Black Monday: the Great Solar Eclipse of 1652.' *The Collation: Research and Exploration at the Folger*, 22 August 2017, https://collation.folger.edu/ 2017/08/black-monday-great-solar-eclipse-1652/ (accessed 21 July 2021). Translation from Latin by Sarah Powell, EMMO palaeographer.

[12] Anthony Grafton, 'Some Uses of Eclipses in Early Modern Chronology.' *Journal of the History of Ideas* 64(2), 2003, pp. 213–29, on p. 221.

ore' and lived to a prodigious age, used the 1652 solar eclipse to mark his lifespan. He recalled that he

> went below ground to assist the miners, and had been thus employed for three or four years, when the great Solar eclipse, vulgarly called the Mirk Monday, happened . . . . He, being then at the bottom of the shaft or pit, was desired by the man at the top to call those below to come out, because a black cloud had darkened the sun, so that the birds were falling to the earth. And this, which he always relates with the same circumstances, is the only event by which his age may be ascertained.[13]

Only extraordinary eclipses that occurred outside these chronological cycles, such as the miraculous eclipse that was recorded at the passion of Christ,[14] were considered to be firm prognostications of horrible events, confined to the Bible or to ancient history. In a reassuring sermon for the Lord Mayor and Aldermen delivered a day before Black Monday at St Paul's Cathedral, the preacher Fulk Bellars stressed that the eclipse was an entirely natural occurrence, stressing that panic was often caused by 'men's unaquaintednesse with the naturall causes of them.'[15] And besides, 'Jesus Christ, the Mysticall or Gospell Sun,' was only 'seemingly eclipsed, yet never going down from his people.'[16] Bellars also assured his audience that there was no need to panic for 'had there been any reall danger in firmamentary Eclipses, our God doubtlesse would have left it upon record in the Scripture.'[17] Instead 'Bellars claimed that they should be treated as allegorical aids to religious meditation, in that the removal of the sun's light was analogous to the removal of Christ's grace.'[18]

Indeed, Lady Hester Pulter (1605–78), a seventeenth-century poet, in her work 'The Eclipse' did not see eclipses as portents of the wrath of God, but rather just normal planetary revolutions; her earthly life instead was 'a journey through many such revolutions' according to natural rhythms: 'I my passage make through Revolution / humbly obedient to my Makers Laws.'[19] Pulter, as did Bellars, also allegorically associated the Sun with divinity, seeing its eclipse as akin to the effects of sin on an individual's relationship with God:

> But O, my sins (my sins), and none but those,
> Makes my poor soul o'erflow with sad annoy;
> 'Tis they, and none but they, do interpose
> 'Twixt heaven and me, and doth eclipse my joy[20]

---

[13] 'On the antient Camelon, and the Picts. By Mr. Walker,' *Archaeologia*, 1, 1, second edn (1779), pp. 230–7, on pp. 230–1, note *z*.

[14] For representations of the eclipse at the crucifixion in western art, see Roberta J. M. Olson and Jay M. Pasachoff, *Cosmos: The Art and Science of the Universe*. London, 2019, 90–1.

[15] Fulk Bellars, '*Epistle Dedicatorie*.' In *Jesus Christ: The Mysticall or Gospell Sun*. London, 1652, Thomason E. 665(15), p. 2. See also Burns, 'Black Monday and Astrology,' p. 145.

[16] Bellars, '*Epistle Dedicatorie*,' title page.

[17] Bellars, '*Epistle Dedicatorie*,' p. 1.

[18] Burns, 'Black Monday and Astrology,' p. 145.

[19] Sarah Hutton, 'Hester Pulter (*c*.1596–1678): A Woman Poet and the New Astronomy.' *Sciences et littérature* 14, 2008, https://doi.org/10.4000/episteme.729. See also Laura Dodds, 'Hester Pulter Observes the Eclipse: Or, the Poetics of the Astronomical Event.' *Journal for Early Modern Cultural Studies* 20(2), spring 2020, pp. 144–68.

[20] Hester Pulter, 'The Eclipse,' stanza 4, line sixteen. The Pulter Project, Northwestern University, https://pulterproject.northwestern.edu/poems/ee/the-eclipse/ (accessed 22 July 2021).

As monarchs in royalist circles thought to be ordained by God, the Sun was of course associated with monarchy, and the astrological identification of the monarch with the Sun was pervasive in early modern England.[21] Solar eclipses were also commonly regarded as allegories for the eclipse of an eminent individual. Elizabeth I's *Rainbow Portrait* by Isaac Oliver (*c*.1600) in Hatfield House shows her holding a rainbow and inscribed above is the motto, *Non sine sole iris* (no rainbow without the sun), an allusion to Elizabeth as the divinely anointed monarch who brings light and virtue to her kingdom. Because 'the equation of the sun as metaphor for a king [or queen] was also always intelligible and effective' for all audiences, royalist propaganda tracts in the Civil War picked up on this association, portraying Charles I as the Sun, ruling his country as the Sun ruled the heavens.[22] 'One month after Charles's defeat at Marston Moor in 1644, the symbol of the sun eclipsed was utilised by one publication, *The Great Eclipse of the Sun OR Charles His Waine*.'[23] Just as the Sun had been eclipsed by the Moon, Charles had been supposedly 'eclipsed by the destructive perswasions of His Queen,' Henrietta Maria.[24] The author feared that the Queen's Catholicism had clouded the king's light of reason, for just as 'Ordinary women, can in the Nighttime perswade their husbands to give them new Gowns, Petticotes, and make them grant their desire,' Henrietta Maria had persuaded Charles to 'protect the Papists.'[25]

However, during the Civil War era, 'sectarian pamphlets took this metaphor more literally, using solar eclipses to predict astrologically the overthrow of monarchy and government. Although these works, like other astrological tracts, often provided the rationale of planetary interposition for a solar eclipse, it was their message of doom that the public remembered.'[26] The metaphorical use of eclipses was seen especially in the effects of prognostications about 'Black Monday,' inspiring broadsheets, sermons, almanacs, astrological pamphlets, and scientific treatises. Over a quarter of the publications collected by the bookseller George Thomason for the month of March related to the significance of the eclipse. 'In the emotionally charged atmosphere of the unstable Rump Parliament, sectarian astrologers forecast not only the fall of Presbyterianism, the reform of the law, and the second coming of Christ but also sickness, flood, and famine.'[27]

These sectarian pamphlets were written in response to an earlier piece entitled *Black Monday*, a 'veiled piece of Royalist propaganda' which had nothing to do with eclipses at all. It appeared in December 1651, and made dire remarks about the Tender of Union of 28 October 1651, the Parliamentary Declaration that England and

---

[21] Geneva, *Astrology and the Seventeenth-Century Mind*, p. 267.

[22] Geneva, *Astrology and the Seventeenth-Century Mind*, p. 267. Roos, *Luminaries in the Natural World*, p. 174.

[23] G. B., *The Great Eclipse of the Sun OR Charles His Waine overclouded by the evil Influences of the Moon . . . Otherwise, Great CHARLES, our Gracious KING, Eclipsed by the destructive perswasions of His Queen*. London, printed according to Order, by G. B., 30 August 1644, Thomason E7 (30), frontispiece. Roos, *Luminaries in the Natural World*, p. 174.

[24] G. B., *The Great Eclipse of the Sun*, frontispiece. Roos, *Luminaries in the Natural World*, p. 174.

[25] G. B., *The Great Eclipse of the Sun*, fol. A2 recto. Roos, *Luminaries in the Natural World*, p. 174.

[26] Roos, *Luminaries in the Natural World*, p. 175.

[27] Keith Thomas, *Religion and the Decline of Magic*. London, 1971; ebook reprint, 1991), pp. 299–300. Some of this material is taken from Roos, *Luminaries in the Natural World*, p. 175.

Scotland should be in a single commonwealth.[28] For context, earlier in the year, on 1 January 1651, the exiled Charles II was crowned King of Scotland at Scone; on 3 September he was defeated at the Battle of Worcester leading a large Scottish army and began his escape northwards; and on 15–16 October, Charles escaped to Fécamp in France from Shoreham and went into exile. He was effectively displaced as King of England and Scotland by the Rump Parliament. Using the common trope of the monarch as the Sun, the anonymous royalist pamphleteer wrote:

> Some begin to be ashamed of their Scotch Covenant, which bound them to promote the honor and dignity of the King and his Successors, *subintelligitur*, if they could not sell him for money to promote their own.
>
> *Sol* (the Prince regent of heaven) who hath naturally the signification of regality, worldly pomp and glory, is ecclipsed in the very degree of his exaltation, and in the regall house of heaven likewise . . . . In a word, this Eclipse, premonstrates a generall madnesse and confusion to all such Kingdoms, Common-wealths, Countries, Cities and Towns as are under the division of *Aries, Libra, Cancer,* and *Capricorn, viz.* England, Scotland.[29]

In response the next year, Nicholas Culpepper (1616–54), herbalist, astrologer, and Fifth Monarchy Man published a sensationalistic account of the terrible effects to follow from the 1652 solar eclipse. His publication was entitled *Catastrophe Magnatum: The Fall of Monarchy*.[30] Culpepper was the son of a clergyman, went to Cambridge in 1634 for a short time, and was apprenticed to an apothecary.[31] His medical training led him to establish his own practice as an astrological physician in Red Lion Street, Spitalfields, London in 1640.[32] He often volunteered his services to the poor, and his political sympathies for the Fifth Monarchy Men led to his enlistment in the Parliamentary Army as a field-surgeon in 1643, where he was seriously wounded in the chest.[33] He retired to his medical practice and by 1649 was publishing translations from Latin to English of leading medical works, such as the *Pharmacopeia Londinesis* of the Royal College of Physicians, because he believed in making such knowledge available to the poor.[34] Culpepper's advocacy of Paracelsian medicine led to his becoming embroiled in further controversy with the College, which followed the methods and humoral medicine of Galen and Hippocrates.[35] Culpepper ultimately died of consumption in 1654.

---

[28] Anonymous, *Black Munday: Or, a full and exact description of that great and terrible Eclipse of the Sun which will happen on the 29. Day of March 1652.* London, 1651; Thomas, *Religion and the Decline of Magic*, p. 344.

[29] *Black Munday: Or, a full and exact description of that great and terrible Eclipse of the Sun*, pp. 3, 5.

[30] Nicholas Culpepper, *Catastrophe Magnatum: Or, The Fall of Monarchie. A Caveat to Magistrates, Deduced from the Eclipse of the Sunne, March 29. 1652. With a Probable Conjecture of the Determination of the Effects.* London, 1651. Thomason dated his receipt of this publication as 31 March 1651.

[31] Patrick Curry, 'Culpeper, Nicholas (1616–1654), physician and astrologer.' *Oxford Dictionary of National Biography.* 23 September 2004 (accessed 28 July 2021), https://www.oxforddnb.com/view/10.1093/ref:odnb/9780198614128.001.0001/odnb-9780198614128-e-6882.

[32] Ibid.

[33] Ibid.

[34] Hill, *World Turned Upside Down*, p. 299.

[35] Ibid. In addition to his medical work, Paracelsus was also more generally known as an astrological prognosticator, his 'disconcerting prophecies based primarily on conjunctions, eclipses, and comets,' of appeal to Culpepper. See Charles Webster, 'Paracelsus: medicine as popular protest.' In Ole Peter Grell

As a Fifth Monarchist, Culpepper desired reform for the Commonwealth, stating that its liberty 'was most impaired by three sorts of men, priests, physicians, lawyers,' who had a monopoly on their professions.[36] He thought this reform of the Commonwealth and indeed of all the European governments would come about due to the effects of Black Monday, predicting that at the beginning of 1655, 'the Government will come into the hands of the People, and everlasting peace we shall enjoy.'[37] After all as the Sun 'is in the heavens, so are Magistrates in a Commonwealth: if the one be afflicted, why may not the other?'[38] Culpepper also had an animistic view of the universe, and argued that the Sun was a 'creature which gives life, light, and motion to the creation: by moving about his own body upon his Axis, he moves the whole creation.'[39] Therefore, because the sun was 'the life of the creation' and the controller of the *anima mundi* or the world soul, its eclipse was 'dismal' to the Earth, leaving it as a 'dead body without a spirit and motion.'[40]

Culpepper's claims concerning the subsequent effect of the lack of sunlight on the Earth and the body was explained by the traditional connection between the vital animal spirits and the Sun. Aristotle's linkage of life with the generative Sun and the motion inherent in the soul-principle—the pneuma—meant that he considered the motion of the heart as indicative of the seat of the soul.[41] The Sun was thus often associated by early modern English writers with the animal spirits and the circulatory system.[42] Physician William Harvey, known for his work on the circulation of the blood, believed that the blood had a celestial nature for it was 'analogous to the element of the stars,' and the heart in the body was as the Sun in the solar system, in the midst of the body and giving life.[43] In this manner, as astrological physician Thomas Tryon claimed, there is 'an Astrology within Man.'[44]

Culpepper then claimed that lack of sunlight in an eclipse moreover caused people to 'conceive strange, ridiculous thoughts of the Divil' resulting in 'an *Epidemicall*

---

and Andrew Cunningham (eds), *Medicine and the Reformation*. Abingdon, Oxon, 2001, pp. 57–77, on p. 61.

[36] Nicholas Culpepper, *A physicall directory, or A translation of the London dispensatory/made by the Colledge of Physicians in London . . . .* London, 1649, Wing #C7540, sig. A; quoted in Hill, *World Turned Upside Down,* p. 298. For material about Culpepper and eclipses, see Roos, *Luminaries in the Natural World,* pp. 175–6.

[37] Culpepper, *Catastrophe Magnatum*, pp. 45 and 68.

[38] Ibid., p. 3.

[39] Ibid., p. 2.

[40] Ibid., p. 2.

[41] Aristotle, *On the Heavens*, trans. W. K. C. Guthrie. Loeb Classical Library, Cambridge, 1960, II.7.289a 20–1, 31–3. Some of this material in this section about solar and lunar effects in medicine is taken from one of the author's earlier publications: Anna Marie Roos, 'Luminaries in Medicine: Richard Mead, James Gibbs, and Solar and Lunar Effects on the Human Body in Early Modern England.' *Bulletin of the History of Medicine* 74(2), autumn 2000, pp. 433–57, on p. 451.

[42] Walter Pagel, 'Medieval and Renaissance Contributions to Knowledge of the Brain and its Functions.' In F. N. L. Poynter (ed.), *The History and Philosophy of Knowledge of the Brain and Its Functions: An Anglo-American Symposium, London, July 15–17 1957*. Oxford, 1958, p. 108.

[43] William Harvey, *Disputations Touching the Generation of Animals*, trans. G. Whitteridge. Oxford, 1981, p. 379. For more on Harvey's connection between the heart and Sun, see Walter Pagel, *William Harvey's Biological Ideas: Selected Aspects and Historical Background*. New York, 1967, esp. pp. 82–124; John S. White, 'William Harvey and the Primacy of the Blood.' *Annals of Science* 43, 1986, pp. 239–55; Roger French, *William Harvey's Natural Philosophy*. Cambridge, 1994.

[44] Thomas Tryon, *Some memoirs of the Life of Mr. Thomas Tryon, late of London, merchant* (London, 1705), pp. 23–4.

desease called *madnesse*.'[45] Insanity, as well as epileptic fits, was considered a lunar disease, the nervous fluids in the body regulated by planetary cycles, an application of astrological belief, and a tenacious humoral physiology and Galenic medicine practised in the early modern period. The Galenic corpus advised that the waxing and waning of the Moon controlled not only the volume of blood in the body, but also the amount of cold and moist phlegm in the brain.[46] Concomitantly, a lack of sunlight, such as that caused by an eclipse, could cause the same symptoms.

Cures for a disease caused by a particular morbificant planet could be healed herbally or with astrological medicine.[47] The medical doctrine of signatures was an extra-Galenic principle that had been popularized by Paracelsus, and by subsequent English publications such as Culpepper's own *English Physitian Enlarged* (1653) or Robert Turner's *Botanologia* (1664).[48] Paracelsian herbal remedies, via the system of astral parallels, went by the principle that each organ and each herb is bound with its own planet, and maladies could be cured sympathetically by employing plants belonging to the planets causing the disease.[49] Each plant had a signature of its medical application, usually resembling the part of the body or the ailment that it could cure—for instance, lentils and rapeseed were thought sympathetically to cure the smallpox, a lunar disease, because the seeds were similar to the spots of the moon.[50] Alternatively, some cures for a disease caused by a particular morbificant planet could be healed antipathetically by a herb of the opposing planet. For example, lunar diseases with their abundance of cold and moist humours could be cured via herbs or food thought to be governed by the Sun such as saffron or lemons, or a gold amulet struck with a picture of the sun (usually when it was at its strongest influence during the vernal equinox).[51] The Sun was associated with gold and the Moon with silver, which were also ancient alchemical analogies.

According to Culpepper, this insanity from the lack of solar beams would eventually 'possess the brains of the Princes thereof,' and, in a neat juxtaposition of astrological medicine and wishful thinking, the incapacitation of the monarchs would

---

[45] Culpepper, *Catastrophe Magnatum*, p. 68.

[46] Roos, 'Luminaries in Medicine,' p. 444; Galen, *De diebus decretoris, in Claudii Galeni Opera Omnia*, 20 vols., ed. C. G. Kuhn. Leipzig, 1821–33; Hildesheim: Olms, 1964–5, 9: 903. Oswei Temkin, *The Falling Sickness: A History of Epilepsy from the Greeks to the Beginnings of Modern Neurology*, 2nd edn. Baltimore, MD, 1971, p. 26.

[47] Here 'morbificant' refers more generally to any planet that causes disease, dependent on season and sometimes on humoral complexion.

[48] Nicholas Culpepper, *The English Physitian enlarged. By Nicholas Culpeper*. London, 1653; Robert Turner, *Botanologia* . . . . London, 1664.

[49] Israel Hiebner, *Mysterium sigillorum . . .*, trans. B. Clayton. London, 1698, pp. 106–7.

[50] Joseph Blagrave, *Blagrave's Astrological Practice of Physick* . . . London, 1671, p. 78; Walter Pagel, *Paracelsus: An Introduction to Philosophical Medicine in the Era of the Renaissance*. Basel, 1958, p. 144.

[51] See Ernest A. Wallis Budge, *Amulets and Superstitions: The Original Texts with Translations and Descriptions of a Long Series of Egyptian, Sumerian, Assyrian, Hebrew, Christian, Gnostic, and Muslim Amulets and Talismans and Magical Figures, with Chapters on the Evil Eye, the Origin of the Amulet, the Pentagon, the Swastika, the Cross (Pagan and Christian), the Properties of Stones, Rings, Divination, Numbers, the Kabbalah, Ancient Astrology, etc*. London, 1930. See also Martha R. Baldwin, 'Toads and Plague: Amulet Therapy in Seventeenth-Century Medicine.' *Bulletin of the History of Medicine* 67, 1993, pp. 227–47; Anna Marie Roos, 'Magic coins' and 'magic squares': the discovery of astrological sigils in the Oldenburg Letters.' *Notes and Records of the Royal Society* 62, 2008, pp. 271–88.

then make way for the 'Fifth Monarchy of the World.'[52] As Culpepper put it in another work, *An Ephemeris for the Year 1652*:

> The Sun's eclips'd in's Throne, and cries aloud,
> O Kings, why, being moral, are ye proud?
> Your Scepter's gone; Democracy takes place:
> Your Hoasts shall fall by sword before your face.[53]

Before this utopia could be established, however, there would be 'nothing but trouble seen under the moon' (the sublunary region of the earth).[54] Pestilence and famine would occur and, if that were not enough, 'violent storms, and unnaturall, if not unheard of hail will be a great prejudice unto the earth.'[55]

Culpepper's tract seized the imagination of other astrological writers, sectarians and non-sectarians alike, who published their own lurid predictions of what would happen due to the effects of Black Monday. One anonymous sectarian wrote *A Year of Wonders*, which prognosticated, like Culpepper's tract, that the eclipse would cause 'The Glorious Rising of the fifth Monarch, Shewing the greatness of that free-born Prince, who shall Reign and govern, and what shall happen upon his Coronation.'[56] Another work published in September 1651 showed on its frontispiece a darkened sun surrounded by a death-head, a sword, fire, and crossbones, accompanied by a warning that the eclipse presaged 'cruel wards and Bloodshed, House-burnings, great Robberies, Thefts, Plundering and Pillaging, Rapes, Depopulation, violent and unexpected Deaths, Famine, Plague.'[57] This author, along with many others, claimed that Black Monday was as 'miraculous an Eclipse, as that was at the outcome of our Saviour's Passion.'[58] The *Shepherd's Prognostication*, published in January 1652, remarked that

> it is no fable, for whosoever lives to see the 29 of March shall surely find the matter here written too true, except the world be brought to an end before that day comes, *Mat. 24.29* [Matthew 24.29] *and immediately after the tribulation of those daies, shall the Sun be darkened, and the Moon shall not give her light.*[59]

---

[52] Culpepper, *Catastrophe Magnatum*, p. 68.

[53] Nicholas Culpepper, *An Ephemeris for the Year, 1652. Being Leap Year, and a Year of Wonders. Prognosticating The Ruine of Monarchy throughout Europe; and a Change of the Law. Manifested by Rational Predictions: From the Eclipses of the Moon. From that most Terrible Eclipse of the Sun .... London, 21 August 1651, Thomason E. 1349(6), p. 20.

[54] Culpepper, *An Ephemeris*, pp. 65–6.

[55] Ibid., p. 67. Roos, *Luminaries in the Natural World*, p. 176.

[56] Anonymous, *The Year of Wonders or, The glorious Rising of the fifth Monarch*. London, 1652, Thomason E. 656 (22), frontispiece. George Thomason dated the reception of his copy of the pamphlet as 21 March.

[57] Anonymous, *Black Munday, or a full and exact description of that great and terrible Eclipse of the Sun which will happen on the 29th day of March 1652. Also an astrologicall conjecture of the terrible effects that will probably follow thereupon*. London, 1651, Wing #3044, and Thomason E. 650(5), frontispiece. Thomason dated this pamphlet 5 September.

[58] Anonymous, *Black Munday*, p. 2.

[59] L. P. [Lawrence Price], 'Praeface' to *The Shepherds Prognostication, Fore-telling, The sad and strange Eclipse of the Sun, which wil happen on the 29 of March this present year 1652*. London, 1652, Thomason E. 668 (1), fol. A2 verso. Thomason dates his copy as 16 January 1652.

Never one to miss an opportunity, William Lilly (1602–81), the foremost Parliamentarian 'judicial' astrologer and propagandist of the era, published *An Easie and Familiar Method whereby to Iudge the effects depending on Eclipses, Either of the Sun or Moon.*[60] Although he claimed that he had 'intended in 1648, to have Printed this small discourse for judging of Eclipses, (for so long time it hath layne by me;), how or by what means it was not then done, I doe not now remember,' it is likely he created the tract to capitalise on Black Monday.[61] Lilly was a notorious opportunist, famous (or infamous depending on political persuasion) for his correct prediction of the outcome of battles in his 1644 *Merlinus Anglicus Junior.*[62] As Blackledge has noted, Lilly's 'judgment was that, overall England's scheme of heaven for 1644 was "averse to monarchy"—Parliament would have a better year than the King.'[63] Lilly reported this year saw a partial solar eclipse (7 May 1644, OS) and an annular solar eclipse (22 August OS) 'of long duration and great, though not visible in our Horizon.'[64] The eclipses of that year argued a tragic 'end and discomfiture of some certain King,' and in July 1644 the Royalists lost a definitive battle at Marston Moor.[65] In his astrological chart forecasting the year, Lilly noted, 'when I speak of the tenth house, I intend somewhat of Kings: when mention is made of the first house, Ascendant or Horoscope, I intend the Commonality in generall'; a reader subsequently wrote in the chart 'Charles' for Charles I and 'Parliament' for the second, so Lilly's allegory was clear to his readers.[66] Having by coincidence made the correct prognostication, he continued to get lucky:

> Lilly's almanac for the following year, 1645, entitled *Anglicus, Peace or No Peace,* made his name by suggesting for June—based on an unfortunate aspect from Mars to the king's Ascendant—that 'If now we fight, a Victory stealeth upon us.' The outcome of the battle of Naseby that month spectacularly confirmed Lilly's pre-eminence over the unfortunate royalist almanac-writer, now his chief rival, Sir George Wharton.[67]

Lilly's almanac published the year before Black Monday, the appropriately named *Annus Tenebrosus* (The Dark Year) predictably foresaw harm to Dutch shipping, a reflection of the First Anglo-Dutch War between the navies of the Commonwealth of England and the United Provinces of the Netherlands.[68] Lilly also proclaimed that the 'Pope . . . seems to be eclipsed by some act or another, and in danger

---

[60] William Lilly, *An Easie and Familiar Method whereby to Iudge the effects depending on Eclipses, Either of the Sun or Moon.* London, 1652.

[61] Lilly, *An Easie and Familiar Method,* sig. A2 recto.

[62] See Harry Rusche, 'Merlini Anglici: Astrology and Propaganda from 1644 to 1651.' *English Historical Review* 80(315), April 1965, pp. 322–33.

[63] Blackledge, *The man who saw the future,* p. 20.

[64] William Lilly, *Merlinus Anglicus Junior: The English Merlin Revived,* London, 1644, p. 6

[65] Ibid.

[66] Ibid., p. 1; Catherine Blackledge, *The Man Who Saw the Future: The 17th-Century Astrologer who Changed the Course of the English Civil War.* London, 2015, p. 20.

[67] *The Oxford Dictionary of National Biography,* 2004 online edition, s.v. 'William Lilly,' https://www.oxforddnb.com/display/10.1093/ref:odnb/9780198614128.001.0001/odnb-9780198614128-e-1004336?rskey=7fUdII&result=1 (accessed 26 July 2021).

[68] William Lilly, *Annus Tenebrosus, or the Dark Year.* London, 1651, p. 4.

either of death, or some private misfortune,' good news for an anti-papal Protestant Nation.[69]

When 'Black Monday' first appeared, there was some panic. Scotland was in the track of eclipse totality, and so there was a real darkness there. As a result, some of the poor in Scotland were 'throwing away all, casting themselves on their backs, and their eyes towards Heaven, and praying most passionately, that Christ would let them see the Sun again, and save them.'[70] In England, which was not in the eclipse's track of totality, there was no noticeable reduction of sunlight, other than a subdued dimness that one would see in a partial eclipse. Thus in Plymouth, a sailor recounted on 'the 29 March 1652, somewhat more than ten in the morning, I was under the Barbers hands in our Ship a trimming, the Barber was inforced by reason of the great Darkness to light a Candle to make an end of his work.'[71] Nonetheless, in preparation for the event, London citizens drank 'Saffron'd Wine' as saffron was a solar herb that substituted for sunshine, and wore Rosemary to protect them 'from Air' generated by the eclipse, staying inside.[72] Another pamphlet writer claimed that the rich fled London and that 'generally through the whole City, journeys, marriages, contracts, bargaines of all sorts, were put off, in consideration of it; such a predominant power has this sacred Art [astrology] over the minds of the people.'[73]

'Black Munday's' actual appearance, at least in England, was thus not so 'black' after all. The eclipse, which was predicted by astrologers to produce an 'Egyptian darkenes over the face of the earth,' in the end, was fairly short.[74] Diarist John Evelyn stated:

> tho' the morning were very Murky, yet was the obscurity no greater than on other clowdy days . . . was that celebrated eclips of the sun, so much threated by the astrologers, and which had so exceedingly alarmed the whole natuion that hardly any one would work, nor stir out of their houses. So ridiculously were they used by knavish and ignorant star gazers!'[75]

John Palmer (1612–79), a rector and later Archdeacon living in Northamptonshire who republished John Blagrave's work on the planisphere, recalled:

> At Ecton Anno Dom. 1652. on Munday March 29. before Noon, I observed the great Eclipse of the Sun by a Telescope and a minute-watch Rectified by the Azimuth of

[69] Ibid., p. 35.
[70] William Lilly, *Merlini Anglici Ephemeris or Astrologicall Predictions for the Year 1653*. London, 1653, Wing #A1885, fol. A3 verso.
[71] Ibid., fol. A4 recto.
[72] On Bugbear Black-Monday, March 29. 1652 or, *the London-Fright at the Eclipse proceeding from a Natural Cause*. London, 1652, Thomason 669 (f.16); Galbrion Albumazar, *Mercurious Phreneticus: Shewing the effect of the Terrible Ecclipse, March the 29. 1652*. London, 1652, Thomason E. 658 (15), p. 2. Saffron and Rosemary were 'solar herbs' that could 'strengthen the Heart, and comfort the Vitals,' and were used to 'resist Poyson, or to dissolve any Witchery, or Malignant Planetary Influences.' (William Lilly, *Christian Astrology: Modestly Treated of in Three Books*. London, 1647, Wing #L2215, 1:71.) Apparently, some of the almanacs further predicted that 'men and women should be suddenly stricken and fall down dead as they went along the streets.' See L.P., *The Astrologers Bugg-beare*. London, 1652, Thomason E. 1301 (2), fol. A3 verso.
[73] Albumazar, *Mercurius Phreneticus*, p. 3.
[74] Ibid., p. 8.
[75] John Evelyn, *Diary. Now first printed in full from the manuscripts belonging to Mr. John Evelyn*, ed. E. S. de Beer. 6 vols. Oxford, 1955, vol. 5, p. 354.

the Sun, taken both before and after, in the company of half a score Gentlemen and Ministers my Neighbours . . . . [T]hough this Eclipse was so great, yet we could read in the time of the greatest darkness within Dores, notwithstanding that the Window was covered with a Blanket.[76]

Indeed, after the eclipse was over, the Council of State put out a paper explaining that eclipses were natural events which could have no political effects. Black Monday ironically made the predictive failures of English astrologers that much more evident. They correctly predicted the 'Black Monday' eclipse all right, but so greatly exaggerated its supposed effects that their profession experienced a darkness all its own.

Many authors subsequently lambasted astrologers for causing such unnecessary social tumult.[77] On 9 April 1652, one 'L.P.' (ballad-writer Laurence Price) published *The Astrologer's Bugg-beare* which showed an astrologer rat contemplating the solar eclipse in the sky.[78] *Black Munday turn'd white, or, The astrologers knavery epitomized*, published about a week after the eclipse, noted Lilly had not described the appearance of the eclipse accurately:

I shall not need to quote any more of his ridiculous absurdities; but conclude with his gross Predictions concerning the Eclipse on March 29. which (according to his Calculation) should have been the greatest that ever eyes beheld in this latter age. Certainly, this argues a great want of faith, and a spiritual darkness; for although there appeared enough to satisfie rational men that there was an eclipse; yet we may observe, that he made the two great Luminaries, and ordereth their course sitteth in the Circle of the Heavens, and will not give His honour unto any other; but drew back the Clouds like a Curtain, and caused the Sun to shew his pleasant Rays and comfortable Beames during the whole time of the eclipse, to the confutation of the great Astrologers, who by the help of Tycho were able to guess at the time of the eclipse, yet could not tell whether the day would be cleer or cloudy.[79]

William Lilly ranted 'for no lesse than a round dozen of Vinegar Pamphlets have been writ against me . . . The Ballad-makers did me the Honour to sing me up and down the streets in their squirting ugly Songs.'[80] Others satirized both Lilly and

---

[76] John Palmer, *The catholique planisphaer. Which Mr Blagrave calleth the mathematical jewel; briefly and plainly discribed, in five books*. London, 1658, pp. 210–11; Black Monday: Eclipses in the Folger Collections, https://folgerpedia.folger.edu/Black_Monday:_Eclipses_in_the_Folger_Collection (accessed 28 July 2021).

[77] Thomas, *Religion and the Decline of Magic*, 300.

[78] L.P. (Laurence Price), *The Astrologers Bugg-beare: being a briefe discription of many pitthy passages, which were brought to passe upon that day which the astrologers painted out for Black-Monday: whereby wee may all see and know that God's power is beyond mans expectation. Mark well and take notice, it is worth your observation. Written by L.P.* London, 1652, frontispiece.

[79] *Black Munday turn'd white, or, The astrologers knavery epitomized: being an answer to the great prognosticks and gross prediciton of Mr. Lillie, Mr. Culpeper, and the rest of the society of astrologers concerning the eclipse of the sun on Munday last* . . . . London, 1652, pp. 7–8. Black Monday: Eclipses in the Folger Collections, https://folgerpedia.folger.edu/Black_Monday:_Eclipses_in_the_Folger_Collection (accessed 28 July 2021); The author referenced Tycho Brahe, the great naked-eye astronomer of the early modern period.

[80] Lilly, *Merlini Anglici Ephemeris 1653*, fol. A2 recto.

Culpepper for predicting the fall of monarchy in general throughout the world, one of these Ballads called 'William Li-Lie.'[81] Another 'On Bugbear Black Monday,' featured doggerel verses such as:

> And why (ye frighted Women!) do ye shake?
> Must an Eclipse needs make the Earth to quake?
> The Heavens their order, and due motion keep.
> Why you disord'red? startle? sigh? and weep?
> Why load ye Coaches and forsake the place?
> As ye confest, its sin were ev'n past grace?[82]

The broadsheet, however, belied some intellectual sophistication. It queried of the astrologer:

> Can all your Art tell certain for a year,
> What fair or foul shall in each day appear?
> If not: your high Prognosticks Ganzaes fail,
> That from your fancies to the Orbes did fall.

Using the wordplay of 'foul' and 'fowl,' the 'Ganzaes' or 'geese' were not only a reference to divinatory birds, but to Francis Godwin's *The Man in the Moone* (1638), a tale about Domingo Gonzales's search for his fortune in the New World. During his journey, he became stranded on a small desert island in the Bahamas after his ship was attacked by English pirates. While on the island, Domingo discovered a species of giant swans or *ganzas* and harnessed them to a chariot of pulleys to escape.[83] However, he did not bargain for the fact that this particular species of swan went on an annual migration to the Moon.[84] The rest of the book was devoted to a description of Domingo's journey through the heavens, his arrival on the inhabited Moon, and his return to Earth with the swans.

After taking off in his chariot of swans, Domingo described the progress of his birds, relating events in their trip to the principles of Gilbertian magnetism. At first the *ganzas* struggled to get off of the Earth, 'panting and blowing, gasping for breath, as if they all presently would have died.'[85] However, after flying for about an hour, 'the

---

[81] *Against William Li-lie (alias) Lillie, that Most Audacious Atheisticall Rayling Rabsheca, that Impious Witch Or Wizzard, and Most Abhominable Sorcerer Or Star-gazer of London, and All His Odious Almanacks, and Others. Written by J. Viccars.* London, 1652.

[82] *On Bugbear Black-Monday, March 29. 1652.* London, 1652.

[83] Roos, *Luminaries in the Natural World*, p. 131.

[84] Such theories about bird migration were thoroughly developed in the seventeenth century, reflecting the new interest in the heavens. Charles Morton (1627–98), best known for his work in the *Compendium Physicae*, compiled a treatise in 1686 in which he hypothesized that birds migrated to the Moon and used Godwin's work as a guide. See T. P. Harrison, 'Birds in the Moon.' *Isis* XLV, 1954, pp. 323–30.

[85] Francis Godwin, *The Man in the Moone: Or a Discovrse of a Voyage thither by Domingo Gonzales, the Speedy Messenger, The English Experience*, no. 459, London, 1638, STC #11943; reprint, New York, 1972, p. 45. Another edition has been prepared by William Poole, ed., *The Man in the Moone*, Peterborough, ON, 2009.

Lines slacked' between the chariot seat and the swans, and 'forcing themselves ever so little,' they were carried towards the Moon with 'swiftnesse and celeritie.'[86] Godwin, apparently familiar with Gilbert's *De Magnete*, explained that this was because the Earth and the Moon exerted a magnetic field, 'in like sort as the Loadstone draweth Iron.'[87] Once the swan chariot had broken free of the 'compasse of the beames attractive' of the Earth, then flying was easy as their vehicle entered the magnetic field of the Moon. In addition, Godwin stated that because the Moon was much smaller than the Earth, its 'attractive power' was 'so farre weaker than that of the earth, as if a man doe but spring upward with all his force, he shall be able to mount 50 or 60 foote high.'[88] At that point, beyond the magnetic sphere of the Moon's attraction, the lunar peoples could in fact 'conveigh themselves in the Ayre' with the help of feathered fans to guide them through the atmosphere.[89] Though Godwin knew nothing of gravity, he utilized Gilbert's theory of magnetism to postulate correctly the weak attractive power of the Moon.[90]

As Domingo approached the Moon, the Earth looked to him 'like another Moone.'[91] The Earth 'did mask it selfe with a kind of brightness' and 'even as in the Moone we discerned certain spots or Clouds.'[92] Domingo also noted those earthly spots changing, realizing it was because the Earth rotated and those spots were continents: 'The Earth according to her natural motion (for that such a motion she hath, I am now constrained to joyne in opinion with *Copernicus*), turneth round upon her owne Axe every 24. Howers from the *West* unto the *East*.'[93] Domingo continued by providing one of the earliest portrayals of seeing an earthrise from the Moon:

> Then should I perceive a great shining brightness to occupy that roome, during the like time (which was undoubtedly none other than the great *Atlantick* Ocean). After that succeeded a spot almost of an Ovall form, even just such as we see *America* to have in our Mapps .... So that it seemed unto me no other then a huge Mathematicall Globe, leasurely turned before me, wherein successively, all the Countries of our earthly world within the compasse of 24 howers were represented to my sight.[94]

From his new perspective, the 'spots' on the Earth were not threatening, but enthralling, just like the umbra of the 'Black Monday' eclipse was a reassurance for some that the natural rhythms of the cosmos were working very well. Eclipses no longer were purely apocalyptic. Fair and 'fowl,' Godwin's tale was a work of fancy, but also one that accurately predicted future technological achievement in an inspirational and visionary description. Astrology itself could also be accurate and predictive

[86] Godwin, *The Man in the Moone*, p. 47.
[87] Ibid.
[88] Ibid., p. 80.
[89] Ibid.
[90] This paragraph draws upon Roos, *Luminaries in the Natural World*, p. 136.
[91] Godwin, *The Man in the Moone*, p. 56.
[92] Ibid.
[93] Ibid.
[94] Ibid., pp. 57–8. Roos, *Luminaries in the Natural World*, pp. 136–7.

of the cycles of the planets and the movement of the stars, and eclipses like 'Black Monday' could be metaphors for the steady rhythm of heavens and inspire faith in the Creation. But the eclipses' temporality of portent could lead to panic and hysteria shaped by opportunism and the quest for power, revealing the tensions between predictability and unpredictability that, in the seventeenth century, were heightened in a time of war and political upheaval.

8

Signs and Portents

Reflections on the History of Solar Eclipses

David Bentley Hart

Philosophical Theology, Philosophy of Mind, Asian Religion

In a certain continuity with both the experience of the twelfth-century 'Wretched, Horrifying, Black, Remarkable' eclipse and that of the seventeenth-century 'Black Monday,' David Bentley Hart contrasts the terror with which eclipses were greeted in pre-modern times with his own experience of beauty and awe when he witnessed the 2017 event. Revisiting cultural environments that we have explored previously in Chapter 3 *(Greco-Roman antiquity),* Chapter 5 *(early Asia), and* Chapter 7 *(early modern England), the author takes us on a reading journey through the ancient Chinese tradition to Lucretius and back up to John Milton and William Shakespeare, to the philosophies of René Descartes (1596–1650), Arthur Owen Barfield (1898–1997), and Martin Heidegger (1889–1976), among others — spanning an arch from pasts unknown to the present day.*

Barfield, Owen Arthur Owen Barfield (1898–1997), British philosopher, author, poet, critic, and member of the Inklings.

cerulean deep sky blue.

Descartes, René (1596–1650) French philosopher, mathematician, and early modern scientist who famously concluded that everything was open to doubt except conscious experience and existence as a necessary condition of this: 'Cogito, ergo sum' ('I think, therefore I am'). In mathematics, he developed the use of coordinates to locate a point in two or three dimensions.

Heidegger, Martin (1889–1976) German philosopher. In *Being and Time* (1927), he examined the ontology of 'Being' (human existence as involvement with a world of objects or *Dasein*.)

heliotrope A plant that turns toward the Sun; a light reddish purple color.

Lucretius (*c*.94–*c*.55 BC) Roman poet and philosopher; full name Titus Lucretius Carus, author of the didactic epic poem *On the Nature of Things* (an exposition of the materialist atomist physics of Epicurus).

Milton, John (1608–74) English poet and author of *Paradise Lost* (1667; revised 1674), *Paradise Regained* (1671), and *Samson Agonistes* (1671), completed with the help of his daughter after he had gone blind in 1652.

Shakespeare, William (1564–1616) English playwright.

sylvan Pastoral; pleasantly rural; wooded, consisting of (or associated with) woods (from mid-sixteenth-century French *sylvain* or Latin *Silvanus* 'woodland deity,' from *silva* 'a wood').

Eclipse, Experience, Effect

I had never observed a full solar eclipse before the one that I, my wife, and my son travelled to see in 2017. It took place on 21 August that year, and we were able to watch it from beginning to end through a cloudless sky from a broad, open prospect atop a mountain outside Nashville, Tennessee. In the city itself, rooftops and balconies and public parks were thronged with spectators, many of them deep in their cups well before the celestial spectacle began to unfold itself; but somehow, quite by chance, we had on the previous day discovered an ideal observation site: secluded, sylvan, but affording an unobstructed prospect of the heavens and wide open to the encircling horizon. We had taken all the proper precautions, needless to say. All three of us were garishly bespectacled with deeply tinted lenses set in lime-green cardboard frames (though I also know that each of us peered out from beneath them at constant intervals as the Moon made its progress across the face of the Sun). I was not at all prepared, I discovered, for the effect the event in its entirety would have on me. At first, I was merely stirred by the fantastic incursion of the lunar disk—now utterly black— upon the solar, and by the ways in which the colours of the world became increasingly unusual (the sunlight on the leaves and grass, for instance, becoming a kind of sullen amber, and the shadows cast by that light a kind of otherworldy heliotrope). It was at about the point when two-thirds of the Sun were obscured, however, that the true uncanniness of the experience began to take hold of me. It was then, almost in an instant, that the entire sky began to change its aspect; visibly darkening, it acquired a strangely glassy quality, and then continued to deepen in hue till it had become a translucent cerulean high overhead but a gauzy violet down at the world's edges. Then, as Sun and Moon reached perfect alignment, and the former was reduced to no more than a thin, glittering, slightly prismatic nimbus around the latter, the dome of the sky began to become diaphanous and I realized with a start that the dozen or so soft gleams of light I saw faintly hovering in its depths were in fact stars. In an instant, the canopy of the lower atmosphere had been withdrawn; for the first time in my life I thought I had some sense of what the *Qur'an* means when it speaks of the sky being rolled up like a parchment at the end of days; and it was a genuinely tremendous and dizzying sight. How could anyone look at such thing, I remember thinking, and not feel, at least fleetingly, as if the whole world were dissolving? In other ages, neither so disenchanted nor so sensually surfeited as ours, it must have been a truly terrifying experience. Even I felt a tremor or two of existential vertigo as I stood there gazing into the abyss of space through a suddenly crystalline sky, although I certainly had no fear that the frame of things was in any imminent danger of disintegrating. Then, after several minutes, the vision began to fade, the pallid stars melted away again into the brightening sky, the full circle of the sun gradually emerged again from behind the moon, and the ordinary fashion of the world was restored.

It took some time for me to decide what the entire effect of the eclipse on me had been—or, rather, the effects, since my reactions continued to develop and fluctuate for some time. Certainly the experience had been of something majestic and mysterious; I had been overwhelmed, astonished, and even vaguely alarmed when the stars above had become visible through the attenuated daylight. Why this should have come as a surprise to me I cannot say, since it is an amply attested phenomenon of which I had been apprised many times in the past. True, several friends who had themselves witnessed full solar eclipses had all told me that they had seen nothing of the sort, and that for them the sky had remained obstinately opaque even when the Sun had been fully obscured. Still, it was all there in the literature, again and again. Diodorus Siculus had included a mention of the stars' visibility in his brief account of the eclipse of 15 August 310 BC that had preceded the escape of the tyrant Agathocles with sixty ships from a Carthaginian blockade of Syracuse;[1] both Plutarch (in 'On the Face of the Moon's Disk') and Phlegon of Tralles (in the *Olympiads*) had recorded the same detail, the former in regard to one on 20 March 71 AD,[2] the latter in regard to one on 24 November 29 AD;[3] so had Marinus of Neapolis about one that occurred on 14 January 484 AD, in a passage from his *Life of Proclus*[4] that I myself had once translated as a school exercise. Only a week before our sojourn in Nashville, moreover, I had begun skimming through a remarkably comprehensive compendium of eyewitness accounts, *Historical Eclipses and Earth's Rotation* by F. Richard Stephenson,[5] and I had noticed that the same detail had recurred there with a fairly pervasive regularity throughout its pages. The very first account I had come upon in the book, in fact, was a brief mention of a solar eclipse (dated now as having occurred on 24 October 444 BC) in the classic Chinese text *Records of the Grand Historian* (or *Shǐjì*), which had prominently mentioned the stars on high shining out in the middle of the day.[6] Excerpts from the *Anglo-Saxon Chronicles*, the *Chronicon Scotorum*, the *Chronicon* of Andreas of Bergamo, the *Historiae Byzantiae* of Leo the Deacon, and any number of other annals and histories from antiquity and the Middle Ages had included frequent references to the phenomenon. There were, for example, an unusually large number of reports from across northern Europe of the solar eclipse of 2 August 1133, and not a single one of them, as far as I can recall, omitted mention of the starry firmament.[7] The same is true of the various accounts by the Muslim and Christian observers of an eclipse of 11 April 1176 in Asia Minor,[8] as well as of all three of the extant reports of another that occurred on 9 August 975 in Japan.[9] And so forth and so on. And yet, even though I had acquainted myself with these facts just a few days earlier, I had still managed to depress my expectations to the degree that, when the stars in fact appeared, it came to me as a revelation.

It was only a week or so later, however, when I had further acquainted myself with old accounts of eclipses past, that I realized something rather curious. While my own experience of the stars becoming visible during the eclipse had in most obvious respects matched the descriptions of various ancient, medieval, or even early modern chroniclers, it had also differed from theirs in one very conspicuous way. Whereas all of them had remarked upon it, and some had noted the wonder and terror the sight had evoked among its witnesses, absolutely none of them had said anything at all

---

[1] Diodorus Siculus, *Bibliotheca Historica*, XX.5–6.
[2] Plutarch, *De facie in orbe lunae*, 931D–E.
[3] Phlegon of Tralles, *Olympiades*, fr. 17 (preserved in Eusebius, *Chronicon*).
[4] Marinus Neapolitanus, *De vita Procli* (also Περὶ Εὐδαιμονίας or *De felicitate*). XXXVII.
[5] F. Richard Stephenson, *Historical Eclipses and Earth's Rotation*. Cambridge, 1997.
[6] 史記 (*Shiji*), 15.
[7] See Francis Richard Stephenson, *Historical Eclipses and Earth's Rotation*. Cambridge, 1997, pp. 392–3.
[8] Ibid., pp. 394–5, 439–45.
[9] Ibid., pp. 267–8.

regarding what to me seemed most immediately obvious about it: to wit, its ravishing beauty. As I say, I had felt a momentary sense of disorientation at the apparition of the stars, and that aforementioned vertiginous moment of amazement; but, more persistently and far more affectingly, I had been aware of the mysterious loveliness of it all—the stars like pale white opals, the sky like liquid glass, the faint iridescence surrounding the Sun's corona, the strange, shadowy, pearl-hued twilight that suddenly hung above the world. I had felt a genuine sense of gratitude for and delight in all of it, as well as a desire to dwell in the moment as long as I could. And *that* was what it would have first occurred to me to say about the experience if asked about it, and probably what I would have said last as well, and what I would have placed the greatest stress upon throughout. By contrast, all the reports that I read from the more distant past had been notable either for their dry tone of indifference or for the palpable air of horror pervading them. Those who mentioned the stars generally described their appearance as terrifying, or at least bizarre, and nearly all of them exaggerated the gloom through which the stars had shone out.

Now, needless to say, those documents come to us from a world considerably less, if not *enlightened*, at least *illuminated* than our own, and from centuries when the dark really was as often as not something to be dreaded, and when the night was often very dark indeed, and when the Sun's departure each day regularly disclosed a dazzling and unfathomable ocean of stars, far too sublime to inspire mere admiration unmixed with some element of fear. We who live in the glow of electric lights and who rarely ever see the naked sky at night can scarcely imagine what the ordinary difference of night from day felt like before gaslight and then its electric sequel began to dispel the shadows; but it only makes sense that in such a world the darkening of the Sun and sudden appearance of the stars during the day should have seemed more ghastly than glorious. Whereas, to me, this shimmering glimpse beneath a veiled Sun into the depths of space came as a rare and wonderful blessing, to most observers in previous ages the same experience must almost necessarily have seemed like something of a curse—something 'disastrous' in the etymologically precise sense of that word—something, that is, 'ill-starred.'

Pre-modern Portent and Pestilence

The earliest extant account of a solar eclipse (at least, so all the classical authorities and anthologists appear to agree, if not all the current scholarship) comes from China and is preserved in the *Shūjīng* or *Book of Documentation*. The text is difficult to date, however, and may come from as much as a millennium and a half after the event supposedly described; so one can make only an educated guess as to which eclipse it describes (if any particular one as such); and the most likely, it seems, is one of 22 October 2134 BC (though others from 2136 BC or 2159 BC are also plausible candidates).[10] In any event, whenever it occurred, if it occurred, it was said to be a grave misfortune for at least two persons. The court astronomers Hi and Ho were

---

[10] 書經 (*Shūjīng*), III.1.

supposedly put to death for having failed to predict the event; it was an unpardonable dereliction on their part for the simple reason that it was widely believed in China in their day that solar eclipses were caused by a celestial dragon who must be driven away with apotropaic ceremonies (arrows fired into the sky, the clamour of beating drums and clashing cymbals, songs and imprecations) if the world was to be rescued from the fall of eternal night. It is a curious tale, however, inasmuch as it encourages us to think (rightly or wrongly) that even four millennia ago Chinese astronomy recognized eclipses as natural phenomena whose occurrences could be predicted by proper observation and calculation, and yet also tells us that, for the court and culture of China many centuries after the 'fact', this reality in no way weakened the conviction that eclipses are moments of cosmic crisis. And such would seem to have been the general pattern in much of the ancient and medieval world. Cultures that were able to foresee the advent of eclipses—Greek, Babylonian, Chinese, and so on—were no less disposed than others to fear them and take them as portents auguring only ill fortune. Whether predictable or not—whether natural or not—they were in every case events that apparently no one much enjoyed or awaited with any great enthusiasm.

There are, admittedly, reports of eclipses from those centuries that are little more than bare notations, devoid of any pronouncements on their significance, and marked by neither alarm nor elation. And there are treatments of eclipses of a purely rationalist and empirical kind, such as those of Aristotle and Lucretius. On the whole, however, when a meaning was assigned to eclipses, almost no one seems to have expected much in the way of good news. Even the few instances of an eclipse being read as a favourable portent are very much, so to speak, matters of perspective. There is one ancient Babylonian[11] text by one Rašil the Elder that treats an eclipse from 27 May 669 BC as a clear omen of military success for Rašil's Assyrian king;[12] but that, of course, means it was also a baneful omen for the forces on the other side; and, even for the Assyrians, it was not seen as representing an unmixed blessing. Similarly, the eclipse of 17 November 1183 in Japan, as recorded in the *Genpei Jōsuiki*, was good news for the forces of clan Taira only because it had so devastating an effect on the nerves of the forces of clan Minamoto.[13] There are also, it must be granted, a few anecdotes in the ancient sources about eclipses that produced happy results. According to Herodotus, for example, an eclipse on 28 May 585 BC—one that, supposedly, had been predicted by Thales of Miletus—so surprised the armies of the Lydians and the Medes on the battlefield that not only did they at once lay down their arms, but their nations very quickly agreed upon a lasting peace.[14] The *Nihon Kiryaku* reports that an eclipse seen in Japan on 9 August 975 AD led to a general amnesty.[15] But, even in these cases, the benign consequences of the phenomenon nevertheless resulted from the terror it inspired and the malignity it was taken to represent. They were not the consequences of anything like a grateful response to the splendour and beauty of the event. As far as the pre-modern and early modern accounts are concerned, eclipses

---

[11] See Chapter 5 by John Steele for further background to Babylonian eclipse records.
[12] Quoted in Hermann Hunger, *Astrological Reports to Assyrian Kings*. Helsinki, 1992, p. 220.
[13] 源平盛衰記 (*Genpei Jōsuiki*), 46.
[14] Herodotus, Histories, I.74.
[15] 日本紀略 (*Nihon Kiryaku*), Zenpen. 6.

may be described as terrible, sad, grim, even 'hideous' (as Gregory of Tours said of the one that occurred on 4 October 590 AD);[16] or, at best, they may be described with clinical detachment; they are never praised for their loveliness or their enchanting strangeness. They seem never to have evoked rapt wonder, awe inflected with reverence, or delight. Rather, their effect in general—as Archilocus said in response to the eclipse of 6 April 648 BC—was one of 'profound fear.'[17] They were accounted to be, as the *Book of Odes* says, the most evil omens of all.[18] Or, as the book of Joel puts it, whenever 'the sun shall be turned into darkness and the moon into blood,' one can be certain that 'the great and terrible day of YHWH is coming.'[19] In Shakespeare's *King Lear*, Gloucester speaks of nature being scourged by the 'sequent effects' of solar and lunar eclipses: 'love cools, friendship falls off, brothers divide: in cities, mutinies; in countries, discord; in palaces, treason; and the bond cracked 'twixt son and father.'[20] Milton used solar eclipses and their dire implications as apt images of Samson chained and 'eyeless in Gaza,' or of Satan's ruined glory in *Paradise Lost*:

> As when the Sun, new risen,
> Looks through the horizontal misty air,
> Shorn of his beams, or from behind the Moon,
> In dim eclipse, disastrous twilight sheds
> On half the nations and with fear of change
> Perplexes monarchs.[21]

In short, every association drawn in the pre-modern or early modern sources between the prodigy on high and events here below is a 'monstrous' one (in the most technical sense of the term); whatever the nature of that occult sympathy that relates the celestial conjunction to the terrestrial course of affairs may be, it brings nothing but calamity: a violent insurrection in the city of Ashur, for instance (the eclipse of 15 June 763 BC as recorded in the *Assyrian Chronicles*);[22] warfare, ruined harvests, blizzards, violent political divisions, floods, devastating cold or heat or rains (as Pindar says of the eclipse of 30 April 463 BC);[23] the death of a Dowager Empress (as on 4 March 181 BC, according to the *Shūjīng*,[24] and on 18 January 120 AD, according to the *Hòu Hànshū*);[25] the rampages of Attila's armies in Italy (as on 24 February 453 AD, according to Gregory of Tours);[26] the death of a great philosopher (as on 14 January 484 AD, according to Marinus);[27] the deaths of kings (as on 1 May 664 AD, according to *The Anglo-Saxon Chronicle*,[28] and on 20 March 1140, according to William of

---

[16] Gregory of Tours, *Historiae*, X.23.
[17] Archilocus, fr. 122.
[18] 詩經 (*Shijing*), 'At the Conjunction in the Tenth Month.'
[19] Joel 2:31.
[20] Act I, scene 2.
[21] Book I, ll. 594–9.
[22] Quoted in Alan R. Millard, *The Eponyms of the Assyrian Empire 910–612 BC*. Helsinki, 1994, p. 58.
[23] *Paean* IX.
[24] 書經 (*Shūjīng*), V.18.
[25] 後漢書 (*Hòu Hànshū*), 28.
[26] *Historia Francorum*, II.3 ff.
[27] Marinus Neapolitanus, *De vita Procli*, XXXVII.
[28] Sole entry for 664.

Malmesbury);[29] the deaths of Byzantine Christians at the hands of Turkish invaders (as on 25 May 1267, according to Gregoras Nicephoros);[30] the end of the Song Dynasty (as on 25 June 1275, according to the *Sòng Shǐ*);[31] and so on and so forth. It is impossible, moreover, not to notice the regular association in many of these sources between eclipse and earthquake, occasionally because of some chance conjunction of the two at some particular time or place, but chiefly because of some vague but prevalent sense that the two phenomena were naturally related and shared in a common malignancy. One finds the connection drawn by Thucydides, by Phlegon, and by countless other writers. It is such a rhetorical commonplace, in fact, that in *The Revolt of Islam* Shelley naturally employs it as a single fused metaphor:

> With hue like that which some great painter dips
> His pencil in the gloom of earthquake and eclipse.[32]

And then too, of course, there is the testimony of folklore from every quarter of the globe. Around the world, understandably enough, eclipses have often been understood—or at least poetically described—as assaults upon the Sun by some great celestial monster seeking to devour it: in China, as already mentioned, tradition said it was a dragon, in Vietnam a great toad, among the Norse peoples a ravenous celestial wolf, in India the demon Rahu (or, at least, his severed head), in Korea 'fire dogs,' and so on. The traditional view of the Kwakiutl is that it is the sky itself that is trying to devour the Sun. Moreover, folk beliefs regarding the pestilential effects of eclipses, and regarding especially the danger they pose for pregnant women and the babies in their wombs, recur with remarkable frequency across a range of cultures far removed from one another in space and time. In India, it was once widely believed that a solar eclipse is a source of dangerous contamination and contagion, and that one should avoid eating or drinking anything its shadow has touched, and should bathe and change clothes once it has passed, praying all the while for protection from its noxious influences.

I could go on, but it seems rather pointless. Suffice it to say, for most ages before our own and in every culture of which we have records, the sight of a full solar eclipse, as far as the explicit testimonies of the chroniclers are concerned, brought joy to no one, typically evoked not so much wonder as horror, and on the whole was regarded as something menacing and perhaps catastrophic. Curiously enough, and after long reflection, I find something enviable in this.

Modern Mechanism and Mindlessness

Obviously, we should rejoice that today we can watch an eclipse not only without terror, but with genuine delight. Because it is simply a brute event for us, which we would

---

[29] *Historia Novella*, II.
[30] *Byzantinae Historiae*, IV.8.
[31] 宋詩 (*Sòng Shǐ*), 67.
[32] Canto V, stanza 23.

not even know *how* to see as a portent of imminent doom or an *omen pestiferum* or a harbinger of civil calamity, we are free to exult in it as something rare and beautiful. But, granting this, we should not deceive ourselves that in this matter we enjoy a privilege denied our ancestors on account of our superior knowledge or our keener powers of reasoning. As I have already noted, many ancient cultures were well aware of the mechanics—or, at the very least, the calculably mechanical nature—of eclipses, and entertained no doubt that they were entirely natural phenomena; and yet this in no way robbed eclipses of their prophetic significance for those cultures. What sets us apart from many ancient peoples in this respect has nothing to do with our sciences or our more 'enlightened' view of things but everything to do with a set of metaphysical prejudices that we have absorbed from the culture around us. We may not be able to view the natural order as a realm of invisible sympathies and vital spiritual intelligences, but this is solely because our vision of reality is the product of four centuries of mechanistic dogma, which tells us that matter is something essentially dead, nature a machine, and the mind a resident alien within (or merely an emergent property of) an intrinsically mindless universe. These, however, are not truths of reason or deliverances of the sciences; they have neither empirical nor theoretical warrant; they are merely the prevailing biases of our times, not only no more intellectually respectable than a belief in celestial signs and portents, but indeed a good deal less so, inasmuch as they are so wholly unreflective. This is not to deny that it is our great good fortune to have been delivered from the terror of the night. The dark is our most ancient adversary, it seems fair to say, and throughout much of the Earth we have vanquished it utterly. And yet, even so, it is worth recalling that our capacity for terror is more or less directly proportionate to our capacity for awe, reverence, or prudence; fear and piety, dread and elation, horror of the unfamiliar and moral restraint before the unknown are simply varying intensities along a single emotional continuum. If we were still able to believe that nature is intrinsically communicative, intrinsically eloquent—that she addresses us with warnings, admonitions, and auguries—perhaps we would be less able to violate, despoil, pollute, and destroy the natural world with quite the gay abandon that we do. It is hardly to our credit—and probably not at all to our benefit—that we no longer experience the cosmos as charged with intentional meaning, and no longer perceive either nature's regularities or her anomalies as embassies to us from the mystery that lies within all things. Now nothing is inviolable to us, nothing sacrosanct; everything natural is merely a resource to be exploited or an obstacle to be surmounted or a burden to be shed. Not only are we spiritually impoverished as a result; we are rapidly murdering our world.

There was, however, a time when the world *did* speak, or when we believed that it spoke, and, because we believed, it did in fact have a meaning of its own for us, already there resident within it before any we might choose to impose upon it. The natural order appeared to us as a system of intelligible signs, a language declaring more than merely itself, the overwhelming eloquence of some intelligent agency behind the visible aspect of things. Throughout human history, most peoples have assumed that, when we gazed out upon the world, something looked back and met our gaze with its own, and that between us and that numinous other there was a real—if infinitely incomprehensible—communion. Practically every sane soul presumed a 'vitalist' and 'panpsychist' model of the world. And, frankly, every truly sane soul still does, at least implicitly. Explicitly, though, most of us have succumbed to the mythology visited upon us by the mechanical philosophy at the beginning of the modern period, and have learned to think of life as no more than a fortuitous and functional arrangement of intrinsically lifeless matter, and have grown ever more accustomed to living according to the logic of mechanism, within a machine of our own contrivance: philosophical, technological, economic, political, social. Having rendered the world mute—or, at least, having deafened ourselves to its entreaties and admonitions—we have reached a condition as a culture in which we feel no compunction in forcing it to serve whatever ends we will for it. Along the way, in order to preserve our vision of a dead world generating illusory meaning and beauty, we have had to extinguish the occasional counterrevolution—Romanticism, principally, and then each of its ever fainter, dying echoes—but in our late modern moment we really have achieved a cultural consensus, principally because our ubiquitously 'technologized' form of life and the market economy it serves have now exceeded our power to resist it, except in the most trivially local and elective ways.

How we arrived at this point is not difficult to recount. The first principle of the new organon that took shape in early modernity was a negative one: the exclusion of any consideration of formal and final causes, and even of any distinct principle of 'life,' in favour of a method purged of metaphysical prejudices, allowing all natural systems to be conceived as mere machine processes, and all real causality as an exchange of energy through antecedent forces working upon material mass. Everything physical became, in a sense, reducible to the mechanics of local motion; even complex organic order came to be understood as the emergent result of physical forces moving through time from past to future as if through Newtonian space, producing consequences that were all mathematically calculable, with all discrete physical causes ultimately reducible to the most basic level of material existence. And while, at first, many of the thinkers of early modernity were content to draw brackets around physical nature, and to allow for the existence of realities beyond the physical—mind, soul, disembodied spirits, God—they necessarily imagined these latter as being essentially extrinsic to the purely mechanical order that they animated, inhabited, or created. Thus, in place of classical theism's metaphysics of participation in a God of infinite being and rationality, those thinkers granted room only for the adventitious and finite Supreme Being of Deism or (as it is called today) intelligent design theory. But, of course, this ontological liberality was unsustainable. Reason abhors a dualism. Any ultimate ground of explanation must be one that unites all dimensions of being in a simpler, more conceptually parsimonious principle. Thus, inevitably, what began as method soon metastasized into a metaphysics, almost by inadvertence. For a truly scientific view of reality, it came to be believed, everything—even mind—must be reducible to one and the same mechanics of motion. Those methodological brackets that had been so helpfully drawn around the physical order now became the very shape of reality itself.

It is always difficult to tell which comes first, of course—the ideological revolution or the evolution of new material conditions. In all likelihood, they are only two sides of a single indivisible process. There is no point in wondering, for instance, whether it was the rise of the mechanical philosophy or the rise of the newly enfranchised and prosperous mercantile class that was the deepest rationale of the early modern project of transforming our vision of the world into one of pure material efficiency. All we can say is that those who created the machine built better than they knew. Certainly, they could not have foreseen how rapidly and inexorably it would come to displace the world that was once our home, or how comprehensively it would alter the very frame of the reality we had till then inhabited, from its most unfathomable foundations to its most inaccessible heights. What began as a dim intuition that nature might be reimagined as a mechanism, the better to penetrate its secrets and marshal its forces, in time became the practical reality of an immeasurably multifarious technology that would relentlessly consume nature in order to enlarge and perpetuate itself. And, far from requiring the connivance of any calculating intellect, this process became more systematic and intricate the more it broke free from the conscious intentions of any single mind. The now hackneyed science fiction trope of computers waking to self-awareness and turning our own inventions against us is positively quaint in its naïve belatedness. The conquest has for the most part already come to pass, not by some conscious entity suddenly emerging from some particularly dense integration of computer code, but rather by an impersonal law of exponential development whose existence we could scarcely have conceived before it had already escaped our control. Consciousness, it turns out, is not only unnecessary for this process; it soon becomes a hindrance, which must be overcome *by* and incorporated *into* this process (with the rise of the internet, that too has more or less already been accomplished).

Eclipse, Revelation, and Recovery of Participation

In the weeks following my experience of the eclipse of 2017, I have to admit I found myself constantly retreating—reluctantly but inexorably—to the late thought of Heidegger. For him, modernity was simply the time of realized nihilism, the age in which the will to power has become the ground of all our values; as a consequence it is all but impossible for humanity to inhabit the world as anything other than its master. As a cultural reality, it is the perilous situation of a people that has thoroughly forgotten the mystery of being, or forgotten (as Heidegger would have it) the mystery of the difference between beings and being as such. This is simply nihilism, in the simplest, most exact sense: a way of seeing the world that acknowledges no truth other than what the human intellect can impose upon things, according to an excruciatingly limited calculus of utility, or of the barest mechanical laws of cause and effect. It is a 'rationality' of the narrowest kind, so obsessed with *what* things are and how they might be used that it is no longer seized by wonder when it stands in the light of the dazzling truth *that* things are. It is a rationality that no longer knows how to hesitate before this greater mystery, or even to see that it is there. As for Heidegger's narrative of Western culture's descent toward this nihilism, it is by turns astute and absurd, the latter being the case the further back into intellectual history he attempted to project its sources. Where he was certainly correct, however, was in recognizing that a crucial boundary had been crossed at that moment when the ground of truth came to be situated in the self that perceives rather than in the world that shows itself; and in philosophy this inversion of the order of truth probably achieved its first fully emblematic expression in the decision of Descartes to begin his philosophical project as a rational reconstruction of reality from the vantage of the subject, erected upon the *fundamentum inconcussum* of his own certainty *of* himself: 'I will close my eyes,' he says in the third *Meditation*, '. . . stop up my ears . . . avert my senses from their objects . . . erase from my consciousness all images . . . .' But, if one begins thus, what world can possibly appear when one opens one's eyes and ears again? Only a fabrication and brute assertion of the human will, an inert thing lying wholly within the power of the reductive intellect. The thinker is then no longer answerable to being; being is now subject to him.

It is thus, for Heidegger, that our age can be understood simply as the age of technology, which is to say an age in which reasoning has become no more than a narrow and calculative rationalism that no longer sees the world about us as the home in which we dwell, where it is possible to draw near to being's mystery and respond to it; rather, it sees the world now as mere mechanism, and as a 'standing reserve' of material resources awaiting exploitation by the projects of the human will. Now the world cannot speak to us, or we cannot hear it. Being's mystery no longer wakens wonder in us; indeed, we habitually mistake that mystery for just another technological question: What physical forces produce which physical states? As the mystery of the world is not a thing we can manipulate, we have forgotten it. Not only do we not try to answer the question of being; we cannot even understand what the question is. The world is now what we can 'enframe,' something whose meaning we alone establish, according to the degree of *usefulness* we find in it. The regime of subjectivity has confined all reality within the limits of our power to propose and dispose. More and more, our culture has become incapable of reverence before being's mystery, and therefore incapable of reverent hesitation. And it is the effect of this pervasive, largely unthinking impiety that underlies most of the special barbarisms of our time. It is certainly what continues to drive us towards ecological destruction. But, again, we may well have passed that last moment when as a culture we still had any discretion regarding how the future will unfold. The machine perpetuates itself.

At the same time that I found my thoughts turning to Heidegger, however, they were also turning towards Owen Barfield. He too constructed a genealogy of our lost awareness of the world as anything more than a brute mechanical event—a genealogy that at certain junctures is as grim as Heidegger's, but also at other junctures more hopeful. For him, there was a kind of spiritual destiny for humanity that necessitates our current moment of dialectical separation from what he calls 'original participation'—the immediate and unreflective communion with the mystery that declares itself in nature's signs and portents—but only for the sake of an ultimate reconciliation with the world under the form of a more reflective, objective, and yet very real 'final participation.'[33] Presumably that final state would be more speculatively aware than the first, but would also involve a recovery of the reverence, restraint, wisdom, and love—though not the terror, perhaps—that once characterized our openness to the world. I suspect that this last phase of the story is in fact a vain expectation, and Barfield simply had no notion how absolute our estrangement from nature had become. But it is consoling to think otherwise: to think, that is, that an escape from the economic and cultural machine in which we have imprisoned

[33] See especially Owen Barfield, *Saving the Appearances: A Study in Idolatry*. London, 1957.

ourselves is still possible, and that we could yet make our technology (which is, after all, very much a part of our essence as human beings) a benign and healing presence within nature, and that we could even find our way back to the living world, once again attentive to its voice. It is pleasant to think, that is, that we may one day recover the wisdom to experience every eclipse as a kind of portent—if not of some impending catastrophe or epochal transition, at least of that perennial mystery that it is a destructive madness to forget. If we are not able to do this, however, then in a somewhat ironically inverted sense every eclipse will still serve as an omen after all, whether we acknowledge it or not: and that an omen of the greatest disaster imaginable.

III
ARTS AND LITERATURE

9

Dante's Total Eclipses

Alison Cornish

Medieval Italian Literature, Dante Studies

Across history, artists and writers have tried to capture and employ total solar eclipses in their works. In the cultural moment of Dante Alighieri—Italy in the thirteenth and fourteenth century—the wonder, awe, or marvel inspired by eclipses is regularly invoked not as contrary to the scientific pursuit, but as its motivation. The author of the Divine Comedy *uses eclipses less as omens of current and impending catastrophe and more as occasions for insight, along the lines of the apocalyptic unveiling. For Dante, eclipses are both obstructions and obfuscations, indicating spiritual darkness, failing, or corruption (as in Dante's likening of the dislocation of the papacy to Avignon to an eclipse of Rome's Sun), and, on the other hand, 'wondrous' events that trigger new knowledge and insight. Dante uses eclipses to exemplify spiritual realities, such as the qualitative (not quantitative) diversity of goodness in the universe, and the free choice of intelligent beings to wait for illuminating grace.*

admiratio (Latin) marvel, wonder.

Albertus Magnus (*c.*1200–80) Dominican theologian, philosopher, and scientist; known as Doctor Universalis. A teacher of St Thomas Aquinas, he was a pioneer in the study of Aristotle.

angels Immaterial free beings created simultaneously with the physical cosmos, some of whom move the visible celestial spheres.

Argo Mythological ship built and commanded by Jason to retrieve the Golden Fleece.

Aquinas, St Thomas (1225–74) Italian philosopher, theologian, and Dominican friar; known as Angelic Doctor and regarded as the greatest figure of scholasticism (see his commentaries on Aristotle as well as the *Summa Contra Gentiles* and *Summa Theologiae*).

Beatrice Dante's love and guide through *Paradise*.

Boethius (*c.*480–524) Roman Saint, martyr, philosopher, and author of *De consolatione philosophiae* (*On the Consolation of Philosophy*).

canticle One of three parts of Dante's *Divine Comedy*: *Inferno*, *Purgatorio*, or *Paradiso*.

crux (Latin) cross; puzzling or difficult problem, or vital, basic, decisive, or pivotal point.

defectus (Latin) failing (as of light of Sun or Moon in an eclipse).

Donation of Constantine medieval document purporting to prove the Emperor Constantine's gift of the entire Roman empire, territory and authority, to the pope.

empire Office and institution of single temporal leader which Dante thought should be distinct from the spiritual authority of the Church.

horizon (Celestial) great circle on the celestial sphere midway between the zenith and the nadir.

kairos (Greek) significant opportune moment or turning point.

Latona Mythical mother of twin gods Apollo (Sun) and Diana (Moon).

Neptune Latin name of Greek god of the sea, Poseidon.

papacy Office and institution of the pope, head of the Latin Christian Church.

passio (Latin) suffering (as in Christ's Passion; also used to refer to eclipses).

primum mobile (Latin) outermost transparent sphere in Ptolemaic, geocentric universe, imparting simple daily rotation to the whole rest of the celestial system.

'Sun of the angels' God.

Virgil (b. 70 BC) Ancient Roman poet of Augustan era and Dante's guide through *Hell* and *Purgatory*.

zenith Point directly overhead.

In the history of their observation and interpretation, scientific knowledge and predictability of eclipses do not diminish the awe they inspire, or the pregnant significance they seem to herald. As C. Philipp E. Nothaft put it in Chapter 3, eclipses were 'the most complex scientific problem that pre-modern cultures found successful ways to address,' and yet they continued to be seen as portents. What is at stake is meaning—meaning in the cosmos in general. Dante enters the conversation after the transformation of Latin astronomical knowledge over the course of the twelfth and thirteenth centuries (traced by Giles Gasper in Chapter 5). The sheer quantity of astronomical references in Dante's works inspired one woman who began participating in eclipse expeditions in the late nineteenth century (mentioned by Roberta J. M. Olson), the astronomer Mary Acworth Evershed (under her maiden name, M. A. Orr), to write and illustrate a whole book explaining them: *Dante and the Early Astronomers*. Evershed would become the first director of the British Astronomical Association's historical section, a position currently held by Mike Frost. Tracy Daugherty played on the title of her monograph in his own recent biography of her: *Dante and the Early Astronomer*. Evershed was particularly keen on assessing the accuracy of Dante's astronomical descriptions, notoriously erroneous for the traditionally accepted fictional date of the journey (Easter 1300). The importance of the date of the journey, like the importance of eclipses and other astronomical phenomena of which Dante makes poetic use in his works, is that it bears significance beyond the literal, measurable, and datable. In a rational, authored, and legible universe, Easter is a new beginning, a passover, and an exodus. An eclipse is both an ominous obstruction and a clarifying revelation.

The three canticles of the *Divine Comedy* rhyme on their last word—*stelle* ('stars'). Dante Alighieri, a political exile from his native Florence in the early fourteenth century, styled himself a citizen of the world: 'To me,' he wrote in a letter to a Florentine friend, 'the whole world is a homeland, like the sea to fish.' Declining an offer of return to the city on ignominious conditions that would in essence require apology for crimes of which he was innocent, he consoled himself with the fact that he could, from anywhere, see the stars.

What! can I not anywhere gaze upon the face of the sun and the stars? can I not under any sky contemplate the most precious truths, without I first return to Florence, disgraced, nay dishonoured, in the eyes of my fellow citizens? Assuredly bread will not fail me![1]

The stars, a term that refers to all celestial lights, make a triple rhyme of Dante's three-part epic, as a sign of hope on re-emerging from Hell (*riveder le stelle*), as a goal for the final climb at the top of the purgatorial mountain (*salire alle stelle*), and as a model for spiritual alignment as a result of this journey and this vision: when one's desire and will are moved by the same love that moves the Sun and other stars (*l'Amor che move il sole e le altre stelle*).[2] When seeking direction on the unfamiliar terrain of Purgatory, Dante's guide Virgil addresses the Sun's 'sweet light' as a reliable guide, 'if other reason does not compel to the contrary.' 'Sun' is virtually equivalent to 'authoritative guide' in the political speech laid out at the centre of the poem, where we are told that Rome used to have two Suns to illumine two roads, one 'of the world' and the other 'of God,' whose institutional representatives are emperor and pope. This brightest of celestial lights provides hope from the very beginning of the poem when the lost traveller emerges from the dark wood to see the 'planet that leads people straight on every path.' An eclipse, especially a total eclipse of the Sun, would thus seem to signify a bad thing, suggesting we are lost, in the dark, off track, without a guide, incapable of seeing what is the case and, more importantly, which way we should go. Yet at the same time, Dante associates eclipses with knowledge and with insight, even as the result of blindness. Eclipses are sources of wonder that trigger the search for hidden causes and they are an obstruction that makes other things visible, including and especially things beyond the immediate, the material, and the contingent. From frightening omen to originator of rational science to metaphysical symbol of the Crucifixion, this particular dramatic natural phenomenon brims with significance across Dante's works that access all these traditions.

Eclipses, inspiring both wonder and terror, as David Bentley Hart observes, are also associated with the beginnings of knowledge, of philosophy, of what Aristotle called the pursuit of science 'for its own sake' in the *Metaphysics*.[3] The desire for knowledge begins in wonder—*admiratio* in the Latin translations—and ready examples of marvellous things that first ignited philosophical inquiry are the 'changes' of Moon and Sun, which medieval commentators translated as 'passions' and, more explicitly, as eclipses. For example, Albertus Magnus states:

They were doubtful about greater things whose cause was not so evident, such as the passions of the moon according to its mansions, increases, and eclipses [*passionibus*

---

[1] Dante Alighieri, *De vulgari eloquentia* I.6.3, trans. Steven Botterill. Cambridge, 1996. *The Letters of Dante*, trans. Paget Toynbee. Oxford, 1966.

[2] Quotations from Dante's *Commedia* are from Dante Alighieri, *La Commedia secondo l'antica vulgata*, ed. Giorgio Petrocchi, Milan, 1966–7. Translations are mine unless otherwise indicated. On the triple rhyme of 'stelle,' see John Ahern, 'Dante's last word: the *Comedy* as a *liber coelestis*.' *Dante Studies* 102, 1984, pp. 1–14.

[3] Aristotle, *Metaphysica* I.2 (982a29), In M.-R. Cathala and R. M. Spiazzi (eds), Thomas Aquinas, *In duodecim libros metaphysicorum Aristotelis expositio*. Turin, 1950, p. 17: 'sola libera est scientiarum; sola namque haec suimet causa est.'

lunae secundum mansiones, accessiones, et eclipses] and about those regarding the sun and the stars ... as when there is an eclipse [*et de his quae sunt circa solem ed astra, sicut quod ... eclipsatur*].[4]

Thomas Aquinas invokes Aristotle's statement on wonder, using the solar eclipse as the precise example, to answer the question of whether human happiness consists in seeing the divine essence. Since human beings naturally desire to know, they are not perfectly happy as long as something is left for them to desire and seek to know ('man is not perfectly happy, so long as something remains for him to desire and seek'). The object of all human understanding is to know 'what a thing is.' If they can see an effect, they will naturally want to know the cause.

> And this desire is one of wonder (*illud desiderium est admirationis*), and causes inquiry, as is stated in the beginning of the Metaphysics (i, 2). For instance, if a man, knowing the eclipse of the sun, consider that it must be due to some cause, and know not what that cause is, he wonders about it, and from wondering proceeds to inquire (*admiratur, et admirando inquirit*). Nor does this inquiry cease until he arrives at a knowledge of the essence of the cause.[5]

The natural desire to seek causes, exemplified in the curiosity inspired by a solar eclipse, will never desist until it discovers the first cause, which can be seen only in God.

Boethius has Lady Philosophy sing about celestial phenomena ('Si quis Arcturi sidera nescit') whose causes are hidden from those who, ignorant of astronomy, will be disturbed by an eclipse of the Moon, beating gongs until it reappears. The regular and frequent movements of the stars will astound someone ignorant of 'high heaven's law.' While no one marvels at the melting of the snow under the growing heat of the Sun or how the north-west wind causes the sea to crash against the shore, because the causes are evident, in rarer events, like eclipses, the uneducated are troubled, because the causes 'are hidden and disturb men's hearts' and 'unexpected things astound the excitable mob.'[6] Here too we see the association of eclipses with wonder

---

[4] Albertus Magnus, *Metaphysica* I.2.6. In Bernard Geyer (ed.) *Alberti Magni Opera Omnia*, tome 16.1, Monasterii Westfalorum: In Aedibus Aschendorff, 1960, p. 23: '*De maioribus dubitantes erant*, quorum causa non erat adeo parata, *sicut de passionibus lunae* secundum mansiones, accessiones, et eclipses, et de his *quae sunt circa solem ed astra*, sicut quod ... eclipsatur.' For a discussion of this passage in relation to Dante, see Patrick Boyde, *Dante: Philomythes and Philosopher*. Cambridge, 1981, pp. 49–51.

[5] Thomas Aquinas, *Summa Theologiae*, ed. P. Caramello. Turin, 1952–6, Iª–IIae q. 3 a. 8, p. 17: 'Et ideo remanet naturaliter homini desiderium, cum cognoscit effectum, et scit eum habere causam, ut etiam sciat de causa quid est. Et illud desiderium est admirationis, et causat inquisitionem, ut dicitur in principio Metaphys. Puta si aliquis cognoscens eclipsim solis, considerat quod ex aliqua causa procedit, de qua, quia nescit quid sit, admiratur, et admirando inquirit.' *The Summa Theologiæ of St Thomas Aquinas* (1920), translated by Fathers of the English Dominican Province. Online edition © 2017 by Kevin Knight.

[6] Boethius, *Consolation* IV.5, *The Theological Tractates*, ed. and trans. H. F. Stewart, E. K. Rand, S. J. Tester. Cambridge, MA, 1973, pp. 354–7: 'Si quis Arcturi sidera nescit / Propinqua summo cardine labi, / Cur legat tardus plaustra Bootes / Mergatque seras aequore flammas, / Cum nimil celeres explicet ortus, / Legem stupebit aetheris alti. / Palleant plenae cornua lunae / Infecta metis noctis opacae / Quaeque fulgenti texerat ore / Confusa Phoebe detegat astra: / Commouet gentes publicus error / Lassantque crebris pulsibus aera./ Nemo miratur flamina Cori / Litus frementi tundere fluctu / Nec niuis duram frigore molem

(*legem stupebit aetheris alti*) and emotional disturbance (*illic latentes pectora turbant*). Knowledge removes the cloud of ignorance and puts an end to the marvels (*cessent profecto mira videri*). Here, already in late antiquity, in a text of wide diffusion in the Middle Ages, we have the eclipse invoked both as ominous portent, in which people feel they are implicated and on which they can have some effect, and as spur to a purely rational and scientific search for natural causes of observable phenomena.

Eclipses give knowledge about material facts of the universe. In the unfinished philosophical treatise he called 'Banquet' (*Convivio*), in which Dante teaches some fundamentals of science to those ignorant of Latin in the context of commenting on a set of his own lyric poems, an eclipse is mentioned as one of the ways in which the number, relative positions, and movements of the celestial spheres are known. That the heavens are nine in number is known from observation, from the sciences of perspective, arithmetic, and geometry, but also from other 'sensible experiences', such as the eclipses of the Sun, by which it is apparent to the senses that the Moon is beneath the Sun.[7] In Dante's last work, *Questio de aqua et terra*, a natural scientific discourse on why some earth (the heaviest element) rises above water, the extent of habitable land, from the straits of Gibraltar at its westernmost extremity to the mouth of the river Ganges in the East, was determined, he says, by means of eclipses of the Moon.[8] Following Aristotle's methodology in the *Physics*, Dante acknowledges that the scope of *Questio de aqua et terra* is limited to the realm of natural and 'mobile' elements—to wit, water and earth—which can only confer a moderate degree of certainty. This method is empirical, not mathematical, proceeding from 'what is better known to us but less known to nature' to 'what is more certain and better known to nature', which is to say from discernible effects to unknown causes.[9] The example Dante chooses to use of such a scientific procedure is how the eclipse of the Sun led to the recognition of the interposition of the Moon, so that, as seen in the *Metaphysics*, men began to philosophize because of their wonder (*unde propter admirari cepere phylosophari*).

/ Feruente Phoebi soluier aestu. / Hic enim causas cernere promptum est, / Illic latentes pectora turbant. / Cuncta quae rara prouehit aetas / Stupetque subitis mobile uulgus, / Cedat inscitiae nubilus error, / Cessent profecto mira uideri!'

[7] Dante, *Convivio* II.iii.6, ed. and trans. Andrew Frisardi, Cambridge, 2018, pp. 64–5: 'Sì che secondo lui, secondo quello che si tiene in astrologia ed in filosofia poi che quelli movimenti furon veduti, sono nove li cieli mobili; lo sito delli quali è manifesto e diterminato, secondo che per un'arte che si chiama perspettiva, e [per] arismetrica e geometria, sensibilmente e ragionevolmente è veduto, e per altre esperienze sensibili: sì come ne lo eclipsi del sole appare sensibilmente la luna essere sotto lo sole.'

[8] Dante Alighieri, *Questio de aqua et terra 54*, ed. Marco Baglio, in *Dante Alighieri, Nuova edizione commentata delle opere di Dante*. Vol. V. Rome, 2016, pp. 728–30: 'Nam, ut comuniter ab omnibus habetur, hec habitabilis extenditur per lineam longitudinis a Gadibus, que supra terminos occidentales ab Hercule positos ponitur, usque ad hostia fluminis Ganges, ut scribit Orosius. Que quidem longitudo tanta est, ut occidente sole in equinoctiali existente illis qui sunt in altero terminorum, oritur illis qui sunt in altero, sicut per eclipsim lune compertum est ab astrologis. Igitur oportet terminos predicte longitudinis distare per clxxx gradus, que est dimidia distantia totius circumferentie.'

[9] *Questio 59*: 'Propter causam vero efficientem investigandam, prenotandum est quod tractatus presens non est extra materiam naturalem, quia inter ens mobile, scilicet aquam et terram, que sunt corpora naturalia; et propter hec querenda est certitudo secundum materiam naturalem, que est hic materia subiecta.'

In such matters effects are better known to us than causes, for it is by them that we are led to the knowledge of causes, as is manifest (for it was the eclipse of the sun that led to the recognition of the interposition of the moon; so that men began to philosophise because of their wonder), the path of investigation in the things of nature must needs be from effects to causes; this method, though it may yield adequate certainty, yet cannot yield such certainty as the way of investigation in mathematics, which is from causes, or the higher, to effects, or the lower. And so we are to look for such certainty as may be had in this style of demonstration.[10]

Eclipses are thus the exemplary empirical question, the one that first produces wonder and amazement, but whose causes have been extensively and accurately uncovered by science and philosophy. Yet eclipses are perhaps even more important for their moral meanings. In his *Confessions*, written at the end of the fourth century AD, Augustine characterizes the ignorant as 'amazed' by the precision with which the philosophers can accurately foretell the 'failing' (*defectus*) of the two great luminaries, down to the year, the month, the day, the hour, and the degree. Yet he immediately follows this acknowledgment of scientific accomplishment by reducing its relative importance. While those who do not understand causes are left to marvel, the arrogant scientists who understand these things, are themselves eclipsed: 'They depart from your great light and fall away (*recedentes et deficientes a lumine tuo*).'[11] This is what happens when knowing the 'periods of the world' takes precedence over seeking knowledge of the Lord. Augustine uses the very insights of this science to draw his moral lessons: 'The wise man is fixed like the sun; but the fool changes like the moon.' An understanding of how the light of the moon increases 'to our eyes' as it distances itself from the sun serves as a lesson that 'the soul of man, receding from the sun of righteousness ... turns all its strength toward external things, and becomes more and more darkened in its deeper and nobler powers.' The increased light of the Moon is, paradoxically, a sign of its distance. So too, 'when the soul begins to return to that unchangeable wisdom ... all that light of the soul which was inclining to things that are beneath is turned to the things that are above, and is thus withdrawn from the things of earth; so that it dies more and more to this world, and its life is hid with Christ in God.'[12] This 'eclipse' of a lesser light by a greater one is similar to what happens frequently in Dante's *Paradiso*. And this particular exegesis, drawing the opposite moral conclusion from the apparent light of the celestial bodies, demonstrates the possibility that the significance of visible astronomical events lies precisely in what we cannot see with our eyes.

---

[10] *The Latin Works of Dante*, trans. Philip H. Wicksteed, New York, 1969, *Questio de aqua et terra* 61: 'Propter causam vero efficientem investigandam, prenotandum est quod tractatus presens non est extra materiam naturalem, quia inter ens mobile, scilicet aquam et terram, que sunt corpora naturalia; et propter hec querenda est certitudo secundum materiam naturalem, que est hic materia subiecta [...] quia quia eclipsis solis duxit in cognitionem interpositionis lune, unde propter admirari cepere phylosophari—, viam inquisitionis in naturalibus oportet esse ab effectibus ad causas.'

[11] Augustine, *Confessions* V.iii.4, trans. Carolyn Hammond. Cambridge, MA, 2014, pp. 190–91.

[12] Augustine, *Letter* 55, to Januarius, trans. J. G. Cunningham, in *Nicene and Post-Nicene Fathers*, First Series, Vol. 1, ed. Philip Schaff, Buffalo, NY, 1887. Revised and edited for New Advent by Kevin Knight. <http://www.newadvent.org/fathers/1102.htm>.

In *Paradiso*, where Dante is on a journey of knowledge, the solar eclipse is used to describe an excess of intellectual curiosity. Just before the final part of his examination on the three theological virtues in the heaven of the fixed stars, the examinee tries too hard to peer into the dancing light that he knows is the Apostle John. Dante stares more fixedly because of a pious belief that the beloved disciple had ascended bodily into heaven.[13] In his effort to see whether or not this is the case, Dante blinds himself, just as when, out of a desire for knowledge of a phenomenon, someone looks directly into an eclipse of the Sun. Such a person, Dante says, becomes non-seeing through trying to see.

> Like the person who stares and endeavors to see the sun eclipse a little and who, through seeing, becomes non-seeing, such I became at that last fire, when it was said to me: 'Why do you dazzle yourself in order to see something that has no place here?'[14]

In *Metaphysics* knowledge progresses from a visible effect that inspires inquiry into its hidden—that is, invisible—causes. Here, by straining to see beneath the brilliant light of John's (normally invisible) soul, Dante seeks a material and visible body. It is an attempt to give material cause to a metaphysical reality, to see the body of the soul that is speaking to him, the dark thing behind the light. John's gentle rebuke is in fact to say that his body is earth and remains down there, in the earth (*in terra è terra il mio corpo*). The examination of love then takes on an eclipse-induced blindness, since not seeing does not prevent speaking and reasoning (*ragionando*) and knowledge can be sought in the abstract, in the absence of visible phenomena.[15]

Because eclipses are relatively rare, an apparent anomaly that 'disturbs men's hearts,' as Boethius put it (*Illic latentes pectora turbant*), they have been associated with something amiss in the otherwise reliable regularity of the celestial spheres. Science explains the darkness as a mere appearance, due to relative orientation of physical bodies that are in fact in no way affected or damaged by what to uninformed humans looks ominous, or indeed, apocalyptic. Knowing the causes of things dispels the wonder and, more importantly, the dread. Yet Dante also retains the notion of an eclipse as a sign of something gone wrong. In a letter to the Italian cardinals, he laments the removal of the papacy from Rome to Avignon as 'so unwonted an eclipse of Rome or rather of her sun.' This 'present misery' has brought grief compounded by shame to the rest of the inhabitants of Italy.[16] In this case the 'sun' suffering eclipse is the papacy. According to the hierocratic theory of papal supremacy, the two lights

---

[13] Aquinas, *Summa theologiae*, III Supplementum 77.1.ob 2: 'But the resurrection of certain members that desire nobility from their being closely connected with the Head was not delayed till the end of the world, but followed immediately after Christ's resurrection, as is piously believed concerning the Blessed Virgin and John the Evangelist.' Rachel Jacoff, 'Dante and the Legend(s) of St John.' *Dante Studies* 117, 1999, pp. 45–57.

[14] *Paradiso* 25.118–23: 'Qual è colui ch'adocchia e s'argomenta / di vedere eclissar lo sole un poco, / che, per veder, non vedente diventa / tal mi fec' ïo a quell' ultimo foco / mentre che detto fu: 'Perché t'abbagli / per veder cosa che qui non ha loco?'

[15] *Paradiso* 26.4–6: 'dicendo: Intanto che tu ti risense / de la vista che haï in me consunta, / ben è che ragionando la compense.'

[16] Dante, *Epistole* XI.23, ed. Baglio, p. 212: 'Et si ceteros Ytalos in presens miseria dolore confecit et rubore confudit, erubescendum esse vobis dolendumque quis dubitet, qui tam insolite sui vel Solis eclipsis

created in heaven to illuminate the day and night were traditionally glossed as papacy (Sun) and empire (Moon).[17] Dante will contest this theory in his political treatise on monarchy—a work that was posthumously condemned and publicly burned. In that work, he concedes that there are indeed two institutions analogous to Sun and Moon, but the lesser light is not dependent on the greater for its existence, only for the light that it reflects. He explains the need for both papacy and empire on the basis of humans' dual nature, material and spiritual, which entail two distinct goals. One is happiness in this life, consisting in what we can do for ourselves, and figured in the earthly paradise. The other goal is the happiness that consists in the eternal enjoyment of the vision of God, figured in the celestial paradise, for which our own resources are insufficient. Happiness of either type can only be found by means of a lighted path, and therefore two different lights are needed for the pursuit of these twin goals. Reasoning and philosophy can inform the practice of moral and intellectual virtues conducive to human flourishing on Earth. Spiritual teachings that transcend human reason can instil the practice of the three theological virtues that lead to eternal life: faith, hope, and charity. Since knowledge of these goals is no longer natural to fallen creatures, perpetually vulnerable to acquisitiveness, human beings need also to be educated and disciplined towards them with what Dante calls 'bit and bridle.' It is therefore not just a question of a light to illuminate a path, but also an authority to govern, as laws and their enforcement both teach and form individuals and society. This is the reason two guides are necessary: 'the supreme Pontiff, to lead mankind to eternal life in conformity with revealed truth, and the Emperor, to guide mankind to temporal happiness in conformity with the teachings of philosophy.' While all roads should lead to that 'Rome of which Christ is a Roman,' that is, to the celestial kingdom, there is an ineluctable role for the secular ruler. The pope's 'jurisdiction,' so to speak, is ultimately limited to the things that pertain to the happiness of the other life, the celestial paradise.[18]

The poet goes quite a bit further in the *Purgatorio* and blows up the astronomical trope altogether when he has a chivalrous Lombard lament that Rome—'that made the world good'—used to have not *a* Sun and *a* Moon, but two Suns. The terrestrial world needs its own laws, its own law-giver, who should nonetheless be able at least to glimpse the highest point, the bell tower of the true city, the celestial one, of which everyone is a citizen.[19] The secular ruler is therefore bound to know something of the eternal order of things, even if it is not his jurisdiction. But that does not make the emperor a Moon in relation to the papal Sun. Rather there is a science and a wisdom specifically appropriate to the governance of worldly affairs, which can be learned, as Dante says in *Monarchia*, by reason as discovered by the philosophers. In *Purgatorio* Dante upgrades this other, secular knowledge, from Moon to Sun, and declares that there are, or rather *were*, two Suns. In ancient, unspecified times, Rome,

causa fuistis?' English version from Paget Toynbee, *The Letters of Dante*, Oxford, 1966. Available at: https://www.danteonline.it/opere/index.php.

[17] Anthony Cassell, "'Luna est Ecclesia": Dante and the "Two Great Lights". *Dante Studies* 119, 2001, pp. 1–26.

[18] Dante, *Monarchia* III.xvi.7–11, trans. Prue Shaw. Cambridge and New York, 1996.

[19] *Purgatorio* 16.94–6: 'Onde convenne legge per fren porre; / convenne rege aver, che discernesse / de la vera cittade almen la torre.'

which is to say, the world, once had both these Suns to light up both roads to human flourishing, one earthly and one divine.[20] The point is that *both* are necessary, and the current disastrous state of affairs stems from the one extinguishing the other.[21] The papacy, starting with the ill-advised Donation of Constantine—where the first Christian emperor allegedly gave all temporal wealth, power, and territory to the Church—has laboured mightily against any would-be emperors in its environs, and has taken on the prerogatives of warfare and material gain not in its purview. The spiritual rulers do not fear the secular ones; and secular rule, now in the hand of clerics, has no one to correct or enlighten it. Allen Mandelbaum translates the statement that the one (Sun) has put out the other (*l'un l'altro ha spento*) as each has *eclipsed* the other.

In the final canticle of the *Commedia*, Suns can in fact be multiple and they can eclipse one another by virtue of greater brightness. A category of exceptionally wise souls, *spiriti sapienti*, who come to greet Dante and Beatrice in the solar heaven appear to stand out against the body of the Sun, not as darker entities, but by virtue of their higher degree of luminosity (*non per color, ma per lume parvente*).[22] At this point Beatrice exhorts Dante to give thanks to the 'Sun of the angels' for having raised him to the height of 'this sensible one,' meaning the literal celestial body, the largest and brightest planet in the geocentric system. When he does so, Dante's obedient reorientation of love toward God eclipses his beloved escort into oblivion (*che Bëatrice eclissò ne l'oblio*), which makes her laugh.[23] Beatrice's final disappearance from the story, and from Dante's poetry, is accompanied by a similar kind of obliteration of a bright light by an even brighter one when the poet depicts how, at dawn, the stars wink out, one by one, up to and including even the most beautiful (*infino alla più bella*), at the approach of a greater light, the 'bright handmaiden of the sun.'[24]

Even so, the corruption of the papacy, resulting from its eclipse of the empire, ruffles feathers even quite high up in Paradise. After a successful examination on the three theological virtues in the heaven of the fixed stars, Dante's principal examiner, St Peter, begins to fulminate about what is going on down here, on Earth, in Rome, which he calls—three times, as if tripling his indignation—'my place, my place, my place.' The entire heaven flushes red, as if Jupiter and Mars were birds and exchanged their plumage.[25] Peter's anger is reflected as 'shame for another's fault' in all the blessed, including Dante's very own Beatrice, who change colour, reddening at the thought of such scandal. So, Dante says, Beatrice 'transmuted her semblance' and just such an eclipse, he believes, was in heaven, when the highest power suffered,

---

[20] *Purgatorio* 16.106–8: 'Soleva Roma, che 'l buon mondo feo, / due soli aver, che l'una e l'altra strada / facean vedere, e del mondo e di Deo.'

[21] *Purgatorio* 16.109–12: 'L'un l'altro ha spento; ed è giunta la spada / col pasturale, e l'un con l'altro insieme / per viva forza mal convien che vada; / però che, giunti, l'un l'altro non teme.'

[22] *Paradiso* 10.40–2: 'Quant' esser convenia da sé lucente/ quel ch'era dentro al sol dov' io entra'mi, non per color, ma per lume parvente!'

[23] *Paradiso* 10.58–61: 'come a quelle parole mi fec' io; / e sì tutto 'l mio amore in lui si mise, / che Bëatrice eclissò ne l'oblio. / Non le dispiacque, ma sì se ne rise.'

[24] *Paradiso* 30.4–9: 'quando 'l mezzo del cielo, a noi profondo, / comincia a farsi tal, ch'alcuna stella perde il parere infino a questo fondo; / e come vien la chiarissima ancella / del sol più oltre, così 'l ciel si chiude / di vista in vista infino a la più bella.'

[25] *Paradiso* 27.13–15: 'e tal ne la sembianza sua divenne, / qual diverrebbe Iove, s'elli e Marte / fossero augelli e cambiassersi penne.'

which is to say, when God was crucified. He compares her righteous discoloration, for no fault of her own, to the total and universal eclipse of the Sun at the Passion of Christ.[26] Two cantos later Dante will moreover have Beatrice pointedly refute the idea that the darkness at the Crucifixion was a natural and predictable celestial phenomenon.

All three synoptic gospels recount that the Sun was darkened for three hours during the Passion of Christ on Good Friday. In the Gospel of Luke, darkness came over all the Earth (*tenebrae factae Sunt in universam terram*) and the Sun was obscured (*obscuratus est sol*).[27] In a canto devoted to the immaterial creatures called angels, Beatrice will make a stern scientific point about this passage. A number of eminent authorities, such as Dionysius the Areopagite, Vincent of Beauvais, Michel Scot, John of Sacrobosco, Pierre d'Ailly, and even Thomas Aquinas, opined that the obfuscation at the Crucifixion was due to a solar eclipse—that is, to the imposition of Moon between earth and Sun. Without naming names, Beatrice blithely accuses all such people of *lying*.

> One says that the moon turned back at the passion of Christ and interposed itself so that the light of the sun did not shine; and he lies, because the light hid itself of its own accord, since such an eclipse was visible to both the Spaniards and the Indians.[28]

Apart from the ungainly fact that the Moon, which we know was full at Passover and therefore at the Passion, would have had to speedily back-track half its monthly cycle in an instant in order to interpose itself between Sun and Earth, a solar eclipse is never seen all over the world (*in universam terram*). In fact, it is only seen by a privileged few, as eclipse-hunters well know. At Christ's crucifixion, to the contrary, the light of the Sun was hidden to both the Spaniards and the Indians, Dante says, which in his geography means from the westernmost to the easternmost ends of the dry land that emerges from the ocean. Beatrice's arguments about the solar obfuscation are scientific, but her purpose is to conclude that that particular phenomenon was not in the ordinary course of things; it was, rather, a miracle: an event whose causes will always remain hidden to the investigations of the philosophers and thus retains the legitimate response of *admiratio*.

Beatrice is no stranger to science. Early on in the celestial journey, she proposes a material experiment to disprove a scientific hypothesis about the nature of the Moon. In that lowest sphere, Dante asks his beloved guide why the closest and most plainly visible celestial body seems to be covered in dark spots. As a good pedagogue, she first asks her pupil what he thinks might be the cause. His answer is that the stuff

---

[26] *Paradiso* 27.34–6: 'così Beatrice trasmutò sembianza; / e tale eclissi credo che 'n ciel fue / quando patì la supprema possanza.'

[27] Luke 23:44–5: 'Erat autem fere hora sexta, et tenebrae factae sunt in universam terram usque ad horam nonam. Et obscuratus est sol.' For representations of the eclipse at the Crucifixion in Western art, see Roberta J. M. Olson and Jay M. Pasachoff, *Cosmos: The Art and Science of the Universe*. London, 2019, pp. 90–1.

[28] *Paradiso* 29.97–102: 'Un dice che la luna si ritorse / ne la passion di Cristo e s'interpuose, / per che 'l lume del sol giù non si porse; /e mente, ché la luce si nascose / da sé: però a li Spani e a l'Indi /come a' Giudei tale eclissi rispuose.' For a summary of the debate about these opinions, see the commentary of Robert Hollander in Dante, *Paradiso*, trans. R. Hollander and Jean Hollander, New York, 2007, *ad loc.*

of which the Moon is made is thicker in parts and thinner in others; it is essentially lumpy. If thinness were the cause of that darkness you ask about, she responds, then either it would be starved of its matter in some part (it would be worn through) or its substance would be distributed unevenly throughout, alternating like fat and thin pages in a book (books were of course typically made of animal flesh). Beatrice observes first of all that if the Moon had holes in it (like Swiss cheese), this would be evident in a solar eclipse, when the light of the occluded Sun would shine through (Figure 9.1).[29]

The solar eclipse disproves the first possibility; for the second she proposes an experiment to do at home. Set up three mirrors, two equidistant from you and the third further away. Turning towards the mirrors, set a light behind your back. The point of the experiment seems to be that all three mirrors will reflect the light back with equal brightness, even if the light will appear smaller from the more distant one. This is supposed to demonstrate that dark spots would not result from light being reflected from portions of the Moon's surface that are set farther back.[30] What at least one critic has pointed out is that the experiment would not work because the body of the observer, interposed between the source of light and the three reflectors, would effect an eclipse. The experiment's very observation would render the phenomenon invisible. John Kleiner further noted that the arrangement of the three mirrors (of which one is 'superfluous') puts the observer at the centre of a cross, making him

Figure 9.1 In Giovanni di Paolo's illustration of Beatrice's lesson on the Moon spots, he includes a depiction of a lunar eclipse, even though she specifically references a solar one (*Paradiso* 2.79–81). 'The Heaven of the Moon.' Dante Alighieri, *The Divine Comedy*. London, 36, f. 132r, Italy (Tuscany) *c*.1450.
Source: British Library.

[29] *Paradiso* 2.76–81: 'Ancor, se raro fosse di quel bruno / cagion che tu dimandi, o d'oltre in parte / fora di sua materia sì digiuno / esto pianeto, o, sì come comparte / lo grasso e 'l magro un corpo, così questo / nel suo volume cangerebbe carte. / Se 'l primo fosse, fora manifesto / ne l'eclissi del sol, per trasparere / lo lume come in altro raro ingesto.'

[30] *Paradiso* 2.97–102: 'Tre specchi prenderai; e i due rimovi / da te d'un modo, e l'altro, più rimosso, / tr'ambo li primi li occhi tuoi ritrovi. / Rivolto ad essi, fa che dopo 'l dosso / ti stea un lume che i tre specchi accenda / e torni a te da tutti ripercosso.'

in effect crucified there.[31] Indeed Kleiner observed that every eclipse in the *Paradiso* is related to some kind of cruciform, crucifixion, or crux. Each of these eclipses he called 'threats' to the poem's intelligibility, 'genuine obstacles to interpretive process,' but that find analogue in the notion of the Cross itself as crux, that is, an obstacle in understanding. The 'crucial' sign is thus not just an emblem for the rational order of the cosmos, the Greek letter *chi* laid out by the demiurge in Plato's *Timaeus*, but also St Paul's 'scandal' or 'stumbling-block,' an 'offense to reason.' This unintelligibility, or wandering into error, is necessary in order to trigger the 'Pauline drama of renunciation and revision.'[32]

Christian Moevs took Kleiner's idea to another level, proposing that the mirror experiment could indeed be carried out, but only if the body of the observer were transparent or, even, absent; in other words, if it were on an abstract plane, rather than a physical one. A physical observer, with the light behind his back, would see no light at all in the central, more distant mirror, but only the reflection of his own body. Moevs argues that 'to see only the body is to fail to see it as the self-manifestation, in the world, of the ground of all being.' This failure would in fact be a crucifixion, as indeed the observer is inscribed in a cross, and just as an eclipse is a failure to see the light that exists, so every crucifixion is a failure to recognize Christ. Moevs writes that 'to sin is to eclipse the light of intellect in oneself,' and 'to consider oneself an ephemeral product of nature is to eclipse the light of being in oneself, which is to crucify Christ.' This is reminiscent of Lady Hester Pulter's awareness of sins that 'interpose / 'twixt heaven and me, and doth eclipse my joy,' even as she retains 'reassurance that the natural cycles of the universe continued in regular, self-sustaining order' (see Roos, Chapter 7, p. 131). According to Moevs, the observer who accepts his own crucifixion at the centre of the cross traced by the position of the three mirrors would become, he says, transparent to the light.[33]

Dante's eclipses can be, as in the epistemological model of Aristotle's *Metaphysics*, sources of wonder that inspire the desire for knowledge. They can indicate a negative, shameful, or disastrous obfuscation, as in the eclipse of Rome's 'sun,' the loss of sight due to an excess of curiosity, the blush of the blessed at the behaviour of modern clerics, or indeed the catastrophe of the Crucifixion itself. At the same time, eclipses trigger investigation into realities beyond physical phenomena, such as the nature of love, the recognition of self and others, and the existence of miracles—whose marvellous exceptionality by definition exceeds the scope of natural investigations. The journey recounted in the *Paradiso* through the spheres—from Moon to outermost, crystalline 'first-moved' container of the universe, the Primum Mobile—is a journey of knowledge, fuelled by doubts that only multiply as each one is resolved by patient, smiling, and well-informed Beatrice. The Primum Mobile is the extremity

---

[31] John Kleiner, 'The Eclipses in the *Paradiso*.' *Stanford Italian Review* 9, 1990, 5–32, 19–20: 'Since the observer stands between the mirror and the lamp, the light spreading from the lamp toward the further mirror will be blocked before reaching the further mirror; the observer's body will cast an obscuring shadow over the mirror's reflective surface. [...] By including a superfluous third mirror and arranging the mirrors to form a cross with the observer "crucified" at the center, Beatrice assures the experiment's failure. Crucifixion is once again followed by eclipse.'

[32] 1 Cor 1:23: 'nos autem praedicamus Christum crucifixum: Judaeis quidem scandalum, gentibus autem stultitiam.' Kleiner, p. 25.

[33] Christian Moevs, *The Metaphysics of Dante's Comedy*. New York, 2005, pp. 125–58.

of the physical universe and therefore also of physical science; in other words, of what we can know of material reality. It is where scientific hypotheses to 'save the appearances' bump up against a metaphysical limit. This culmination of cosmological investigation is marked by a grand image of planetary alignment at the start of the same canto where the naturalistic explanation of cosmic darkness at the Crucifixion (a solar eclipse) is refuted along scientific lines.

Paradiso 29 begins with Sun and Moon pictured in perfect equilibrium, a syzygy, on opposite ends of the horizon as if suspended from a point directly overhead, balanced like the two dishes in the scales of Justice.[34] The two great lights of heaven are here described as 'both Latona's children,' the twin gods Apollo and Diana, balanced as if hanging from the zenith (a point directly overhead) and belted by the horizon on which they also oppose each other diametrically when located in the equinoctial constellations of Aries (the Ram) and Libra (the Scales). These coordinates describing the relative positions of three bodies (Moon, Sun, Earth) are not absolutes but are rather contingent on the perspective of an observer standing in a particular terrestrial location, surrounded by the circle of the horizon and vertically aligned with a point directly overhead (the zenith). Such a configuration, as contributors to and readers of the present volume will easily discern, could diagram an eclipse—a lunar eclipse, where the Earth stands directly in the way of the light of the Sun that would otherwise illumine the Moon.

The exact tenor of the elaborate planetary simile is not spatial, but temporal. Dante claims that Beatrice paused in her teaching, between one long lesson and the next, her face painted with a smile, for just as long as it would take these moving planets to 'unbalance' themselves from the line of the celestial horizon. What the simile explicitly compares is quantitative: we are dealing here with geometry, with lines and points and right angles. Because planets are always in motion, spinning daily around the Earth (according to the medieval geocentric cosmos), apart from their other various major revolutions and epicycles, the perfect balance of the tableau cannot last. In fact, it cannot last more than a moment, since the line of equilibrium through which the two moving bodies will pass as they 'exchange hemispheres,' one headed below the horizon and the other rising above it, is a mathematical line. Mathematical lines have no thickness; there is no extent of time needed to pass through what is essentially a single point, because points have no dimensions at all. The value indicated by the elaborate simile is in fact zero. Which is to say, Beatrice never stops talking. The image describes a moment, an instant, a temporal point of geometrical equilibrium, when the planets are balanced as on the perfectly horizontal arms of a scale, at right angles to the plumb line descending from the zenith, and then another moment when that perfect balance is lost. The scales will immediately tip, not because one planet is bigger or heavier than the other, but because these are celestial bodies in constant motion. In relation to the chosen horizon, one will be rising as the other one sets.

---

[34] *Paradiso* 29.1–9: 'Quando ambedue li figli di Latona, / coperti del Montone e de la Libra, / fanno de l'orizzonte insieme zona, / quant'è dal punto che 'l cenìt inlibra / infin che l'uno e l'altro da quel cinto, / cambiando l'emisperio, si dilibra, / tanto, col volto di riso dipinto, / si tacque Bëatrice, riguardando / fiso nel punto che m'avëa vinto.'

The image is splendidly ambiguous, as well as completely relative. Zenith and horizon depend entirely on where we posit an observer. Moreover, if the Sun is in Aries, it would be spring (leaving aside the slippage between astrological signs and astronomical constellations that had already occurred by this time). If it is rising, it is morning. On the other hand, if the Moon is in Aries, then the Sun is in Libra, and the scene is autumnal. Moreover, if the Sun is on the west end of that horizon, it is setting, the Moon is rising, and the uncertain twilight actually portends the coming night. Either way, the image manages to elicit a destabilizing scene out of a totally common astronomical situation: first the celestial lights are balanced; then they are unbalanced. They pass from perfect equilibrium to a growing state of imbalance or, we might say, inequity. Inequity and iniquity stem from the same root of 'equality.' Measured against the perfect right angle, the perfect cross, formed by the plumb line descending from the zenith to intersect the line of the horizon, the two planets moving in opposite directions will immediately tilt the balance and the instantaneous picture of symmetrical justice is suddenly skewed.

As I have argued elsewhere, this planetary configuration, which also entails an eclipse, is a sketch of the world as it might have been created, at a pregnant moment of perfect balance.[35] It is fitting that these imagined planets in motion introduce a discussion of angels since it is angels who move the planets. Celestial motion—which Dante elsewhere refers to as 'circular nature'—is the whole, immense, physical, visible, knowable, and regular system that is sustained by an intermediate metaphysical cause: the various and disproportionate intensity of angels' attention as they look upon what they desire and love what they know. In the Primum Mobile, Dante depicts their nine whirling hierarchies as spinning fiery circles orbiting a single intense point of light that represents the Prime Mover. While still within the physical cosmos, Beatrice will discuss the nature and history of its movers: when, where, and how they were created, and when, where, and how they either fell or began their revolutions. The opening eclipse is a poetic image of how the angels were created—in balance, in justice, and—according to a long tradition—in 'twilight'; which is to say, in a moment preceding their own moral choice. It was Augustine who explained that the literal meaning of the first lines of Genesis was to be found not in the material formation of the created universe, but in those spiritual creatures that Dante describes as myriad shards of mirrors reflecting the Creator's single and unique light.[36] These first intellectual creatures, created in twilight, would decide either to remain in themselves and become night or to await 'morning knowledge' and become 'day.' Whether the moment described is one of spring or autumn, morning or evening, depends—like the whole depicted planetary balance itself—on the moral subject who observes the scene and has a choice. It is not a measure of chronology, but rather an image of kairos—the single momentous moment when a free being decides what it is it really wants. Even the choice of human beings who, unlike angels, take time to deliberate

[35] Alison Cornish, 'Planets and Angels in *Paradiso* XXIX: The First Moment.' *Dante Studies* 108, 1990, pp. 1–28. Alison Cornish, *Reading Dante's Stars*. New Haven, CT, 2000, pp. 119–41.
[36] *Paradiso* 29.142–5: 'Vedi l'eccelso omai e la larghezza / de l'etterno valor, poscia che tanti / speculi fatti s'ha in che si spezza, / uno manendo in sé come davanti.'

their options, takes no time at all; a decision occurs in an instant marking a division between before and after.[37]

This cosmic metaphor of choice, origin of an imperfect world, occurs in the Primum Mobile, limit of the physical universe, and prefaces the story of the angels, who occupy a liminal status between the human and the divine. This eclipse, like all eclipses, depends on the position of the observer. Even if a lunar eclipse, unlike a solar one, is visible all over the world, only an imagined 'monarchic' view can depict the eclipse from the outside, from an external viewpoint where there is no eclipse, only a universe perpetually suffused in light. As Giles Gasper noted, even Sacrobosco, in his astronomical handbook the *Sphere*, was clear that an eclipse is not really a failure of the light: 'not that it ceases to shine but that it fails us because of the interposition of the moon between our sight and the sun.' The angels themselves, according to Dante, have no need of memory, since their vision of the source of all light has never been interrupted, or eclipsed, by a 'new object.'[38]

The last eclipse of the *Paradiso* is, I would suggest, in one of the final images of the whole poem: that of Neptune, god of the sea, looking up to marvel (*ammirar*) at the shadow of the ship of the Argonauts as they set out on the first sea-going journey.

> A single point is greater forgetfulness to me than the twenty-five centuries since the enterprise that made Neptune marvel at the shadow of the Argo.[39]

In one of the many inversions of the last cantos of the poem, this tercet imagines a god marvelling at the achievements of men, looking *up* at a technological innovation, a bold new human enterprise. Because it is also the first ship to sail across the visual field of the god's consciousness, the shadow of the Argo, eclipsing the light, makes something evident that was not evident before. What was once just the undifferentiated luminosity of the underwater world has now been marked by an event; a border has been traced between the world under the surface of the sea and the equally transparent realm of air and light above and beyond. Just as the solar eclipse demonstrated that the Moon lies beneath the Sun, as Dante observed in his *Question on the Water and the Earth*, so the shade cast by the passage of the Argonauts marks an intermediate territory between above and below, between gods and men, between before and after. The passing shadow makes the light legible.

In conclusion, eclipses are not really encounters between bodies. They depend for their drama upon the perspective of the observer. In blocking the light, they trigger the pursuit of knowledge, which is a desire for light. To see an eclipse as only darkness is to fail to see the light behind the body, or at any rate to know such light remains even as the apparent shadow passes in between. While the ignorant fear what such

---

[37] Thomas Aquinas, *De malo* 16.4, *Quaestiones disputatae*, *Opera omnia*, vol. 8. New York, 1949, p. 401: 'sicut et homo in ipso instanti quo certificatur per consilium, eligit quid est faciendum.'

[38] *Paradiso* 29.76–81: 'Queste sustanze, poi che fur gioconde / de la faccia di Dio, non volser viso / da essa, da cui nulla si nasconde: / però non hanno vedere interciso / da novo obietto, e però non bisogna / rememorar per concetto diviso.'

[39] *Paradiso* 33.94–6: 'Un punto solo m'è maggior letargo / che venticinque secoli a la 'mpresa / che fé Nettuno ammirar l'ombra d'Argo.'

a 'defect' might portend and the philosophers explain the hidden causes, the truly wise also contemplate the significance of such phenomena, what they express about realities not made up of corporeal objects. To see an eclipse as only darkness is to be ignorant of how eclipses work, to 'pluck darkness from light' as Virgil put it in *Purgatorio* (*di vera luce tenebre dispicchi*). This failure to recognize the light behind the things, to recognize God in oneself and in one's neighbour for example, is how an eclipse becomes a crucifixion—an intersection that remains a stumbling block, a crux, rather than the cornerstone and a building block of knowledge.[40]

---

[40] When Virgil responds to Dante's perplexity that, unlike in the sharing of earthly goods where a greater number of people means therewill be less for each, the sharing of celestial goods produces more of it to go around, making all its possessors wealthier, just the way light is multiplied the more mirrors there are, he characterizes Dante as 'plucking darkness from true light,' precisely because he is fixated on earthly things: 'Però che tu rificchi / la mente pur a le cose terrene, / di vera luce tenebre dispicchi.' For the cornerstone, Matthew 21:42: 'Dicit illis Jesus: Numquam legistis in Scripturis: Lapidem quem reprobaverunt aedificantes, hic factus est in caput anguli.'

10

Eclipsed?

The Nineteenth-Century Quest to Capture Solar Eclipses in Art, Science, and Technology

Roberta J. M. Olson

History of Art, Cultural History engaged with the Natural Sciences (Astronomy, Natural History, Ornithology)

From the fourteenth-century literary tradition around Dante, Chapter 10 *leaps forward to the eighteenth and nineteenth centuries—a new era, grounded in the Copernican revolution with the Sun at the centre of the solar system, which impacted the visualization of total solar eclipses and connected to momentous revolutions in science, technology, and journalism before and after the beginning of photography. Until recently, artistic representations of total solar eclipses have not been examined in depth within their scientific and cultural context. The author's articles on the subject, her collaborations with astronomer Jay M. Pasachoff beginning in 1992, and her book* Cosmos with Pasachoff *(2019) have been among the first serious considerations. The following chapter contains new material on the rich context of Western nineteenth-century art history on total solar eclipses.*

Audubon, John James (1785–1851) French–American artist, naturalist, and ornithologist.

Constable, John (1776–1837) English Romantic landscape painter.

eclipsareon A mechanical device to demonstrate eclipses, invented by Scottish astronomer James Ferguson.

Friedrich, Caspar David (1774–1840) German Romantic landscape painter.

Goethe, Johann Wolfgang von (1749–1832) German writer and polymath.

Halley, Edmond (1656–1742) English astronomer, geophysicist, mathematician, and meteorologist.

Linnell, John (1792–1882) English engraver, portrait painter, and landscape painter.

Runge, Philipp Otto (1777–1810) German painter, draughtsman, and colour theorist.

triptych A three-panelled work (e.g. a painting or a sculpture with three panels or parts).

Turner, Joseph Mallord William (1775–1851) English Romantic painter.

Wordsworth, William (1770–1850) English Romantic poet.

Not all the spectacular celestial phenomena visible from Earth occur in the dark skies of night. The Sun, the brightest light in the diurnal firmament, has a singular beauty too dangerous to view without protection except during totality. Nowhere is this magnificence more apparent than during a solar eclipse, as the Moon's shadow gradually obscures the solar disc, and in a total eclipse covers it for minutes, haloed by the stunning corona. Among the most awe-inspiring astronomical events visible without instruments, solar eclipses are an opportunity to witness the powerful motion of celestial bodies. No wonder that they have inspired astronomers, visual artists, and illustrators since time immemorial.[1] Like comets, meteors, meteor showers, and other reported heavenly fireworks, solar eclipses were long viewed superstitiously as portents, as both negative and positive signs, of future events or change, but less so in cultures able to predict their occurrence (see Hart Chapter 8).[2] Even in an early sixteenth-century depiction of a solar eclipse showcasing the solar corona (Figure 10.1), the phenomenon is accompanied by a plague of insects and a 'great dying.'[3] Gradually, as scientific and astronomical advancements increased

Figure 10.1 Solar eclipse of 1483 and the plague of locusts near Mantua. Fol. 87 from the 'Augsburg Book of Miracles' (*Augsburger Wunderzeichenbuch*), 1550, watercolour, gouache, and black ink on paper, 202 x 307 mm, The Cartin Collection.
Source: Photo courtesy of the Cartin Collection.

[1] For solar eclipses in western art, see Roberta J. M. Olson and Jay M. Pasachoff, *Cosmos: The Art and Science of the Universe*, London (2019), pp. 48–85.

[2] See Roberta J. M. Olson, *Fire and Ice: A History of Comets in Art*, New York (1985); Roberta J. M. Olson and Jay M. Pasachoff, *Fire in the Sky: Comets and Meteors, the Decisive Centuries, in British Art and Science*, Cambridge (1998).

[3] Till-Holger Borchert et al., *The Book of Miracles / Das Wunderzeichenbuch / Le Livre des Miracles*, Cologne (2013), vol. 1, p. 127.

the understanding solar eclipses, artists' and illustrators' observations and ability to portray the galvanizing event followed suit.

This chapter briefly surveys trends in western culture from the eighteenth century through to the early twentieth century, during a period which witnessed an increased scientific understanding of the Sun. It explores complex transformations that are preserved in visual material and works of art, including great leaps in technology courtesy of the Industrial Revolution. They facilitated a transition from recording astronomical phenomena via draughtsmanship to photography, a term which was coined in 1839 by the English polymath John Frederick William Herschel and literally means 'drawing with light.'[4]

Astronomer Royal Edmond Halley's broadside predicting the path of the 3 May 1715 solar eclipse (Figures 10.2, 2.1) was a watershed for eclipse observations because it illustrated for the first time the lunar umbra's path across the Earth's surface in a very focused geographic map of England. In the text of the broadside, Halley petitioned for what we now call 'citizen science,' requesting that people send him their observations. Since his predictions were slightly off, he produced a sequel that corrected the path of the 1715 eclipse and illustrated the solar eclipse of 1724 across England to Continental Europe. Equipped with the twenty-first-century, three-dimensional topographic mapping of the lunar surface from the Japan Aerospace Exploration Agency's Kaguya (SELENE) satellite and NASA's Lunar Reconnaissance Orbiter, astronomers have greatly improved on Halley's accuracy. They can now predict the time of an eclipse—including the appearance of Baily's beads and the diamond-ring effect arising from the alignment of valleys on the lunar edge—to a tenth of a second.

Most eclipse images up to Halley's time were unconvincing, especially those purporting to show totality. Only with the German Baroque altarpiece *Vision of St Benedict* by Cosmas Damian Asam (1735)—with its dazzling ray issuing (slightly inaccurately) from the lunar disc—was totality followed by the diamond-ring effect successfully portrayed. Asam painted several solar eclipses, and there is evidence that he was an early eclipse-chaser.[5] In the mid-eighteenth century, when astronomical demonstrations were popular entertainments, Scottish astronomer and instrument maker James Ferguson, who began as an artist, invented the 'eclipsareon.' Like a cometarium that demonstrated the orbit of comets around the Sun, this mechanical device duplicated the syzygy—the lining up of the three celestial bodies—that create a solar eclipse (Figure 10.3).[6]

As science revealed new understandings about the origins of the world and the universe, artists in the Western tradition were inspired and challenged to incorporate them into their subject matter and to investigate new paths in a rapidly changing world in which technology played a decisive role.

---

[4] Larry Schaarf, 'Sir John Herschel's 1839 Royal Society Paper on Photography,' *History of Photography* 3(1) (1979), pp. 47–60.

[5] Roberta J. M. Olson and Jay M. Pasachoff, 'St Benedict Sees the Light: Asam's Solar Eclipses as Metaphor,' *Religion and the Arts* 11 (2007), pp. 299–329; Olson and Pasachoff, *Cosmos*, fig. 74.

[6] See Patricia Rothman, 'By "the Light of His Own Mind": The Story of James Ferguson, Astronomer,' *Notes and Records of the Royal Society of London* 54(1) (2000), p. 39.

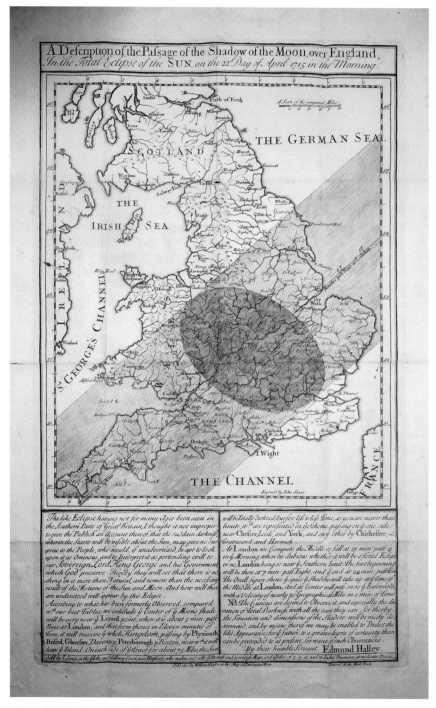

Figure 10.2 Edmond Halley, Broadside with the path of the solar eclipse 3 May 1715, 1715, engraving, 404 x 247 mm. Collection of Jay M. and Naomi Pasachoff.

Source: Photo courtesy of Chapin Library, Williams College, MA.

Figure 10.3 James Ferguson, 'eclipsareon,' pl. XIII from Ferguson and Jeremiah Horrocks, *Astronomy Explained upon Sir Isaac Newton's Principles, and Made Easy to Those Who Have Not Studied Mathematics . . .* (London, 1790).
Source: University of Minnesota Libraries, Minneapolis, MN, Public Domain via HathiTrust.

Nature, Landscape, and Light in the Early Nineteenth Century

Towards the end of the eighteenth century, in the wake of the Enlightenment and its great strides in the sciences, the artistic, literary, musical, and intellectual movement of Romanticism swept the Western world. Emphasizing subjectivity and imagination, the inspiration of the individual, and an exploration of the wonders of Nature, Romanticism generated two artistic responses to solar eclipses. For some visual artists, as for writers and musicians, it was a mystical, quasi-religious relationship with nature; for others it encouraged a more scientific investigation of natural phenomena via studies of meteorology, light, and landscape painting *en plein air*. Unprecedented developments in astronomy, meteorology, and aeronautics produced a new perception of the sky as a layer that separates the Earth from space and supported the exploration of the atmosphere. In addition, the growing sophistication of optics and physics played a role, as did a need to understand the cosmos and the place of human beings in it. A number of Romantic painters applied the new scientific knowledge and their empirical observations to their works. Among the artists who portrayed solar eclipses—both their unearthly light and the syzygy of the heavenly bodies— were English and German painters from cultures with strong landscape traditions and literary ties to nature.

Early in the century, the English painter Joseph Mallord William Turner depicted phases of a solar eclipse in his Eclipse Sketchbook (Figure 10.4). Obsessed with the effects of light, Turner studied aspects of it incessantly in sketchbooks and paintings over his prodigious career. He portrayed sublimely beautiful rainbows, lightning, sunsets, the aurora borealis, moonlight and the Moon through clouds, fireworks, exploding shells, light shining through atmospheric conditions, and studied their relationships to colour. He also included an allegorical comet in a watercolour illustrating Napoleon's fall from power, and used meteors as metaphors in his engraved vignettes for Milton's *Paradise Lost* in which light triumphs over darkness.[7] Moreover, Turner's small watercolour *Galileo's Villa* is related to his thinking about cosmic forms. It includes a telescope, a celestial globe, and a celestial map with orbits of spherical bodies.[8]

John Gage theorizes that Turner acquired his knowledge of cosmology from Mary Fairfax Somerville whose 1831 treatise on astronomy, *Mechanisms of the Heavens*, he had acquired.[9] Somerville, herself an amateur painter, admired Turner's work. Moreover, Turner referred to her early experiments on the magnetizing properties of colour, which for him were a confirmation of the influence of light. Together with Caroline Herschel, Somerville was elected as one of the first honorary female members of the Royal Astronomical Society (she died 44 years before it admitted women). She was also a friend of Caroline's nephew, polymath astronomer John Herschel— who as previously noted coined the term 'photography'—and other scientists.[10] In her memoirs, Somerville comments about Turner: 'I frequently went to Turner's studio, and was always welcomed. No one could imagine that so much poetical feelings existed in so rough an exterior . . . .'[11]

[7] See Olson and Pasachoff, *Fire in the Sky*, pp. 190–6, figs. 105–8.
[8] See A. J. Finburg, *A Complete Inventory of the Drawings of the Turner Bequest*, London (1909), vol. 2, p. 895, no. 87.
[9] John Gage, *J. M. W. Turner. 'A Wonderful Range of Mind'*, New Haven and London (1987), pp. 222–4.
[10] See Kathryn A. Neeley, *Mary Somerville: Science, Illumination, and the Female Mind*, Cambridge and New York (2001).
[11] Mary Somerville, *Personal Recollections, From Early Life to Old Age, of Mary Somerville*, London (1873), p. 269.

Because she followed one of the fashions of the day—not saving correspondence but simply preserving one of Turner's signatures as a specimen—we know little about their relationship.[12] Somerville's experimental work, however, was of particular interest to Turner because it concerned the solar spectrum, and his contacts with Somerville may have fuelled his interest in astronomical phenomena.

The study of the Sun at various times of day and seasons became one of the hallmarks of Turner's work. In his sketchbooks he drew a number of landscapes with sunsets and pulsing solar discs, while in his mature paintings he sometimes built up the solar disc with an underlayer of priming gesso, rendering it as a two-dimensional disc in relief below the impasto of the viscous oil paint and the pulsing white light generated by swathes of pigment from his broad brushstrokes. His most direct impressions of nature are found in nearly 300 sketchbooks, which were not only his visual archive for potential masterpieces, but also reveal his serious dedication to studying nature. Among his earliest depictions of the Sun is a spread in the Wilson Sketchbook (1798–9).[13] In several other studies from the Skies Sketchbook (1818–19), Turner records the Sun as it sinks in the dramatic twilight sky and the crescent Moon punctuates the firmament, as well as a glowing sunset he depicted on folio 42, caused by the post-Tambora volcanic eruption, which also includes a solar halo sixty years before Bishop's Ring became known to science.[14]

In his Eclipse Sketchbook of 1804, Turner documented the onset of a solar eclipse with the buildings at the left framing the transient skyscape (Figure 10.4), which continues on the next folio to the right in the spread.[15]

It is assumed that Turner witnessed this partial eclipse in England and, therefore, the likely date is 11 February 1804.[16] He captured its ethereal light effects on two other folios (folio 2a, which shows more than half of the solar disc obscured, and folio 6, which features the Sun three-quarters eclipsed).[17]

Although English painter John Constable famously stated that 'The sky is . . . the chief organ of sentiment,'[18] he never portrayed a solar eclipse in his hundreds of meteorologically correct sky and cloud studies, although he did represent rainbows. Nevertheless, Constable kept a weather diary and considered his artistic goals in harmony with science.[19] He knew the work of some scientists, including the English chemist and amateur meteorologist Luke Howard, who introduced his seven cloud types in a lecture in 1802 to the Askesian Society, a debating club for scientific thinkers in London. Howard published his ideas in *Essay on the Modification of Clouds* (1804) and *The Climate of London* (1818–20). His watercolour cloud studies

[12] Gage, *J. M. W. Turner*, p. xxv, n.2. The Somerville Papers are on deposit in the Bodleian Library, Special Collections, Oxford.

[13] Ian Warrell, *Turner's Sketchbooks*, London (2014), 33. For the sketchbooks, see ibid., pp. 7–17.

[14] Ibid., 102–3; Natalia Bosko, 'Exploration of the Sky: The *Skies* Sketchbook by J. M. W. Turner,' *Journal of the Royal Astronomical Society of Canada* 113(4) (2019), pp. 141–2.

[15] See Finburg, *A Complete Inventory*, vol. 1, pp. 224–5; for a link to all the pages in the sketchbook (inv. TB LXXXV, 1a–7), see n. 17 below.

[16] Warrell, *Turner's Sketchbooks*, p. 14, which reproduces the full spread fols. 1a, 2. This was an unusual hybrid eclipse, shifting between an annular and total eclipse with a narrow total eclipse band, beginning and ending as an annular eclipse. In England it was partial.

[17] See: https://www.tate.org.uk/art/artworks/turner-landscape-with-the-commencement-of-an-eclipse-d05246/.

[18] Quoted at greater length in Mark Evans, *Constable's Skies: Paintings and Sketches by John Constable*, New York and London (2018), p. 6.

[19] See John E. Thornes, *John Constable's Skies: A Fusion of Art and Science*, Birmingham (1999).

Figure 10.4 Joseph William Mallord Turner, *Landscape with the Commencement of an Eclipse* (left side) from the Eclipse Sketchbook [Finburg LXXXV], 1804, black and white chalk and charcoal on beige paper, 166 x 112 mm, Tate Britain, London.
Source: Tate Britain, London.

are held at the Science Museum, London.[20] In 1836, Constable also remarked in his *Discourses*: 'Painting is a science and should be pursued as an inquiry into the laws of nature. Why, then, may not a landscape be considered a branch of natural philosophy, of which pictures are but an expression.'[21]

Another English painter, John Linnell, created cloud studies in 1811 that predate those of Constable.[22] Like Turner, he tried his hand at poetry,[23] and like his friend the artist and poet William Blake, Linnell was mesmerized by the Great Comet of 1811—the naked-eye comet of 1811–12 (C/1811 F1, formerly comet 1811 I)—first visible in March through telescopes. Afterwards he began to study the solar system, and in his journal he made a sketch of the planets in relation to Earth and another of the comet.[24] He also drew three more developed comet studies that, like Turner's eclipse drawing, include buildings from his point of view in the urban landscape. The trio of independent sheets were bound into an album with three studies of the solar eclipse of 19 November 1816. These annotated drawings, which demonstrate the strong links between science and art—that is, that the fundamental work of observation, intensive interrogative gaze at phenomena, and then its representation are common foundations to both art and science—reveal that Linnell possessed astronomical knowledge from an unknown source. When he died, the *Art Journal* and the *Athenaeum* described Linnell as 'our greatest English naturalist,'[25] but his naturalism increased with his religious conversion in 1811, the year of the comet. This dichotomy led him to understand landscape and the meticulous design of nature as proof of God's existence, so that truth to nature was not just an aesthetic criterion but also a moral obligation, which heightened the realism and the intensity of his works.[26] He believed that awesome astronomical spectacles revealed the divine presence in the universe.

It is not known whether Linnell associated with astronomers, although some of his friends had allied interests. In his unpublished journals, Linnell records in the early days of September 1811 meetings with John Varley and his brother Cornelius from Blake's circle. Cornelius Varley was an artist, inventor, and optical instrument maker who devised and patented the graphic telescope in 1811.[27] He wanted to create a portable instrument for landscape painting that resembled the camera lucida, but with the power of a telescope; it could be used with lenses of different strengths, enabling the user to outline and to enlarge or reduce the image and to bring fine details into focus. Drawings produced with the device offered photographic accuracy thirty years before the first camera. Varley sold Linnell several optical

---

[20] Ibid., pp. 188–91; Richard Hamblyn, *The Invention of Clouds: How an Amateur Meteorologist Forged the Language of the Skies*, London (2001).

[21] John Constable, *Discourses*, edited by Ronald Brymer Beckett, Ipswich (1970), p. 69.

[22] Katharine Crouan, *John Linnell: Truth to Nature (a centennial exhibition)*, London and New York (1982), p. 9.

[23] David Linnell, *Blake, Palmer, Linnell and Co.: The Life of John Linnell*, Lewes (1994), p. 23.

[24] The entries are on fols. 31v and 32r of the artist's *Journals* in the Fitzwilliam Museum, Cambridge; see Olson and Pasachoff, *Fire in the Sky*, fig. 62.

[25] Katharine Crouan, *John Linnell: A Centennial Exhibition*, Cambridge (1982), p. ix.

[26] Ibid., p. xi. See also Alfred Thomas Story, *The Life of John Linnell*, 2 vols., London (1892).

[27] Huon Mallalieu, 'Varley the Optician,' in *Cornelius Varley: The Art of Observation*, London (2005), pp. 25–31. Varley's eldest son, Cornelius John, became an astronomer, publishing watercolours of comets and stars in the *Journal of Astronomical Observations*.

devices: a camera obscura in 1811, a camera lucida in 1816, and a graphic telescope in 1848.[28]

As mentioned, Linnell's trio of eclipse drawings depict the total solar eclipse of 19 November 1816, which was only partial in London.[29] (Totality passed through Scandinavia and Eastern Europe.) From his inscriptions on the second drawing, he observed the eclipse from No. 2 Streatham Street, Bloomsbury, where it began at 8:04 a.m. and ended at 10:19 a.m., with maximum coverage at 9:09 a.m. Sunrise occurred at 7:32 a.m. It is unknown how Linnell viewed the partial phases safely, whether via projection or filtered through a darkened glass, although the accuracy of the size of the crescent may indicate projection. Unfortunately, the volume of his journal for 1816 is lost.

His first solar eclipse drawing in silvery graphite on off-white paper contains three registers of sketches with four spheres representing the Sun in three early, successive stages of the eclipse and two lines of inscriptions.[30] Near the lower edge Linnell wrote: 'Eclipse of Sun Nov' 19 1816'; at the upper left near the most exacting sphere: '5 min. af. 8–' (8:05 a.m.) with only a slight corner occulted; the larger sphere in the middle register has a greater portion obscured, while two smaller spheres below record a later phase.

Linnell's second eclipse drawing, and the most ambitious of the album, is a fold-out sheet (Figure 10.5). Maximum coverage in London was 78% of the diameter of the Sun at 9:09 a.m. The Sun was then at an altitude of 11° and azimuth of 143°.[31] Linnell drew this complex grouping in white, black, yellow, and reddish pink chalks on brown paper, inscribing it at the upper edge: 'Eclipse of the Sun Nov.r 19.1816 see The other sketch' (implying that this was the second). He included rectangular structures at the lower centre divided by a vertical line that is part of a separate box at the lower right within which he drew an orb in reddish pink chalk with a slight occultation. From the inscription below it ('colour of appearance at first'), this pink orb represents the first contact of the Moon with the Sun, the stage in the upper register of the previous sheet. Moving in a counter-clockwise direction, above and to the right are two white crescents, representing the maximum occultation of the eclipse. To the right Linnell inscribed '20 / min [after] / 9.' (9:20). His emphasis here is on the blinding white light emitted by the two crescents. Counterclockwise to the left he drew the Sun in yellow and white chalks emerging from behind the Moon's shadow and emitting a yellow glow. The adjacent inscription at the right ('1/4 to X / or 20 min') indicates the approximate time, 9:40 or 9:45. The final study on the sheet in white chalk is near the completion of the eclipse cycle and is inscribed 'X oclock' (10:00). The sequence Linnell recorded took two hours and five minutes.

His third eclipse drawing is another fold-out sheet, in black and white chalk on grey-green paper, but the artist only drew on its left side (Figure 10.6). It portrays the same eclipse at 9:00 a.m., shortly before maximum coverage, but it is different

[28] Crouan, *John Linnell: Truth to Nature*, p. viii.
[29] Roberta J. M. Olson and Jay M. Pasachoff, 'The 1816 Solar Eclipse and Comet 1811 I in John Linnell's Astronomical Album', *Journal for the History of Astronomy* 23 (1992), pp. 121–33.
[30] Ibid., fig. 4.
[31] Ibid., p. 133 n. 11.

Figure 10.5 John Linnell, *Studies of the Solar Eclipse of 19 November 1816*, 1816, coloured chalks on brown paper, 132 x 720 mm, irregular, the British Museum, London, inv. 1967, 120.3.5.

Source: © the Trustees of the British Museum, London.

from the other two drawings. Instead of focusing on the Sun, the artist has concentrated on the setting and the rooftops of London. It is inscribed below a baseline in graphite: 'J. Linnell Del. / Eclipse of the Sun. Nov.r 19.–186. As it appeared at 9 oclock / Mor[ning?]/ Streatham Street Bloomsbury / Nov 19'; on the right of the verso: 'Looking from The House of No. 2 Streatham St. /Bloomsbury.' The address was that of the family home where he lived with his father until 1817 in order to save money for his marriage.[32] In this drawing Linnell incorporates a distillation of his two more scientific eclipse drawings. Moreover, his inscription includes 'Del.' for *delineavit* (he drew) which, together with the border beneath the image, suggests that he intended the composition for engraving, although he added similar inscriptions to drawings that were never reproduced. Linnell did not continue studying astronomical subjects, although meteorology always played an important role in his naturalistic works.

The German Romantic artist Philipp Otto Runge was a mystical Christian who attempted to express the harmony of nature, humanity, and the divine in his art through the symbolism of colour, form, and numbers. He developed a friendship with polymath Johann Wolfgang von Goethe based on common interests in colour and art. In 1809, Runge finished his manuscript *Farben-Kugel* (*Color Sphere*), which was published in 1810.[33] As was the case with some Romantic artists, Runge was interested in the total work of art (*Gesamtkunstwerk*). He planned such an extravaganza in a series of four huge paintings known as *The Times of the Day*, which were designed

---

[32] Story, *Life of Linnell*, 1, pp. 80–6; Crouan, *John Linnell*, p. xi.
[33] Philipp Otto Runge, *Die Farben-Kugel, oder Construction des Verhaeltnisses aller Farben zueinander*, Hamburg (1810).

Figure 10.6 John Linnell, *Studies of the Solar Eclipse of 19 November 1816*, 1816, black and white chalk on grey-green paper, 132 x 720 mm, irregular, The British Museum, London, inv. 1967, 120.3.6.

Source: © the Trustees of the British Museum, London.

to be seen in a special Gothic chapel, accompanied by music and poetry written by his friend Ludwig Tieck.

Runge painted in oil two versions of *The Morning*, including *The Small Morning* (Figure 10.7), but the other paintings did not advance beyond his drawings before he died of tuberculosis in 1810. In 1803, he had commissioned engravings after his preliminary line drawings in order to share his ideas and to announce the series; he presented a set to Goethe, who fittingly displayed the prints in his music room. Runge issued a second, altered edition in 1807 for commercial distribution. He conceived of

Figure 10.7 Philipp Otto Runge, *The Small Morning (Der Kleine Morgen)*, 1806–8, oil on panel, 109 x 85.5 cm, Hamburger Kunsthalle, Geschenk von Herrn Paul Runge, 1891, inv. HK-1016.

Source: Wikimedia Commons.

the series as a radical, personal vision of landscape with a complex symbolic iconography based on underlying cosmic geometry depicting the arrival and departure of light during four stages of the day. Their gigantic botanicals echo the organic processes of conception, growth, decay, and death,[34] and they are surrounded by eclectically symbolic frames.

In the lower central border of *The Small Morning* Runge represented a solar eclipse just after totality. Supported by the feet of two putti in a manner that resembles the *imago clipeata* on ancient Roman sarcophagi held up by pagan genii, it evokes the reappearance of light echoed in the upper border, where the rays stream over the heads of the cherubim, and in the main panel, where the Sun rises over a sublimely painted landscape below Aurora, goddess of dawn. There is no eclipse in the relatively finished compositional drawings (dated 1803) or the print. Instead, at the centre of the lower border of the painting Runge positioned an ouroboros (an ancient mystical symbol with a serpent eating its own tail symbolizing eternity) crisscrossed by two upside-down burning torches, and at the centre of the upper border a Sun disc with hieroglyphics and rays surrounded by cherubim.[35] Perhaps Runge introduced the solar eclipse after reading about or observing the hybrid (annular/total) eclipse of 11 February 1804, the same one drawn by Turner (Figure 10.4). This event featured a narrow path of totality, beginning and ending as an annular eclipse, which was quite far south of Runge's Hamburg in the north of Germany.

With the growth of Romanticism, the Moon took pride of place as a nearly ubiquitous presence in poetry, music, and art, mostly in nocturnes. This was natural because throughout its cycle the Moon is nearly omnipresent, even periodically during daylight hours, whereas solar eclipses are relatively rare and last for short periods of time. Nevertheless, like the Moon, the Sun and solar eclipses expressed ideas about nature versus industrialization and modernity that involved time and change, as well as the powerful principles operative in the cosmos. A case in point is a watercolour depicting a solar eclipse over the Riesengebirge Mountains by the quintessential German Romantic artist Caspar David Friedrich (Figure 10.8), which heretofore has been misidentified as depicting the Moon.[36]

The work is redolent with sentiments of the Sublime (awesome greatness beyond all calculation in art, aesthetics, and literature) and the seductively mystical power of nature.[37] In this watercolour Friedrich captured the haunting, unearthly light of an annular eclipse and the atmospheric mist at that time and location. Like the artists previously discussed, he was fascinated with light, and he experimented with inserting transparencies into the paper of some watercolours to achieve the desired luminous effects.[38] Although the watercolour is undated, it is now possible to date

---

[34] For the evolution of the series see Hanna Hohl, *Philipp Runge, Caspar David Friedrich: The Passage of Time*, edited by Andreas Blühm, Amsterdam and Zwolle (1996); Markus Bertsch et al., *Kosmos Runge: Der Morgen der Romantik*, Munich (2010), pp. 138–54, 158–218, 386–90.

[35] See Bertsch et al., *Kosmos Runge*, pp. 145, 148, 152.

[36] The work is catalogued at the National Gallery of Art as *New Moon above the Riesengebirge Mountains*.

[37] The concept derives from Edmund Burke's *Philosophical Enquiry into the Origin of Our Ideas of the Sublime and Beautiful* (1757), a treatise that separated the Beautiful (that which is aesthetically pleasing) from the Sublime (whose power is inspiring but also frightening) and was important in Romanticism.

[38] Birgit Verwiebe, 'Erweiterte Wahrnehmung: Lichterscheinungen—Transparentbilder—Synästhesie,' in *Caspar David Friedrich: Die Erfindung der Romantik*, Munich, Essen, and Hamburg (2006), pp. 38–344.

Figure 10.8 Caspar David Friedrich, *Solar Eclipse of 7 September 1820 over the Riesengebirge Mountains*, 1820, watercolour, pen and grey ink, over graphite on paper, 262 x 365 mm, National Gallery of Art, Washington, DC, Wolfgang Ratjen Collection, Purchased as the Gift of Helen Porter and James T. Dyke and Dian Woodner, inv. 2007.111.10.

Source: Public Domain.

it by the eclipse he depicted: the annular eclipse of 7 September 1820, around 1:00 p.m. (Figure 10.9).[39] Therefore, its title should read, as in Figure 10.8, *Solar Eclipse of 7 September 1820 over the Riesengebirge Mountains*.

Romanticism's celebration of natural phenomena also appears in works of art by American artists, an influence undoubtedly stoked by the ideas of Alexander von Humboldt who travelled to North and South America. Take, for example, the singular portrait by artist-naturalist John James Audubon (Figure 10.10). Dated 1821 and having a New Orleans provenance, the work is one of Audubon's bread-and-butter likenesses, drawn in the Crescent City after the artist-naturalist suffered bankruptcy following the Panic of 1819.

---

[39] Gerrit Moll, 'On the Solar Eclipse which took place on September 7, 1820. Communicated in a Letter to J. F. W. Herschel. Esq., Foreign Secretary, from Professor Moll of Utrecht,' *Memoirs of the Royal Astronomical Society* 1 (1822), p. 145.

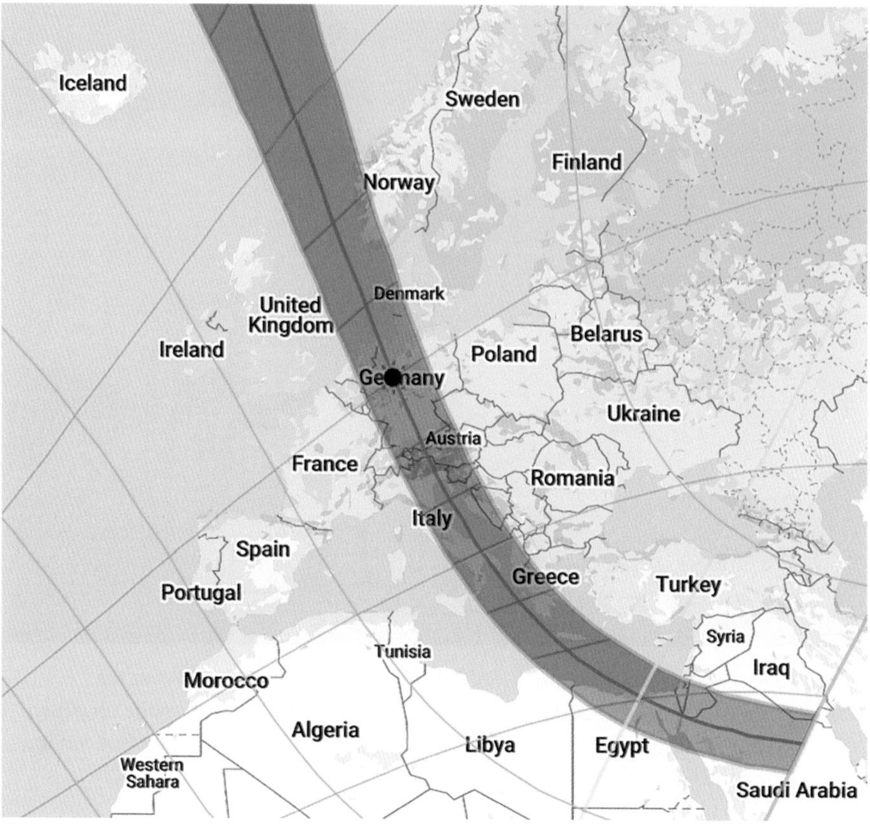

Figure 10.9 Path of the Solar Eclipse of 7 September 1820.
Source: Map by Xavier M. Jubier; map data © 2001 Google.

In the past, the portrait has mistakenly been identified as portraying the Marquis de Lafayette and is only known from a damaged photograph.[40] It represents an older man who has suffered a stroke (note the drooping right side of his mouth). His glance engages the viewer's, and his craggy features and weary mien communicate strength of character. At the upper left corner Audubon depicted a solar eclipse—a sliver of the partially occulted Sun and its corona—as the solar rays break through after maximum coverage. The surrounding clouds underline the unusual darkness and dramatic nature of the celestial event (see Harrison Chapter 14). Most likely, it represents the annular solar eclipse of 27 August 1821, which in New Orleans enjoyed 95% coverage,[41] and/or it may be emblematic. As a symbol, the eclipse could reference the sitter's recovery after a stroke, like the Sun's emergence from behind the

[40] Edward H. Dwight, *A National Exhibition of the Works of John J. Audubon*, Philadelphia (1938), no. 38 (the only known reproduction of the work). According to the Frick Art Reference Library, its provenance is Simon J. Schwartz, New Orleans in 1926; Erskine Hewitt, New York; Parke-Bernet, New York, 18–22 October 1938, lot 1076; John Scott Linsall, Lexington, KY.
[41] See: https://Moonblink.info/Eclipse/eclipse/1821_08_27/.

Figure 10.10 John James Audubon, *Portrait of an Unknown Man with a Solar Eclipse*, 1821, black chalk, Conté crayon, and black ink wash on paper, 305 x 222 mm, whereabouts unknown.

Source: Edward H. Dwight, *A National Exhibition of the Works of John J. Audubon,* exh. cat., Philadelphia (1938), no. 38, repr.

Moon's shadow. On the day of this annular eclipse, Audubon was 150 miles north of New Orleans at Bayou Sara, totally intoxicated with representing birds (there is a gap in his 'Mississippi River Journal' from 25 August until 10 October without a single entry), which does not mean that later in October, when he returned to New Orleans, he could have been commissioned to draw this portrait. Perhaps someone furnished him with a print featuring a solar eclipse or described it to him. At the time, there was a cultural awareness of solar eclipses, as seen in William Wordsworth's poem 'The Eclipse of the Sun' (1820):

> No vapour stretched its wings; no cloud
> Cast far or near a murky shroud;
> The sky an azure field displayed;
> 'Twas Sunlight sheathed and gently charmed,
> Of all its sparkling rays disarmed,
> And as in slumber laid:—...
>
> High on her speculative Tower
> Stood Science waiting for the Hour
> When Sol was destined to endure
> That darkening of his radiant face
> Which Superstition strove to chase...[42]

Blinded by the Light: A Different Tack on Representing Solar Eclipses

Beginning in 1827, the so-called father of Danish painting, Christoffer Wilhelm Eckersberg, and the mathematician and astronomer Georg Frederick Ursin were colleagues at the Royal Academy of the Fine Arts in Copenhagen. They collaborated on a number of projects during Denmark's Golden Age and shared a passion for solar eclipses and the natural sciences.[43] In 1819, while Ursin was working on his doctorate from Copenhagen University, he was appointed an observer at the Observatory. His dissertation topic was a solar eclipse in 1820 (*De eclipsi solari VII Sept. MDCCCXX Dissertatio . . .*) that enjoyed five editions between 1820 and 1870 and marked the inception of his interest in solar astronomy. Visible that year not far from Copenhagen, this annular solar eclipse occurred on 7 September 1820—the same one that Friedrich observed (Figures 10.8 and 10.9). The Danish capital enjoyed a partial eclipse of 90% obscuration, but it is not known whether Eckersberg observed it because his journal entries at the time were irregular.[44]

At the Academy, Ursin made his most important pedagogical contributions by promoting the more practical teaching of mathematics, geometry, and science, thereby spreading that knowledge to a wider audience. His textbooks enjoyed multiple editions and were widely used and translated into other languages.[45] In addition, Ursin collaborated with Eckersberg on two highly technical treatises about perspective

[42] William Wordsworth, *Wordsworth's Poetical Works*, edited by E. de Selincourt and Helen Darbishire, Oxford (1954), vol. 3, pp. 184–6.

[43] See Roberta J. M. Olson and Jay M. Pasachoff, 'Shedding Light on Danish Astronomy during the "Golden Age", *Artibus et Historiae* 85 (2022), pp. 277–303.

[44] Christoffer Wilhelm Eckersberg, *C. W. Eckersbergs dagbøger*, vol. 1, edited by Villads Villadsen, Copenhagen (2009).

[45] WorldCat.org lists 130 works by Ursin in 343 publications: http://worldcat.org/identities/lccn-no2006059401/.

that are noted below.[46] Underlining their friendship, Eckersberg painted Ursin's sympathetic occupational portrait in 1836 (Figure 10.11).[47]

He portrayed his colleague as a teacher, standing at a blackboard with a white chalk diagram illustrating the geometry of a solar eclipse, his fingers marking the location where the lunar umbra (shadow) hits Earth. It footnotes not only the general topic of Ursin's dissertation, but also the two men's joint interest in the phenomenon.

Like the English artists discussed above, Eckersberg was interested in studying nature, painting *en plein air*, and capturing ephemeral light and meteorological conditions at specific moments. To that end, he invented a device called the 'perspective octant' that offered mechanical assistance, like the camera obscura but simpler, to rationalize the dimensions of objects in nature that are otherwise fugitive, suffusing his works with a timeless mathematical precision that perfected nature.[48] He painted the Moon and rainbows in exquisite landscapes, and he notes comets and other astronomical phenomena in the cloudy skies over maritime Denmark in his journal, frustrated about cloudy days when there were eclipses. His first recording of observing a solar eclipse dates from 15 May 1836, when there was an annular eclipse with an 89.4% coverage in Copenhagen. 'The solar eclipse, which began after 3 o'clock and lasted until 5 o'clock. So wonderful when the weather was favourable, although the air was not quite clear when it was at its highest, and a strangely dark day prevailed.'[49] Adjacent to his written description in the manuscript, he sketched a thumbnail diagram of the eclipse with an off-centre annulus.[50] This was the very year he painted Ursin's portrait.

Many of Eckersberg's works with an emphasis on perspective and meteorological and astronomical phenomena contain enigmatic narrative elements, such as a number of anecdotal scenes, some based on the plates in his 1841 treatise on perspective. They feature hallucinatory light effects, and their mysterious nature is heightened by figures depicted in transitory states of motion and psychological tension. The most inscrutable of the group is the *View Through a Doorway* (Figure 10.12)

It is related to the vignette at the lower left of plate VI (figure III) of *Linear Perspective as Applied to the Art of Painting* (Figure 10.13).[51] The woman in the etching, who has her back to the viewer and wears a shawl, hurries in the dark entry hall towards the

---

[46] Ursin read and commented on the manuscript of Eckersberg's first—*Forsøg til en Veiledning i Anvendelsen af Perspektivlæren for unge Malere* (*Attempt at a Guide for the Application of the Theory of Perspective for Young Painters*) of 1833—and wrote from the artist's notes the text for the second with engravings by Eckersberg—*Linearperspectiven, anvendt paa Malerkunsten, en Række af perspektiviske Studier med tilhørende Forklarginer* (*Linear Perspective as Applied to the Art of Painting: A Collection of Studies in Perspective with Accompanying Explanations*) of 1841. For the illustrations in the latter, see Roberta J. M. Olson, 'Balancing the Real and the Ideal: The Preparatory Studies for Eckersberg's Treatise on Perspective,' *Master Drawings* 61(2) (2023), pp. 147–80.

[47] See Philip Weilbach, *Maleren Eckersbergs Levned og Værker*, Copenhagen (1872), p. 245; Emil Hannover, *Maleren C. W. Eckersberg: En Studie i dansk Kunsthistorie*, Copenhagen (1898), pp. 207, 392, fig. 99; Konrad Monrad et al., *C. W. Eckersberg 1783–1853: Artiste Danois à Paris, Rome & Copenhague*, Paris (2016), pp. 248–9.

[48] Patricia G. Berman, *In Another Light: Danish Painting in the Nineteenth Century*, New York (2017), 109–10, fig. 73.

[49] Eckersberg, *Dagbøger*, vol. 1, 689. The manuscript is *C. W. Eckersberg's accounts and diaries notes 1810–1853*, Copenhagen, The Royal Danish Library, Additamenta 1138 quarto, vol. 9, 60v.

[50] Olson and Pasachoff, 'Shedding Light on Danish Astronomy,' fig. 5.

[51] A preparatory drawing for pl. VI, fig. III is in the Statens Museum for Kunst, Copenhagen, inv. KKS2020-7.

Figure 10.11 Christoffer Wilhelm Eckersberg, *Professor Georg Frederik Ursin Explaining a Solar Eclipse*, 1836, oil on canvas, 39 x 29.5 cm, private collection.
Source: Wikimedia Commons.

door and the raking light in the street, presumably drawn by the disturbance outside. By contrast, in the painting Eckersberg represented a woman standing just outside the doorway, bathed in light and looking down the street in the direction towards which people are running. Shading her eyes with her right hand, she holds a corner

Figure 10.12 Christoffer Wilhelm Eckersberg, *View Through a Doorway*, 1845, oil on canvas, 31 x 27 cm, Statens Museum for Kunst, Copenhagen, inv. KMS8847.
Source: SMK Photo / Jakob Skou-Hansen.

of her apron in her left. Both versions of the scene are equally ambiguous and have a disturbing sense of the dramatic action taking place beyond the picture frame. Neither explains the mystery as to why the figures are running towards an unknown location in a setting illuminated by a strange light source. Whereas the painting only has a simple rectangular doorway, the plate vignette features a lunette (demilune window) above the doorway whose metal frame around the segments of glass creates a pattern resembling a setting Sun and its rays. Eckersberg may have adopted this design as a clue to the nature of the event. One of the themes behind Eckersberg's enigmatic works is that people experience the surrounding world only in glimpses and can never

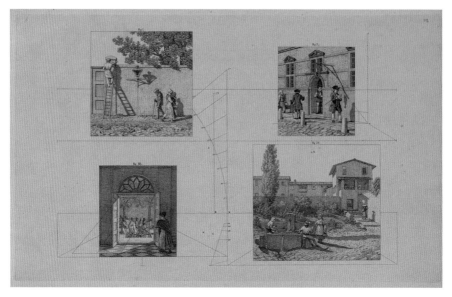

Figure 10.13 Christoffer Wilhelm Eckersberg and Georg Frederik Ursin, *Linearperspectiven, anvendt paa Malerkunsten, en Række af perspectiviske Studier med tilhørende Forklarginer* (Copenhagen, 1841), plate VI, Metropolitan Museum of Art, New York, The Elisha Whittelsey Fund, inv. 2008.519.
Source: Public Domain.

fully apprehend it. The Danish artist's works, therefore, incorporate the elusiveness and fragmentary, unsettling nature of the modern age, a state resulting from a general collapse of political, social, and ideological values on both national and international levels around the turn to the nineteenth century.[52]

Taking into account Eckersberg's interest in solar eclipses, together with the evidence in *View Through a Doorway* of 1845 and the illustration in his treatise, the event that mesmerizes and motivates the figures is a solar eclipse.[53] All the dramatis personae look up at the sky and run towards a place to better view the hypnotic solar display. The woman in the painting wisely, but inadequately, shields her eyes against the harmful rays to view the eclipse, an action which should never have been attempted without protective, blackened glass. Eckersberg also captured the unique light, diminishing with the sharpening of shadows and a slight colour shift during the final partial phases of a total solar eclipse. Rather than focusing on the astronomical bodies of the eclipse, Eckersberg instead represented the characteristic solar light, shadows, and excitement that accompany this celestial spectacle.

The identification finds support in the artist's drawing of the 28 July 1851 solar eclipse (Figure 10.14), which he also documented in his journal.[54] In *View of the Solar*

---

[52] Philip Conisbee et al., *Christoffer Wilhelm Eckersberg, 1783–1853*, Washington, DC (2003), p. 150.

[53] Eckersberg's artistic devices help to create a sense of mystery and incomprehensibility that increase the strangeness and anxiety produced by the solar eclipse itself, as discussed in other chapters of this book.

[54] Eckersberg, *Dagbøger*, vol. 2, p. 1215. For the drawing, see Monrad et al., *C. W. Eckersberg*, p. 318, fig. 119; *The Danish Golden Age: Ten Drawings from Private Collections*, Hamburg (2019), no. 8.

Figure 10.14 Christoffer Wilhelm Eckersberg, *View of the Solar Eclipse of 28th July 1851*, 1851, black ink and gray and brown wash, over graphite on paper, 250 x 160 mm, whereabouts unknown.

Source: Le Claire Kunst, Hamburg.

Eclipse of 28th July 1851, a group of people have gathered to view the total solar eclipse that could be seen in Copenhagen in the afternoon (Figure 10.15).

This event was the first solar eclipse whose corona was captured in a correctly exposed coronal photograph (Figure 10.16), signalling the beginning of a new era and the late arrival of the depiction of the solar corona, as noted by Jay M. Pasachoff in Chapter 2. Credit for the achievement goes to Johann Julius Friedrich Berkowski, a skilled daguerreotypist in the Prussian city of Königsberg (now Kaliningrad, Russia).[55]

To safely view the partial phases of a solar eclipse—before and after totality— various methods were used, including a blackened (often smoked) glass, as Eckersberg represented in his drawing. The artist also recorded it in his journal: 'Around 3 p.m. the solar eclipse started, which was almost total at 4 p.m., and terminated around 5 p.m.'[56] Although his words are dispassionate and scientific, Eckersberg's drawing, like his other works involving eclipses, captures the emotions of the witnesses to this celestial marvel.

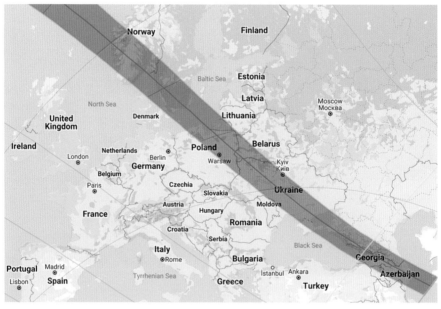

Figure 10.15 Path of totality of the Solar Eclipse of 28 July 1851.

Source: Map by Xavier M. Jubier; map data © 2001 Google.

---

[55] See Reinhard Schielicke and Axel. D. Wittman, 'On the Berkowski daguerreotype (Königsberg, 1851 July 28): The first correctly-exposed photograph of the solar corona,' *Acta Historica Astronomiae* 25 (2005), pp. 128–47. See also Olson and Pasachoff, *Cosmos*, p. 266.

[56] Eckersberg, *Dagbøger*, vol. 2, p. 1215.

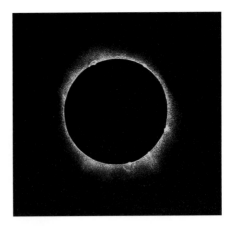

Figure 10.16 Johann Julius Friedrich Berkowski, solar eclipse, Königsberg, 28 July 1851, daguerreotype.
Source: Courtesy of Alex D. Whittmann.

Mid-Century Journalism and the Rise of Photography

Whereas the paintings and drawings discussed above belonged to Romanticism's interest in natural phenomena and symbolism, as well as artists' personal, singular reactions to them, the later nineteenth century produced less complicated images of solar eclipses, which were widely disseminated and understood by many individuals. Rapid technological developments, among them the industrialization of printing and the expansion of ever more sophisticated printed media—in books and prints, then magazines and newspapers—engendered a print culture that went way beyond the newsworthy single-leaf broadsides, the tabloids of earlier centuries, not only in Great Britain, but also in European and North American countries, and, in fact, worldwide. This 'democratic' development went hand in hand with the genesis of journalism and photography, art forms which, like prints, could be produced in multiples.

One indication that scientific knowledge about solar eclipses was filtering down to a wider culture, which was no longer fearful but rather curious about them, is their appearance in humorous contexts in widely circulated printed material. A case in point is *An Eclipse Lately Discovered in the Georgium Sidus* (Figure 10.17), which belongs to the tradition of the eighteenth-century British satirical prints that were displayed in coffeehouses where people gathered to read the news before the wide circulation of papers and magazines. It depicts the frontal head of General Wellington eclipsing the profile of King George IV, a caricature that visualizes Wellington's rise in popularity, which casts a shadow over England but not Ireland, because he favoured Catholic causes. The reference to *Georgium Sidus* is a witty take on English astronomer William Herschel's discovery of a new planet in 1781, which was later renamed Uranus. The French illustrator Jean Ignace Isidore Gérard, who was known by his pseudonym J. J. Grandville, included *A Conjugal Eclipse* in his imaginative satirical masterpiece *Un Autre Monde* (Figure 10.18). It sounds a lighter note, portraying a solar eclipse as a celestial love affair, a kiss between the anthropomorphic

Figure 10.17   John Philips ('Sharpshooter'), *An Eclipse Lately Discovered in the Georgium Sidus, and Quite Unexpected by any of the Astronomers*, 1829, hand-colored engraving, 328 x 232 mm, The British Museum, London, inv. 1935,0552.3.126.

Source: © the Trustees of the British Museum, London.

Figure 10.18 J. J. Grandville (Jean Ignace Isidore Gérard), *Une Éclipse Conjugale*, from *Un Autre Monde*, 1844, hand-colored wood engraving, Bibliothèque nationale de France, Paris.
Source: gallica.bnf.fr / BnF.

Sun and Moon floating above the Earth in classicizing costumes. Below a gaggle of anthropomorphized astronomical instruments cast shadows outside the penumbra and track the voluptuous cosmic drama like fanatical reporters and eclipse-chasers (see also Lange, Chapter 11, Figures 11.11, 11.17, and 11.22).

Also, in a humorous mode, Mark Twain—who was born with Halley's Comet and fulfilled his wish to die with its 1910 apparition—inserted a large solar eclipse illustration on the opening page of the sixth chapter of *A Connecticut Yankee in King Arthur's Court* (1889).

At mid-century there was an explosion of journalism, accompanied by the invention of new reproductive techniques, to illustrate topical events for the masses. It continued the media revolution that had begun in the eighteenth century with cheaper paper and disposable print, changing forever the public's relationship with the dissemination of knowledge that today has multiplied exponentially with the internet. Among the flood of books and newspapers, some of which featured solar eclipses, were *The London Illustrated News, Harper's Weekly* (see Lange Figure 11.13a), and Frank Leslie's various publications, one of which, *Frank Leslie's Chimney Corner*, featured Winslow Homer's wood engraving 'Looking at the Eclipse' on the cover of its 16 December 1865 issue. Naturally, eclipse expeditions, like those of today, became the social events of the season and were also tracked and illustrated in the press.[57]

Following the first daguerreotype of a solar eclipse in 1851 (Figure 10.16), William and Frederick Langenheim, American daguerreotypists and former journalists, attempted to capture sequentially the total solar eclipse of 26 May 1854 visible in North America in seven daguerreotypes, none of which show totality. (Daguerreotype was the first publicly available photographic process, employing an iodine-sensitized silvered plate and mercury vapour, and named after the painter who invented it, Jacques-Louis-Mandé Daguerre.) Although other early photographers documented the eclipse, only these survive. In the northern hemisphere, the Moon always shadows the Sun from right to left during a solar eclipse, but these images, like all uncorrected daguerreotypes, are reversed, as in a mirror.[58] Warren De la Rue, a pioneer in the application of photography to astronomy, took his photoheliograph to Spain to photograph the total solar eclipse of 18 July 1860.[59] He captured totality and proved the solar nature of the prominences or red flames around the limb of the Moon during a solar eclipse. Other photographers followed suit, among them Henry Draper, who photographed the same total eclipse of the Sun on 29 July 1878 that professor of astronomy Maria Mitchell brought a team of graduates—'Vassar girls' as the press called them—2,000 miles to study near Denver. Whereas previous scientific expeditions had mostly been male, during the late nineteenth century women began participating in eclipse expeditions.[60]

Aiming for even loftier adventures and as an aid to the developing technology of photographic techniques and equipment, some astronomers employed hot-air balloons to ascend to new observational heights, among them French astronomer Jules Janssen (Pierre Jules César), who had studied a solar eclipse in India in 1868. During the German occupation of Paris in 1870, he risked his life for patriotism by ascending in a balloon over France in protest, thus missing the eclipse he wanted to observe in

---

[57] Alex Soojung-Kim Pang, 'The Social Event of the Season: Solar Eclipse Expeditions and Victorian Culture,' *Isis* 84(2) (1993), pp. 252–77.

[58] See: https://www.metmuseum.org/art/collection/search/283180/.

[59] David Le Conte, 'Two Guernseymen and Two Eclipses,' *Antiquarian Astronomer* 4 (January 2008), pp. 55–68 (De La Rue and Paul Jacob Naftal, whose watercolour of the 22 December 1870 eclipse was engraved); Le Conte, 'Warren De La Rue—Pioneer astronomical photographer,' *Antiquarian Astronomer* 5 (February 2011), pp. 14–35.

[60] Alex Soojung-Kim Pang, 'Gender, Culture, and Astrophysical Fieldwork: Elizabeth Campbell and the Lick Observatory—Crocker Eclipse Expeditions,' *Osiris* 11 (1996), pp. 17–43.

Algeria. With the English scientist Norman Lockyer, Janssen is credited with discovering the gaseous nature of the solar chromosphere and the existence of helium, as discussed by Pasachoff in Chapter 2. Likewise, Russian chemist and inventor Dimitry Mendeleev, who devised the periodic table of elements, attempted to observe a solar eclipse from a balloon on 7 August 1887, a failed attempt that Russian artist Ilya Repin depicted in a watercolour.[61] In 1881, French scientist and artist Étienne Léopold Trouvelot published his chromolithographic series of 15 astronomical phenomena, including a total solar eclipse (see Lange, Figure 11.12a) and its scintillating protuberances, which he had drawn in pastel in the 1870s just as astronomical photography was beginning to take hold. By the total solar eclipse of 1882, photography of astronomical phenomena was firmly established, and one photograph even captured a rare eclipse comet—a bright comet that passes near or in front of the solar disc during a total eclipse (Figure 10.19).

Because photographic technology was still in its infancy, international eclipse expeditions continued to employ artists to record solar eclipses as aids for their later studies of the event and the solar surface changes. For example, in 1870 John Brett, the amateur astronomer and latter-day Pre-Raphaelite artist who lead a double life in art and science, was hired as a draughtsman on an expedition to Augusta, Sicily, for the solar eclipse of 22 December.[62] Similarly, Paul F. Naftal, known as a landscapist, accompanied another eclipse expedition the same year to Cadiz, Spain, to especially study the corona. Naftal was assigned to concentrate on atmospheric colour and terrestrial illumination and produced watercolours to that effect, lit by the Sun and its corona, in addition to a study of the isolated Sun and its dazzling halo.[63] Chief among these eclipse expedition draughtsmen was the American Howard Russell Butler, a physics-trained artist who the US Naval Observatory hired for its eclipse expedition in 1918 because he was noted for his unique method of making lightning-fast sketches that produced more precise results than photography.[64] Among Butler's finest solar eclipse portrayals is a trio of paintings depicting the eclipses of 8 June 1918, 10 September 1923, and 24 January 1925 (Figure 10.20).

They are based on his own observations, but, to underline the mystical–religious parallel of the awesome celestial alignments, together they suggest the form of a tripartite Renaissance Christian altarpiece known as a triptych. The trio was long displayed in the former art-deco building of the Hayden Planetarium of the American Museum of Natural History. Butler also executed stunning paintings of reddish solar prominences and of the 1932 eclipse.

Nevertheless, photography ensured that cameras would eclipse artists' talents and tools. Eventually, astronomy looked to the increasing accuracy of the camera, and artists were freed to explore solar eclipses for their own purposes: from

[61] See: https://www.sothebys.com/en/buy/auction/2021/russian-pictures-2/the-solar-eclipse-of-1887-mendeleev-in-his-hot-air/.

[62] John Brett, 'Extrait of a Report of the Eclipse Expedition of 1870, at Augusta, Sicily,' *Monthly Notices of the Royal Astronomical Society* 31(5) (1871), pp. 163–6. For his astronomy, see Christiana Payne, *John Brett: Pre-Raphaelite Landscape Painter*, New Haven and London (2010), pp. 11, 110–13, 179–80.

[63] Le Conte, 'Two Guernseymen,' pp. 63–6, figs. 11–13.

[64] Jay M. Pasachoff and Roberta J. M. Olson, 'The Solar Eclipse Mural Series by Howard Russell Butler,' in *Inspiration of Astronomical Phenomena VIII: City of Stars*, edited by Brian P. Abbott, San Francisco (2015), pp. 13–20.

Figure 10.19 W. H. Wesley, Corona of eclipse with Twefik's Comet on 17 May 1882. Colourized image from Arthur Schuster's photograph.

Source: Philosophical Transactions of the Royal Society of London, 175(1), 1884, pl. 13, colourized by Jay M. Pasachoff.

realistic representations to powerful symbolic images with wildly inventive multi-media results. They range from eclipse aficionado Diego Rivera's *Portrait of Ramón Gómez de la Serna* (1915), which features a total solar eclipse for the pupil in one of the poet's eyes, to George Grosz's *Eclipse of the Sun* (1926), where the total eclipse

Figure 10.20 Howard Russell Butler, *Triptych: Solar Eclipses of 1918, 1923, 1925*, 1926, oil on canvas, 172.7 x 242.6; 243.5 x 170; 167.7 x 234 cm, respectively, American Museum of Natural History, New York.

Source: American Museum of Natural History.

encompasses a dollar sign worshipped by warmongering arms dealers, to Wassily Kandinsky's abstract depictions and Roy Lichtenstein's Pop Art images, and beyond. Katie Patterson's *Totality* installation of 2016 gathered images of total solar eclipses from old drawings and photographs and mounted 10,000 of them on a mirrored ball 83 centimetres in diameter; it spun like a disco ball projecting tiny lights on the walls, to only suggest the electrifying and bedazzling experience of an actual total solar eclipse with a kind of elusive synaesthesia. One of the universe's grandest naked-eye spectacles, a total solar eclipse cannot be captured in its entirety or impact by any camera or video, but can only be experienced fully in person under clear skies in nature.[65]

[65] See Olson and Pasachoff, *Cosmos*, pp. 70–85, for a selection of examples.

11

Total Eclipse of the Art

Vision, Occlusion, Representation

Henrike Christiane Lange

History of Art and Architecture, History of Literature

Posing unique challenges in the history of representation, total solar eclipses involve all of the following: phenomena of movement, light, shadow, obfuscation, momentousness, and gradual dimming down and up as a process with a climactic moment at its centre—all happening on a massive scale and across inconceivable distances, requiring the combination of cosmic imagination with scientific understanding. Complicating things further, the central moment merges the visual cosmic spectacle above with its own dramatic effects on sky and land below. Finally, while the event above is predictable, the weather conditions in between are not—and the tension between hyper-predictable and entirely uncertain conditions of visibility add to the interest in the event's historical-visual documentation as much as to the preciousness and awe of the actual experience. Combining perspectives from art and literary history, the following chapter focuses on total solar eclipse perception and representation (one of the common denominators across the visual and textual arts)—on word, image, media specificity, and a range of historical developments in which these representational challenges connect to central interdisciplinary questions of time, perception, and experience.

Alberti, Leon Battista (1404–72) Italian early modern architect, painter, theoretician, and author of the concept of '*storia*' in a pictorial representation.

aniconic A non-iconic representation that avoids the creation of a direct or indexical image or figure.

Arcadia Mountainous region of the central Peloponnese, Greece; ideal setting of Vergil's *Eclogues*; in European Renaissance art, idyllic place; region or scene of simple pleasure and quiet (traditionally associated with a vision of pastoralism and harmony with nature); 1993 science play by Tom Stoppard.

Baxandall, Michael (1933–2008) British art historian active in London and Berkeley, California.

bird's-eye view A perspective offering a view from above, often used in cartographic representation.

cast shadows Sharply delineated dark shadows falling from an object under conditions of a direct and pointed light source.

chiaroscuro Light and dark values in a work of art, usually of three-dimensional objects as they appear under a certain angle and quality of light in a two-dimensional painterly or graphic representation.

Crashaw, Richard (*c.*1612–49) Early modern metaphysical poet and Anglican cleric.

diaphanous Light, delicate, and translucent; a term often used in the context of the analysis of light effects through and around Gothic cathedral walls and windows.

effigy A statue or likeness; often used as a term of materiality in contrast with the immaterial image on the one hand, and the actual body on the other.

ekleipsis (Greek) omission, disappearance, or abandonment.

genre Categorization of different kinds or occasions of image-making such as still life, landscape, group portrait, etc.

Gombrich, Ernst H. (1909–2001) Austrian-born British art historian in the Warburg Circle.

historiography The history and record of a historical discipline's own methods and approaches.

mot juste / moment juste The exact, appropriate word / the exact, appropriate moment.

mysticism Appreciation and practice of contemplation and self-surrender towards a union with the divine or the absolute, or the spiritual apprehension of knowledge inaccessible to the intellect.

Nash, Paul (1889–1946) Surrealist painter of the *Eclipse of the Sunflower*, 1945 (see cover image, Figs. 11.21a and 11.21b, and pp. 246–49).

numinous Quality of a strong religious or spiritual nature; indicating or suggesting the presence of the divine or the absolute.

occlusion Obstruction or blockage. Here, a representational tactic of grouping, organizing, hiding figures and/or objects in a representation of three-dimensional phenomena on a two-dimensional surface.

perspective In early modern Western art: the technique of representing depth on the basis of geometrical perspective between a central point and converging parallel lines in relation to an ideal point of view.

phenomenology Branch of philosophy focusing on the science of phenomena as distinct from that of the nature of being.

Pliny (Gaius Plinius Secundus, called Pliny the Elder) (23/24–79) Roman author, naturalist, natural philosopher, and author of the *Naturalis Historia* (*Natural History*).

political iconography Method and field of study in art history, specifically in the social history of art, founded in Germany by Martin Warnke (1937–2019) in response to the involvement of art historians under the Nazi regime (see Martin Warnke, *Das Kunstwerk zwischen Wissenschaft und Weltanschauung*, Gütersloh 1970; and the index for political iconography in the Warburg-Haus in Hamburg, Germany).

Romanticism Term for different kinds of artistic and literary counterforces within the European Enlightenment, often in contrast to the idea of the rational age.

sepia Brown-tint aesthetic, typical for early photography.

still life Genre of painting mostly inanimate objects.

trompe-l'oeil Eye-deceiving style of representation, often in hyperrealistic painting, at times seemingly crossing the artwork's semiotic border and entering the
beholder's space.

visualization Making something visible; realizing an expression of something
(mostly) invisible in a (mostly) visual medium.

Thomasina: You cannot stir things apart.
Septimus: No more you can, time must needs run backward, and since it
will not, we must stir our way onward mixing as we go, disorder out of
disorder into disorder until pink is complete, unchanging and unchangeable, and we are done with it for ever. This is known as free will or
self-determination.

Tom Stoppard, *Arcadia*

Introduction: The Mystery of the Ordinary

Valentine: The ordinary-sized stuff which is our lives, the things people
write poetry about—clouds—daffodils—waterfalls—what happens in a
cup of coffee when the cream goes in—these things are full of mystery, as
mysterious to us as the heavens were to the Greeks.

Tom Stoppard, *Arcadia*

It is not night nor the usual dusk that helps to mask the break between night and day;
it is a sudden and untethered twilight that undermines all expectation of experience,
startling to animals as well as to those who have learned to expect the sudden breaking
of night into day—that day, that place, that time. To be standing under a shadow cast
by a dramatically large and distant object, rather than observing a shadow thrown by
some object in the viewer's immediate environment, remains unrepresentable even
after the experience is fulfilled and therefore maintains its numinous power. As cosmic clockwork, the event possesses something ordinary, yet to those experiencing
the brief removal of the Sun's light, warmth, and energy in the middle of the day and
the intensity with which the corona shines around the darkness it always seems to be
mysterious and mystical.

Formally, if not figuratively, the challenge should be simple enough for artist
or amateur—a dark, circular centre surrounded by the glow of a corona, maybe a
'diamond ring' appearing from its all-encompassing edge, and the uneven glow of
coronal streamers. But already this minimal description invites an infinite number
of potential eclipses and variations of their appearance, especially since each single
one is not one single visual fact but a fluid phenomenon with too many facets in
too quick a succession to clearly express it fully as a single image, a picture, or any
conventional medial object, and any single image attempt of capture or visualization can look vastly different depending on what specific aspect should be shown
(Figures 11.1 and 11.2).

A composite image or momentary picture focusing on the circular appearance in
the sky has therefore come to be a preferred method, as well as the cartographic

Figure 11.1 Wadesboro Observatory Solar eclipse of 28 May 1900, photograph for the Smithsonian Solar Eclipse Expedition taken during totality by Thomas Smillie.

Source: UC Berkeley Historical Slide Library (est. 1938–2018), Doe Memorial Library, Berkeley, California.

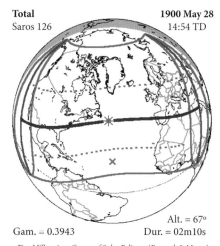

Figure 11.2 NASA, diagrammatic representation of the track of totality of Wadesboro Observatory: Solar eclipse of May 28, 1900.

indication of the mapping of its shadow, or the artistic embellishment of the core form in some other medium, usually engaging either the circular form (e.g. in drawings or relief sculpture), the darkness–light dynamics (e.g. in paintings or photographs), or the disruptive nature of the perception of the event (e.g. in soundscapes, music, film, theatre). In any such case, the specificity of a single eclipse representation seems to be at the centre of interest, forcing the scientist or artist to make decisions as creative and as challengeable as those made by Handel or Pink Floyd (see Chapter 12's consideration of music and sound). Generally, approaches from art history into the sciences have been most fruitful in exploring the interplay between art and science not as illustrations, or matters of influence into the one or the other direction, but rather as 'structural intuitions' shared by artists and scientists when confronting the world, and manifestations of a deeper structure that connects both spheres—as is evident in poignant examples such as those modelled by Martin Kemp in his popular *Nature* journal essays and bound together in *Visualizations: The Nature Book of Art and Science*.[1] But these strategies seem to be more elusive when it comes to total solar eclipses: The event of a total solar eclipse escapes capture by representation since it combines a host of complex visual challenges.[2] Each one of these challenges entrains its own long historiographies in art history and theory, each experimentally approached in practice as well as explored in theory in multiple ways. All combined in the experience of the eclipse, we have to consider at once the specific art histories of quite a range of phenomena: light, darkness, shadow, motion, science/technology versus naked-eye astronomy, scientific expert and citizen-scientist amateur viewership, to name only the most obvious. But then one has also to consider wider conditions such as the extraordinarily wide reach of the experience, that is, how a cosmic event also plays out on Earth, how it motivates and validates animal and human behaviour and knowledge (see Chapter 13, 'Animal Behaviour and Eclipse' and Chapter 14, 'Weather and the Solar Eclipse: Nature's Meteorological Experiment'). Finally, the ultimate challenge of a full visual representation of the event stands in yet another paradoxical, triple demand: somehow, one would have to show the view up to the Sun from Earth as well as the reversed gaze to Earth and the circumspection to other viewers together all at once—since the interplay of the experience between individual and collective is a core part of eclipse narratives (see Chapter 4 on eclipse-chasing).

This chapter will explore what it is that makes the artistic representation and, more specifically, the visualization of total solar eclipses categorically unique and what kinds of decisions have helped producers and viewers of images, maps, and artwork to orient themselves in anticipation or in memory of the happening. Four major directions of the gaze structure the various approaches to representational strategies—looking up, looking down, looking around, and looking inwards into the cosmos of symbolic light and shadow. Depending on the direction of the gaze and the medium of representation, total solar eclipses are shown by lumping together, dividing, merging, condensing, sequencing, and representing the event by representing only its effect (e.g. the all-around horizon darkness) or only the dynamics of its experience (e.g. audiences, instruments of observation).

Whitney Davis opens his study *Visuality and Virtuality: Images and Pictures from Prehistory to Perspective* with the question of whether the eclipse's appearance

---

[1] See the innovative grouping of art–science matters under sections such as 'Microcosms,' 'Spatial visions,' 'Nature on the move,' 'Graphic precision,' 'Space and time,' or 'Process and pattern,' in Martin Kemp, *Visualizations: The Nature Book of Art and Science*. Berkeley and Los Angeles, CA, 2000.

[2] See the survey of eclipse imagery in Ian Blatchford: "Symbolism and Discovery: Eclipses in Art." *Philosophical Transactions of the Royal Society A: Mathematical, Physical and Engineering Sciences* 374 (2016), pp. 1–26.

between fluid phenomena in a cosmic visual space can itself be a representation, an image, a picture, challenging those very definitions.³ A fitting starting point for the millennia of artwork and imagery covered in his book, Davis shows the image of the solar corona from the anonymous photograph of the total eclipse of 28 May 1900 (Figure 11.3a). He includes the plate for analytical definitions on 'The Anthropology of Images and Imaging Pictures,' comparing and contrasting the black-and-white portrait of the corona with a colourful, hyper-realistic trompe-l'oeil by William Michael Harnett from 1888—*Still Life: Violin and Music* (Figure 11.3b, Metropolitan Museum of Art, New York) from which elements stand out pretending to be real things in real space.

Davis describes the definitional nuances in the corner of the field of art history known as visual-cultural studies with a powerful reference to total solar eclipses:

> In art history one hears, for example, not only that a picture is an image, that is, that it depicts—and perhaps that it is a map, a scale model, and so on. One also hears, perhaps, that a *painting* (suppose it is not pictorial) is an image, that an effigy (suppose it is aniconic) is an image, or that an architectural setting for paintings and effigies (suppose it contains no representational elements beyond whatever the paintings and effigies might represent) is an image. Here 'image' simply designates a phenomenon in visual space that has been produced specifically to be registered, at least in part, including visible representations. Indeed, in art history as it has expanded to include visual-cultural studies and *Bildwissenschaft* [the science of the image] one might hear that a solar eclipse (a physical event in real space that does not occur because it is a phenomenon) is an image, that is, that we have an image of it or a way of registering and maybe representing it (suppose that we watch it through special sunglasses, or photograph it through a telescope [. . .]. [In this case,] the term 'image' can be replaced with a more particular and precise label: [. . .] 'solar eclipse photographed through a telescope.'⁴

Davis illustrates this point with an anonymous nineteenth-century photograph of the total eclipse of 28 May 1900 from the observatory in Wadesboro, North Carolina.⁵ With that specific illustration, Davis concludes that the image and its phenomenology can be read in reverse or upside down, as the phenomenon in space and time becomes itself a representation, or picture:

> In this sense, it seems that sometimes the analytic work of the term 'image'—as distinct from the term stating an extension in a given instance—is best limited to cases in which we do not know, or cannot discern, what kind of phenomenon we are dealing with in visual space. Someone might not discern, for example, that a phenomenon in his visual space is a representation, even a picture.⁶

---

³ Whitney Davis, *Visuality and Virtuality: Images and Pictures from Prehistory to Perspective*. Princeton, NJ and Oxford, 2016. See also Davis, *A General Theory of Visual Culture*. Princeton, NJ, and Oxford, 2011.
⁴ Davis, *Visuality and Virtuality*, p. 17.
⁵ See fig. 0.8 in Davis, *Visuality and Virtuality* 'Anonymous, nineteenth century. Solar corona observed at Wadesboro Observatory, North Carolina; photograph of the total eclipse of May 28, 1900.'
⁶ Davis, *Visuality and Virtuality*, p. 17.

Figure 11.3a Solar corona observed at Wadesboro Observatory, North Carolina; photograph of the total eclipse of May 28, 1900.
Source: UC Berkeley Historical Slide Library.

Indeed, Sun and Moon have, as far as we know, no awareness of causing such great and disturbing experience as totality for spectators on Earth. This question regarding the status of any given image or representational reality is lifted onto an entirely different scale when considering the cosmos and its clockwork. Contrasting any more conventional representational challenge of a happening on Earth, cosmic phenomena are not only incomparable to simple objects on Earth by the much larger distances; different conditions of light, atmosphere, and visibility; obvious limitations

Figure 11.3b William Michael Harnett, *Still Life: Violin and Music* (1888).
Source: UC Berkeley Historical Slide Library.

or lack of any tactile connection; and often a much larger necessity of scientific unlocking or deciphering of what one is to find, and where, and when, with the naked eye or through the telescope. Cosmic phenomena then also regularly require much more than the already elaborate skills of a visual artist on Earth working on objects and optical appearances in their realm. The total solar eclipse is also in this context unique and a phenomenon of superlatives—with details such as the sheer dynamic range of brightnesses displayed from the searing intensity of the diamond ring, or, when that is gone, the full-Moon intensity of the inner corona to the faintest tendrils of the outer corona, matched with the techniques behind, for instance, Figure 2.10 that reduce contrast range so all can be seen—demanding the command of many technical skills from the artist and even from the camera or digital technology that aims at producing an image of an experience. In the attempt of producing representation, of creating images of a specific instance or general description of a total solar eclipse, we encounter basic representational parameters—light, shadow, movement, momentousness, perspective, the position or even existence of a spectator. All of these have already their own historiography—their own, separate histories within the history of art. Even traditional terms such as *chiaroscuro* or our neologism of the *moment juste* (expanded from the *mot juste* of the poets) must be reconsidered with the theme of total solar eclipses, but many of them are contradictory in terms to one another, or at least seemingly paradoxical. Of course, a creative, imaginative, and prismatic view of visible and invisible matter is needed to make sense of the cosmos, as Joel R. Primack and Nancy Ellen Abrams put it in *The View from the Center of the Universe: Discovering our extraordinary Place in the Cosmos*:

> Any picture of the expanding universe can only be symbolic, because the universe [cannot] directly be seen. No one can step outside of it to look at it, no one can see all times, and over 99 percent of its contents are invisible. Symbols are the way to grasp the universe. They free us from the limitations of our five senses, which evolved to work in our earthly environment. [. . .] No single symbol can ever represent the universe completely. To get a sense of the whole, we have to somehow absorb the meaning of all the symbols together, and this takes imagination.[7]

As noted in Jay Pasachoff's discussion of the corona in Chapter 2, for the longest time the tradition of eclipse recordings did not include its most impressive visual qualities. A tradition that has now multiplied in uncounted media and visual strategies would begin with forms as aniconic as neo-Assyrian clay tablets such as the one with a letter to the king from Nabu-ahi-irba, in this case relating a lunar eclipse (Figure 11.4).

But eventually, the history of science came to merge and compete with those of technology and art, all integrated in the history of photography. As Roberta J. M. Olson has shown in the previous chapter, once photography overlaps with the study of total solar eclipses, we find the beginnings of greater variety in showing the happening—in either different moments, captured on different carrier film, produced in more-or-less sepia-quality hues or different qualities of coarse or smoother paper, and finally, with the complexities of different framing, mounting in a passepartout, and the potential mirroring of the indexical photographic negative image in the process of development (Figures 11.5a and 11.5b). While there is one exact way for a slide to sit in an art historian's projector (*not* upside-down, and *not* inverted), astronomers do not tend to worry about display orientation of an image such as that shown in Figures 11.5a and 11.5b as there are no cues to disorient casual viewers. Most (though not all) telescopes invert the image anyway.

Chasing optical knowledge and understanding of how an eclipse can manifest or be communicated in visual terms, we are after its paradoxical appearance and meaning.

---

[7] Joel R. Primack and Nancy Ellen Abrams, *The View from the Center of the Universe: Discovering our extraordinary Place in the Cosmos.* New York, 2006, p. 9.

Figure 11.4 Clay tablet recording a lunar eclipse in a letter to the king from Nabu-ahi-irba, 24 lines of inscription, neo-Assyrian.
Source: British Library.

Those, as we will find, often appear intertwined in representation across media, tracing how light and shadow, space, time, and place, image and imagination, experience and interpretation, meaning, knowledge, engagement, and ecstasy all contribute to the intrapictorial and interdisciplinary possibilities of total solar eclipses in the realm of the visual. Additionally, the exchange of collective and individual experience and viewership will be considered in the side-glance to the audiences of eclipses, and finally those are also reflected in the ever-developing relationship between observed, manually produced, technically reproduced, and self-reproducing imagery since the

Figure 11.5a Version 1 of the 1870 total solar eclipse in Sicily ('Die Corona bei der Sonnenfinsternis 1870, von Sizilien aus beobachtet').
Source: UC Berkeley Historical Slide Library.

age of photography as they directly or indirectly imply the audiences as observing and image-producing communities. Any attempt to represent the phenomenon of a total solar eclipse requires some strategies for a visualization of something

Figure 11.5b Version 2 of the 1870 total solar eclipse in Sicily ('Die Corona der Sonnenfinsternis vom 2. Dezember 1870').
Source: UC Berkeley Historical Slide Library.

inherently paradoxical, quasi-invisible, fleeting, and communal in strangely personal–collective–personal ways, and with the addition of human-made meaning in different contexts.

The lines of inquiry in the chapters in this section of the book therefore follow a series of basic questions: What are the main strategies for showing a phenomenon that is fundamentally about the eclipse of any regular conditions of sight and visibility? Are there comparable groups of strategies or tools in visual arts, music, and literature? What does an eclipse look like in the visual arts, sound like in music; what might it look or sound or feel like in a text? Which visual arts are best prepared, which ones most interested or available when it comes to this specific challenge? Which kinds of strategies can deliver the most efficient scientific renditions, and which express the best rendition of an experience, or create a work that delivers the experience? How is it that the same phenomenon can be shown by some as the most dreadful darkening of the sky (Figure 11.6a), by others as the most benign romantic idyll for the amusement-park consumption of curiosities (Fig. 11.6b), and, finally, even with the insertion of a full palette of colours left and right in the context of a public building (Fig. 11.6c)? With points of view, moods of observers, possibilities of the representative medium, and weather constantly changing, there seem to be infinite varieties of potential eclipse renditions—for instance, that of 1724 was probably cloudy across much of Britain, which would help explain the dramatic clouds in Figure 11.6a.

There is something to be said about the numinous, truly overwhelming intensity of the event (see umbraphile Mike Frost's notes in Chapter 4, and the fascinating

The appearance of the Total Solar eclipse from Haradon hill May 11. 1724.

Figure 11.6a Elisha Kirkall (1681/82–1742) and William Stukeley, 'The appearance of the Total Solar eclipse from Haradon Hill [Wiltshire, England,] May 11, 1724' (etching with mezzotint and wood-block printing).

Source: UC Berkeley Historical Slide Library.

Figure 11.6b Anonymous, *Die Sonnenfinsternis am 8. Juli 1842* ('*Guckkastenblatt*').
Source: UC Berkeley Historical Slide Library.

Figure 11.6c George Harding, *Solar Eclipse* (oil painting) in the second-floor rotunda ceiling of the US Custom House in Philadelphia, Pennsylvania (1938).
Source: UC Berkeley Historical Slide Library, Selz Collection.

reports published by Kate Russo[8]). Many eclipse-chasers might just stop speaking after uttering the awe and awesomeness of what they see, or simply break out in cheering and clapping, but I would argue that their relative speechlessness is grounded in the intense and overwhelming combination of a host of multi-sensorial phenomena plus the cultural investment of any kinds of interpretation, as a kind of collective subconscious baggage. These two seem to stop most from searching for more specific words, and it seems to be the clash and amalgamation of the specific, hyperintensified appearances that leave one entirely speechless.

This is why we depend on another interdisciplinary approach (indeed, a dramatic and science–poetic practice) for structuring this chapter's chase of totality: The guiding quotes throughout this chapter come from Tom Stoppard's famous and luminously contradictory play *Arcadia* (1993).[9] Quotes are kept in abstraction as to comment on the main points, but not entirely introduced so to not spoil the play for anyone who has yet to see it. The play is set in two temporalities: the present and the time of Lord Byron (c.1809–12), with actions running concurrently. *Arcadia* does not contain the visualization of an eclipse, but it is all about similar paradoxical themes—order and chaos, present and past, certainty and confusion, and indeed, the poster to its Yale Repertory Theater production in the 2014–15 Season strikingly evokes a quasi-ecliptic clash by combining the mirroring day and night skies in two amalgamated profile faces, male and female (Figure 11.7).

During years of interdisciplinary work on the perplexities of total solar eclipses, it became clear to us how much of *Arcadia*'s commentary on time, ages, historical styles, discovery, creativity, and connection across cosmic and personal events is expressing conditions of ecliptic convergences. The quotes are included as one alternative offering to this art–science challenge and a successful example of creative interdisciplinarity in the arts—in *Arcadia* combining science, costume, stage-setting, music, and drama. Tom Stoppard himself gave a helpful hint on the interdisciplinarity of the arts in an interview. In delightfully searching language, the wordsmith slowly works

---

[8] As Russo puts it, 'During totality, you are standing on a planet, observing another heavenly body cross paths with a third. The scale is unimaginable, yet here it is happening right on top of you and around you.' Kate Russo, *Total Addiction: The Life of an Eclipse Chaser*. Berlin, 2012, p.7. Russo also dedicates attention to the 'intense physiological responses' including the 'fight-or-flight' response of the sympathetic nervous system, activated by an initial surge of adrenaline. See Russo, *Total Addiction*, p. 58.

[9] See Tom Stoppard, *Arcadia*. London, 2009. As a series of dialogues between and across multiple groups, such as the dialogues between art and science, between history and present, between different communities (lay and professional), *Arcadia* gives notes on the classical and the Romantic style, and on Classicism and Romanticism as intellectual and emotional attitudes in an age of reason like the stage-setting rational Enlightenment part of the play, and the message on time—most notably, the collapsing of time (present and past) in the scene of the waltz and the transport of Thomasina's ideas, which are before her time as well as timeless. See also Harold Bloom (ed.), *Tom Stoppard*. New York, 2003, and Liliane Campos, *The Dialogue of Art and Science in Tom Stoppard's Arcadia*. Paris, 2011. For a select bibliography on *Arcadia*'s contexts and intertexts, see Campos, *The Dialogue*, pp. 179–80. See also Michael Roeschlein, 'Theatrical Iteration in Stoppard's "Arcadia": Fractal Mapping, Eternal Recurrence, "Perichoresis."' *Religion & Literature* 44, 2012, pp. 57–85; Enoch Brater, 'Playing for Time (and Playing with Time) in Tom Stoppard's "Arcadia."' *Comparative Drama* 39, 2005, pp. 157–68; William W. Demastes, 'Portrait of an Artist as Proto-Chaotician: Tom Stoppard Working His Way to "Arcadia."' *Narrative* 19, 2011, pp. 229–40.

Figure 11.7 Yale Repertory Theater, 2014–15 Season: Advertising for Tom Stoppard's *Arcadia*.

out how he wants to highlight the role of creativity and practice more than craft, or training, and their happy integration in artistic freedom:

> But more than that, I think the arts come from somewhere else. A meaningless phrase, I know. But. I don't think the arts are something you teach and learn. But, I'm missing the point I'm trying to make—which is that there is something about creativity that escapes the physical world, even escapes mathematical law. Artists aren't subject to that. It's hard to explain.[10]

Taking this important comment seriously, the challenge of total solar eclipse representation collapses into the mysterious core of creativity.[11] It is not necessarily the trained scientific expertise but the *practice* of *perception*, the *practice* of *experience* that determines how an eclipse can be appreciated by viewers. As scientist-communicators such as Sir Martin Rees, Sir Patrick Moore, Sir Brian May (the astrophysicist guitarist in the band Queen, originally with bandmates Roger Taylor, Freddie Mercury, and John Deacon), Jay M. Pasachoff, and Tom McLeish FRS have shown in many events for audiences of citizen scientists and students in universities and schools, children of all ages can put on their stargazers' hats and look out into the cosmos for the experience of knowledge and abandonment. In a historical and global perspective, we must distinguish our and our children's experience in well-lit contemporary cities from that of people in environments before or without such overwhelming and numbing artificial light, and consider also the not-so-long-ago worlds without such a wide distribution and machine-over-human power of portable screens and devices manipulating viewers with their various floods of images. But the main directions of the gaze that determine the major categories of representation remain the same: Looking down, up, around, and inwards.

Looking Down: Mapping Experience and Perception

> *Valentine*: It makes me so happy. To be at the beginning again, knowing almost nothing. . . . A door like this has cracked open five or six times since we got up on our hind legs. It's the best possible time of being alive, when almost everything you thought you knew is wrong.
>
> Tom Stoppard, *Arcadia*

The paradoxical phenomena embedded in the eclipse require special strategies for a transmission of the revelation in eclipse as night breaks into day, a sudden rupture of *vita contemplativa* into *vita activa* forcing everyone to be facing the cosmos in a string of contradictory and confusing experiences: of smallness and connectedness,

---

[10] Tom Stoppard as quoted in Frank Rizzo, 'Stage Notes: *Shakespeare in Love*'s Tom Stoppard To Give Free Yale Lecture,' *Hartford Courant*, 4 September 2014 at 2:18 p.m., see https://www.courant.com/ctnow/arts-theater/stage-notes/hc-rizzo-ticker-0907-20140907-story.html and archive of the author.

[11] See Tom McLeish, *The Poetry and Music of Science*. Illustrated edition, Oxford, 2018, extended edition with a new chapter on poetry, Oxford, 2022.

of the surrender to that which is so much larger, the personal individual need for meaning and interpretation, and finally the friction between the cosmic clockwork as such and the constant of cosmic calendars throughout time of human cultures and civilization looking for these events and ascribing meaning to them, and—almost as a reflex triggered by the union of *activa* and *contemplativa*—the intriguing coexistence of the governing universe with specific human rituals and pointed activities around it (such as the eclipse-chaser's life planning around travels, our writing of this book, and your reading now, in the precise moment in the future at which these words will have reached you in real time). Finally, the ambivalence of inherent predictability and unpredictabilities: some conditions can be calculated (see Chapter 1, 'The Cosmic Clockwork—the How and When of Total Solar Eclipses'), others will not be clear until the moment (see Chapters 2, 'The Unveiling of the Corona' and 4, 'From Science to Story: Testimony of an Eclipse-Chaser').

Then again, this seems to be the point of the unrepresentable moment of totality—to see how the entire atmosphere changes, time collapses, the most reliable rules of daily and nightly structure disrupted, all just to experience that the stars, corona, and prominences are always there—one can just not see them without the revelatory rupture of the eclipse. Much like in the day- and night-time experience of light in the diaphanous structure of Cathedral windows, while we denote total invisibility as looking at the stars when the Sun is shining—one just cannot see them—during a total eclipse, the stars reveal that they are in the sky both day and night, and that their visibility to us is determined by the Sun at day and the darkness of its absence at night. The cultural, spiritual, and religious–emotive investment in the stars, in Sun and in Moon, is likewise heightened and dramatized by the experience. Suddenly the spheres of the powerful, warm, and bright Sun of justice and clarity collapse into that of the yearningly pale Moon, the distant target of longing, dreaming, imagination, and fancifulness (as in Ludovico Ariosto's whimsical account of mad 'Astolfo sulla Luna' in the *Orlando furioso*, XXXIV, 70–87). Imagine the catchy, energetic rhythms of Bobby Hebb's 'Sunny', The Beatles' bright and tingly 'Here Comes the Sun', or Norah Jones's confidently matutinal chords of 'Sunrise' playing over the serene and calming 'Song to the Moon' of Rusalka, Friedrich Hollaender's 'O Mond', or 'Der Mond ist aufgegangen' by Matthias Claudius (see also our Conclusion to this book), the clash of moods inspiring human beings to search for its significance as well as for ways in which the happening can be recorded, replayed, and communicated to others.[12] 'Don't let the sun go down on me', as George Michael and Elton John would sing.

The total eclipse speaks of connection across wide spatial distances as well as great distances across time when it brings into the light of day—into the course of day—into the daylight side of life where there is distraction and hustle and bustle and things to do, the daily half that is *activa* for most of us meets the arcane *contemplativa* of the night, when the sky is wide and silent, when one tends to be in single or intimately shared solitude, and the gaze up inspires reflection. The busyness of the day and contemplativeness of the night suddenly collapse. In traditional Western terms, e.g. in those of Chapters 6 and 7, it brings the arresting twilight of the evening prayer

---

[12] Other examples for invigorating, sunny tunes include 'You Are the Sunshine of My Life', 'Aquarius / Let The Sunshine In', 'Walking on Sunshine'.

right into the middle of the day that stands under the activities following the morning prayer, the planting of the apple tree without doubt or hesitation—disrupting the kinds of certainties and revealing that the guiding stars of the night sky are always there, and could always guide. The eclipse questions the practical knowledge and certainties of the peasant and the scholar, the lawyer and the priest, with the astronomer predicting and explaining its course. Handing then the exegesis and interpretation of the event back to other scholars, to philosophers and to artists, the struggle becomes to imagine, visualize, illustrate, and finally represent it in all such dimensions. Of course, no single medium is able to do so sufficiently, but in each medium there is room for striving for its specific appearances in some shape or form. There is a secret charge, potential, and benefit in selecting any specific medium—drawing, painting, film, music, or another channel—and each of these then engages with the at times ominous, at times simply striking, breaking of the night's mysteries into the plain, mundane predictability of the bright sunlit day.

Mapping total solar eclipses as a projection of the predicted path of totality on the Earth seems to be straightforward enough for orientation, and indeed, the development of such maps goes hand in hand with the histories of cartography, geography, and printmaking. Their point of view is defined as a bird's eye view (as opposed to that of a frog); their point is the shadow. But during an eclipse, the experience is not to look at a shadow, but to be covered by a shadow—to be part of a phenomenon rather than witnessing it as a viewer, having it cast over oneself in the path of totality—a shadow larger, wider, falling from an object more distant than is imaginable, vast, mobile, and predictable. Shadow representation is a central part of the work of Ernst H. Gombrich on perception, since it combines all possible challenges to the field of perception *and* representation that paid attention to psychologies of perception and specific issues such as light in painting.[13] Michael Baxandall, in his 1995 book *Shadows and Enlightenment*, discusses 'shadows and their part in our visual experience'[14] with a focus on the juxtaposition of modern and eighteenth-century notions about shadows: First in philosophical terms as experiences described by John Locke FRS (1632–1704), then as epistemologies investigated by Bishop George Berkeley (1685–1753) and Gottfried Wilhelm Leibniz (1646–1716), and finally, in artistic practice, as the Rococo-empiricist shadow, forging a path of understanding from the physical constitution of shadows to their role in perception and cognitive science to the application of this understanding in works of art. Baxandall's study also covers part of the history of shadows in painting, including notes on shadows by Leonardo da Vinci in the Appendix of his book.

Shadow, Baxandall claims, originates in a 'local and relative deficiency of visible light,'[15] and is fundamental to the perception of the world: 'We do not see the world directly, but through the two-dimensional pattern of light that falls on the retina of the eye. This light is not a simple transparent witness of the world [. . .]. It is compromised

---

[13] See E. H. Gombrich, *Shadows: The Depiction of Cast Shadows in Western Art: A companion volume to an exhibition at The National Gallery*. New Haven, CT and London, 1995. See also Gombrich, *Art and Illusion*. London, 1960.

[14] Michael Baxandall, *Shadows and Enlightenment*. New Haven, CT and London, 1995, p. v.

[15] See Baxandall, *Shadows and Enlightenment*, p. 1.

by varied experience.'[16] Definable as the absence of light, shadows are themselves already elusive, indeed 'Holes in a Flux' as Baxandall titled his introduction.[17] More generally formulated, the themes that apply to eclipses are the precise consideration of light sources, illumination, and surface for their representation—but now on a cosmic scale. Baxandall speaks of the manageable appearance and manipulation of shadows in a drawing and an engraving of a figure study by the Italian Rococo painter Giovanni Battista Piazzetta (1682–1754)—all examples limited to a small-scale work of art. With the eclipse, a cosmic shadow falls over everything. The question of light, shadow, and representation as an issue of experience hence includes an array of other core issues familiar to historians and theoreticians of the visual arts: For representation in general, and in particular for visual representation, this drama poses some fundamental challenges, each of those at the heart of what depiction means in figurative and abstract art. An eclipse is not the unfolding of a dramatic moment between actors in one place, as a usual scenery and depiction by Renaissance architect and theorist Leon Battista Alberti's definition of a *storia* would imply or even demand, but a happening between places, fluid, in motion, with more than just visual consequences (including change in weather and temperature, and changing sounds from birds and crickets), in part predictable on a very long timeline yet with an entirely unpredictable momentum of weather and other circumstances ruling appearance and visibility, and with the age-old investments in the occurrence as something deeply meaningful. Additionally, there is a need for reconciliation of the elements of skill (e.g. the prediction of the event) and chance (e.g. the event as it happens).

The figurative tradition of the eclipse shadow of totality has been changing in eclipse representations from Halley's map of the eclipse of 1715 (Figure 11.8) to the much larger vision and scientific detail while also exploring more fantastical imaginaries of representation, such as to make the conical shadow something that can be observed from some kind of Divine, almighty perspective in the cosmos in an all-embracing vision somewhere between Sun and Moon, able to see both at once (Figure 11.9a) as in a frontispiece for Mabel Loomis Todd's 1900 *Total Eclipses of the Sun*, titled 'Moon's Shadow on the Earth.'[18] This, reconsidering earlier visualizations such as those from the Harley manuscript diagram in our Introduction (see Figure I.8), is a reversal of the conical shadow shown in projection from earlier visualizations, so that one can consider that the principle of 'looking down' is representative of eclipse prediction from the ancient Greeks onwards, where the principle of 'looking up' is one delivering only regularities, forming the basis for the early Babylonian and other ancient Near Eastern traditions. Similarly inspired, Mabel Loomis Todd had added a subtly shadowed and thereby spatially animated planar representation 'To Show How Eclipses Take Place' to her 1894 publication of *Total Eclipses of the Sun* (Fig. 11.9b).

Around the time of the publication of Mabel Loomis Todd's imaginative imagery, Simon Newcomb created various citizen-scientist-oriented publications, with illustrations showing the path of totality as a guide on the map, together with the

---

[16] Ibid., p. 8.
[17] Ibid., 'Introduction,' pp. 1–15.
[18] See Mabel Loomis Todd, *Total Eclipses of the Sun*. Boston, MA, 1900.

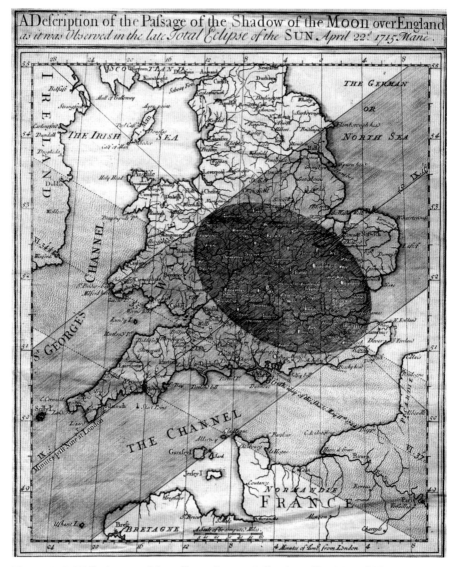

Figure 11.8 Halley's map of the eclipse of 1715. Collection of Jay M. and Naomi Pasachoff.

Source: Courtesy of Chapin Library, Williams College, MA.

timestamps (Fig. 11.10a).[19] Ever searching, in his 1902 *Astronomy for Everybody: A Popular Exposition of the Wonders of the Heavens*, and resorting to Milton's poetry, Newcomb could not decide if a circular or rectangular frame would work best, and published both options, side by side (Figure 11.10b).[20]

---

[19] See Simon Newcomb, *The Coming Total Eclipse of the Sun.* New York, 1900, p. 49.
[20] See also Simon Newcomb, *Astronomy for Everybody: A Popular Exposition of the Wonders of the Heavens.* New York, 1902, p. 141.

MOON'S SHADOW ON THE EARTH

Figure 11.9a Mabel Loomis Todd, 'Moon's Shadow on the Earth,' bird's-eye view of the path of totality combined with God's-eye view on the frontispiece for *Total Eclipses of the Sun* (1900).

Figure 11.9b Mabel Loomis Todd, *Total Eclipses of the Sun* (1894): 'To Show How Eclipses Take Place'.

tra of the stars and nebulæ. For several years it did not seem that much was to be learned in this way about the sun. The year 1868 at length arrived. On August 18th there was to be a remarkable total eclipse of the sun, visible in India. The shadow was 140 miles broad; the duration of the total phase was more than six minutes. The French sent Mr. Janssen, one of their leading spectroscopists, to observe the eclipse in India and see what he could find out. Wonderful was his report. The red prominences which had perplexed scientists for two centuries were found to be immense masses of glowing hydrogen, rising here and there from various parts of the sun, of a size compared with which our earth was a mere speck. This was not all.

of the light reflected from the air might be so diminished that the bright lines from the gases surrounding the sun would be seen. It was anticipated that thus the prominences would be made visible. Both of the investigators we have mentioned endeavored to get a sight of the prominences in this way; but it was not until October 20th, two months after the Indian eclipse, that Mr. Lockyer succeeded in having an instrument of sufficient power completed. Then, at the first opportunity, he found that he could see the prominences without an eclipse! At that time communication with India was by mail, so that for the news of Mr. Janssen's discovery astronomers had to wait until a ship arrived. By a singular coincidence his re-

MAP SHOWING THE PATH OF THE COMING ECLIPSE (MAY 28, 1900), WITH THE EXACT TIME IN THE MORNING AT WHICH THE ECLIPSE WILL OCCUR AT VARIOUS POINTS DESIGNATED.

Figure 11.10a Simon Newcomb, *The Coming Total Eclipse of the Sun* (1900): 'Map showing the path of the coming eclipse (May 28, 1900), with the exact time in the morning at which the eclipse will occur at various points designated.'

As Newcomb's searching in poetry and combinations of mapping and portrait-style photography of the corona show, the technical maps remained somewhat unsatisfying when experts and amateurs were trying to communicate what was to them one of the most impressive sights that nature had to offer. Finally, there is the lack of the story in the sense of Sun and Moon being faceless cosmic actors without human personality, origin story, or interpersonal drama that often is the point of history painting. So much of the traditional iconography in the Western history of the image since the Renaissance age, with Alberti's definition of the *storia*, requires such emotional investment. For eclipse representations, this tradition seems to have actually necessitated a reintegration of familiar motives and storylines into the fantastic imaginary of total solar eclipses. One might say that J. J. Grandville's dalliance of the classically informed bodies of Sun and Moon, united in a kiss in the sky while watched by a set of voyeuristic anthropomorphized telescopes and instruments, is almost a joke about the loss of the dramatic stage that used to structure representation and perception, and a reinvestment of cool scientific appearances with anthropomorphic fantasy (Figure 11.11). Grandville included this blissful 'Conjugal Eclipse' into his collection *Un Autre Monde* (*Another World*), published in 1844. We will return to this merging of love, attraction, desire, and humour at the end of this chapter.

FIG. 76. — Another view of the solar corona.

FIG. 75. — The solar corona during a total eclipse of the sun.

Figure 11.10b Simon Newcomb, *Elements of Astronomy* (1902): 'The solar corona during a total eclipse of the sun' and 'Another view of the solar corona.'

Something else is carried over with the word for the phenomenon: Greek *ekleipsis* meaning omission, disappearance, or abandonment. And as the word comes with a wide array of artistic and poetic possibilities, its visual and otherwise figurative or abstract representations likewise allow a much wider reach into basic psychological, even primal, experience. In addition to his work on definitions and descriptions of shadows, Gombrich has also made much of the recognition of occlusion as a representational tactic.[21] Considering this in an eclipse context, it is interesting that the opportunities of illustrating the phenomenon with coronagraphs are also only possible by occlusion—creating the kinds of images we will focus on next, looking up.

[21] See Ernst Gombrich, 'Mirror and Map: Theories of Pictorial Representation.' *Philosophical Transactions of the Royal Society of London B, Biological Sciences* 270, 1975, pp. 119–49. Gombrich summarizes his discussion of demands for an alternative system of perspectival representation with an emphasis on the difference between appearance and representation of the physical world on one hand, and human reaction and experience on the other: 'We can map the physical world but not its variable and shifting appearance. This conclusion, however, is not intended to discourage artistic attempts to record a visual experience. On the contrary: all experiments on the hoardings, on the screen and in paintings probing our cognitive and emotional response to images should be of interest to the student of human reactions.' See also Barbara Gillam, 'Occlusion issues in early Renaissance art.' *i-Perception* 2, 2011, pp. 1076–97.

Figure 11.11 J. J. Grandville, *Un Autre Monde* (*Another World*): 'A Conjugal Eclipse' (1844).

Source: UC Berkeley Historical Slide Library.

Looking Up: Moment and Focus

Valentine: We're better at predicting events at the edge of the galaxy or inside the nucleus of an atom than whether it'll rain on auntie's garden party three Sundays from now.

Tom Stoppard, *Arcadia*

Looking up from within or underneath the shadow, the corona appears as the most glorious and recognizable pattern of the event (see also Chapter 2), with consequences for the collective imaginary of total solar eclipses, especially since photography and film made it possible to block out the brightest part of the corona for facilitating the careful contrast-processing of the fainter regions, inviting artists to focus on its artistic enhancement. Occlusion is a meaningful principle even for the technology of image production, since a coronagraph must mitigate the brightness to allow the figuration of the Sun. Both the event and the photographic technology— its name derived from the Greek φωτός (*phōtós*), genitive of φῶς (*phōs*), 'light' and γραφή (*graphē*) 'representation by means of lines' or 'drawing,' combined meaning 'drawing with light'—function through light, which makes the corona so attractive in photography as the Sun's halo imprints itself with the help of this indexical technology and medium. Rather counterintuitively, it is in the optically relevant difference of media and delayed image production as much as it is in the framing and cropping of the appearance that the corona-focused idealized snapshot becomes a medium of seemingly authentic portrayal as well as a reminder of the very artificiality of the composite image: From painstakingly enhanced drawings and chromolithographs from circa 1881 (Figure 11.12a) to finely drawn coronal streamers illustrated in a drawing from a photograph from 1898 to the typically focused, narrow-cropped

TOTAL ECLIPSE of the SUN.

Figure 11.12a Total eclipse of the Sun, chromolithograph circa 1881.

Figure 11.12b Coronal streamers illustrated in a drawing by W. H. Wesley from a photograph by Annie Scott Dill Maunder, née Russell (Mrs Walter Maunder), 1898.

photos in 1900 and into the first decades of the twentieth century (Figures 11.12b–11.12f).

While the corona and diamond ring indeed became signs and wonders and the star of the show in the phenomenological representation of total solar eclipses in popular imagery, they can also appear in the chorus line, framed in imagery that documents the phases (Figure 11.13a).

The sequencing imagery from *Harper's Weekly: A Journal of Civilization*, 28 August 1869, compared to the partial eclipse as seen from Austria in 1900 (Figure 11.13b), serves then also to show the fundamental difference to the phases of partial versus total eclipses: without the moment of revelation in the corona, it is a much less dramatic, fluid development of waxing and waning; the interruption of the sequence by the corona's appearance and lingering in the darkness with the fear of it not moving back into normalcy emerges even from this simple phase imagery.

At times, as in the *Harper's Weekly* issue's cover, scientific imagery of the sequence combines those records of the event above with the documentation of the scenery of tents and travelling scientists and amateurs, leading us into the third typical category of total solar eclipse representation through looking around, at bystanding eyewitnesses.

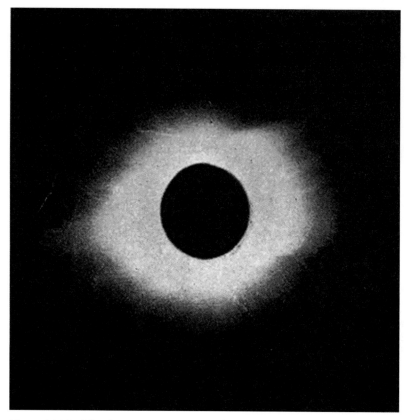

Figure 11.12c The Corona, 28 May 1900, photographed at Algiers by Mr Walter Maunder and party (exposure 0.5 second, taken 5 seconds after mid-totality on Imperial 'fine grain ordinary' plate. Aperture 4 inches, focal length 34 inches.)

Looking Around: Viewers and Audiences

Septimus: We shed as we pick up, like travellers who must carry everything in their arms, and what we let fall will be picked up by those behind. The procession is very long and life is very short. We die on the march. But there is nothing outside the march so nothing can be lost to it. The missing plays of Sophocles will turn up piece by piece, or be written again in another language. Ancient cures for diseases will reveal themselves once more. Mathematical discoveries glimpsed and lost to view will have their time again. You do not suppose, my lady, that if all of Archimedes had been hiding in the great library of Alexandria, we would be at a loss for a corkscrew?

Tom Stoppard, *Arcadia*

In political iconography, the staged total eclipse group portraits of statesmen and scientists form a genre among the staple imagery of journalistic reports on eclipses

Figure 11.12d Second Contact, 28 May 1900, photographed at Wadesboro (Wadesborough) by J. N. Maskelyne with 3.5-inch kinematograph.

(Figure 11.14a). We already know from the historical chapters in this book that the association of power in terms of rulers and political control on Earth has a long and significant history reaching back to the beginnings of societally organized human civilization. In the collective visual subconscious, those total eclipse group portraits are played upon by a strong tradition of power-related satirical imagery by caricaturists such as Robert Cruikshank (1789–1856) in commentaries on 'grand politico-astronomical phenomena,' often using the eclipse of powers as a mocking opposite or challenge to the powers that be—here illustrated with the printed title of the 1820 critical publication *The total eclipse: a grand politico-astronomical phenomenon, which occurred in the year 1820; with a series of engravings, to demonstrate the configuration of the planets. To which is added, an hieroglyphic, adapted to these wonderful times,* a political pamphlet of 26 pages concerning the Queen Caroline affair published in London by Thomas Dolby, here featuring a wig in place of the corona and the word 'VICE' in its dark centre (Figure 11.14b).

The genre of group portraits of scientists, on the other hand, is markedly about different kinds of perceived power. Following the impetus of 'shooting' the Sun (Figure 11.15a), they are shown with maps and calculations in discussion prior or after the event, laying out the preparation that goes into it with as much pride as the subsequent viewing of their success. The latter is documented in the key of group portraiture (Figure 11.15b), culminating in the nestled gazes of viewers looking at a

Figure 11.12e Frederic Dawtrey Drewitt, *Total Eclipse of the Sun* (mezzotint on paper), 'The Sun's Corona – Torreblanca, Spain 30 Aug. 1905' (1907).

small photograph of a group of men looking at a poster-size, framed and hanging aerial photograph of the sky during a total solar eclipse, one man's arm following the path of the opening in the clouds with an energetically extended index finger reminiscent, in its assumption of cosmic power over sky and land, of the famous life-giving hand of God in Michelangelo's *Creation of Adam* scene on the frescoed ceiling of the Sistine Chapel (Figure 11.15c).

Figure 11.12f 'Signs and Wonders 1925,' Diamond ring of the solar eclipse, 24 January 1925.

The pointing finger appears so that it is just so outlined by white clouds behind it, with an opening of the clouds continuing its trajectory. It indicates the division between two pairs of handwritten inscriptions in white on the poster trying to paint the dynamic interaction of the elements of the eclipse, from top to bottom: 'Sunlight in the Sky and on low Clouds beyond Shadow,' 'Low Clouds darkened by the Moon's Shadow,' 'Low clouds in Sunlight in Front of Advancing Shadow,' and 'Woods and Fields visible between Clouds.' The lecturer's dramatic gesture, ruling retrospectively over the sky, the photographed appearance, and the scientific narrative given to the three bystanders as well as the viewer of the photograph, matches the drama of image production. Long before such dramatic sky-high shots from the airplane were thinkable, the nineteenth-century publications with a perennial interest in showing total

Figure 11.13a *Harper's Weekly: A Journal of Civilization*, 28 August 1869.

Figure 11.13b Solar eclipse, 28 May 1900, Sonnenwendstein in Niederösterreich, Austria.

eclipse explorers would likewise dramatize the moment of capture, such as in the caption 'Mrs. Walter Maunder preparing to snap the shutter.' (Figure 11.16a).[22]

Other discourses—colonial and gender in nature—emerge from the imagery of the first female amateur astronomers, a group of them involved with the first filming of a total solar eclipse, in 1900 (Figure 11.16b).[23] True to the present day, the question

---

[22] See Dan Streible, 'Women film eclipses, 1898–1901,' 15 July 2022, https://wp.nyu.edu/orphanfilm/2022/07/15/eclipse/.
[23] Ibid. See also Marilyn Bailey Ogilvie, 'Obligatory Amateurs: Annie Maunder (1868–1947) and British Women Astronomers at the Dawn of Professional Astronomy.' *The British Journal for the History of Science* 33, 2000, pp. 67–84.

Figure 11.14a US President Wilson and Mrs Wilson viewing the eclipse on 24 January 1925.
Source: UC Berkeley Historical Slide Library, Selz Collection.

is; who is privileged to enjoy freedom of travel and information? And how can local communities and eclipse chasers join in the experience together?

To return to the kind of voyeuristic amateur audience in Grandville's 'Conjugal Eclipse,' the satirists' assortment of telescopes and optical machinery, one looking at the eclipse via its reflection in a bucket of water, also illustrate how the moment of the eclipse compresses the focus of viewers in the souvenirs and descriptions, making the viewer of the image yet another viewer of mitigated, channelled experience (Figure 11.17).

Throughout time, this investment of channelled experiences is bound to interpretation and further symbolic readings of scientific reports and images, enthusiastically practiced in the subtle Romanticism that is always embedded in the historical 'rational' age of the Enlightenment and post-enlightened centuries up to the present day.[24]

---

[24] The artistic and cultural struggle between Romantic–sentimental and rational–enlightened urges has since defined Western discourse and society, as Tom Stoppard comments via his characters in *Arcadia* with the example of English landscape architecture specifically on the nestling of stylistic traditions since Antiquity: '[*Hannah:*] English landscape was invented by gardeners imitating foreign painters who were evoking classical authors. The whole thing was brought home in the luggage from the Grand Tour. Here, look—Capability Brown doing Claude, who was doing Virgil. Arcadia! And here, superimposed by Richard Noakes, untamed nature in the style of Salvator Rosa. It's the Gothic novel expressed in landscape. Everything but vampires.'

Figure 11.14b   Robert Cruikshank, *The Total Eclipse: A Grand Politico-Astronomical Phenomenon. . .* (1820).
Source: UC Berkeley Historical Slide Library.

Figure 11.15a To 'shoot' the Sun: Captain Edward T. Pollock (left) and Captain F. B. Littell, scientists of the US Naval Observatory in Washington, 24 January 1925.
Source: UC Berkeley Historical Slide Library, Selz Collection.

Looking Inward: Eclipse as Symbolic–Figurative Medium

Septimus: When we have found all the mysteries and lost all the meaning, we will be alone, on an empty shore.

Tom Stoppard, *Arcadia*

As much as the map is embedded in the history of cartography, the representation of light and shadow has its own art history, and most of it is related to further iconographies and iconologies—coding representations of light and darkness and pointing to vast cultural archives of texts explaining their significance. Light has long found its way in the histories of art as well as, more broadly, in theological and philosophical discourses.[25] Shadows, especially cast shadows, have been a practice and theoretic interest of the discipline from the beginning to the classical late-twentieth-century

---

[25] See Wolfgang Schöne, *Über das Licht in der Malerei*. Berlin, 1954. See also Hans Blumenberg, 'Licht als Metapher der Wahrheit: Im Vorfeld der Philosophischen Begriffsbildung.' *Studium Generale* VII, 1957, pp. 432–47. See also Glory and *Herrlichkeit* in Hans Urs von Balthasar, *Herrlichkeit: Eine theologische Ästhetik. Band 1: Schau der Gestalt*. Einsiedeln, 1961; von Balthasar, *Herrlichkeit: Eine theologische Ästhetik. Band 11: Fächer der Stille, Teil 1: Klerikale Stille*. Einsiedeln, 1984; and von Balthasar, *Herrlichkeit: Eine theologische Ästhetik. Band 11: Fächer der Stille, Teil 2: Laikale Stille*. Einsiedeln, 1984.

Figure 11.15b Group portrait from the early 1950s showing the collective viewing of a textually enhanced large-scale photograph of the sky during the total eclipse of the Sun, 31 August 1932.

Source: UC Berkeley Historical Slide Library, Selz Collection.

Figure 11.15c The hands of God and Adam in Michelangelo's Sistine Chapel's ceiling's *Creation of Adam* (1508–12).

Source: UC Berkeley Historical Slide Library, Baxandall and Partridge Collection.

MRS. WALTER MAUNDER PREPARING TO SNAP THE SHUTTER

Figure 11.16a 'Mrs. Walter Maunder preparing to snap the shutter.'

approaches to perception and cognitive science through the arts, from Michael Baxandall to Ernst H. Gombrich and Victor I. Stoichita.[26] Stoichita's 'provocatively' short history of the shadow quickly lapses from mythologies into the many uncanny occurrences of /shadow/, as the origin of painting and portraiture and dealing with loss in Pliny's *Natural History* XXXV, 15 on outlining the shadow as the origin story of painting and, in *Natural History* XXXV, 43, also for the origin of sculpture with the addition of clay in relief—another dimension of shadow in the history of art. Shadows are then shown to be inherent in the ambiguity of the Platonic shadow stage, as the double and doppelgänger, as the uncanny ghost following in Cruikshank's and Menzel's 'Pursuit of the shadow', to the chapter 'Of Shadow and its Reproducibility during the Photographic Era,' as that which threatens (Nostradamus). How then to imagine, in the line of this tradition, the shadow of the Moon as the double, like the doppelgänger shadow tradition that is attached to the human being? Its doubled complication as uncanny double of the Moon that co-occurs with the eclipse of the Sun—replacing its radiating core with darkness—further complicates the scenario

---

[26] See Baxandall, *Shadows and Enlightenment*, and Gombrich, *Shadows*. See also Victor I. Stoichita, *A Short History of the Shadow*. London, 1997.

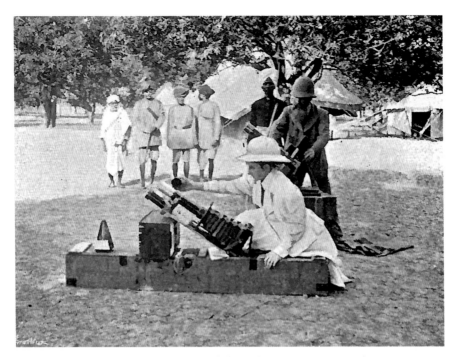

Figure 11.16b Gertrude Bacon and her father John Bacon setting up their cameras in Buxar, India, 22 January 1898.

Figure 11.17 Grandville, 'A Conjugal Eclipse'—Grandville's visual technology audience (detail of Figure 11.11)

and explains why the happening is experienced as so fascinating that it becomes life-changing for the eyewitnesses. It goes beyond any regular visual experience yet keeps all conundrums, paradox, and perplexities of 'normal' visuality within it. The fleeting 'hole in flux' changes the view of the world permanently.

The sudden Holy Friday darkness at the Crucifixion might be the most famous instance of a total solar eclipse in the Western popular imagination, usually restaged in any movies about the Life of Christ as well as in Passion music (see also Chapter 12). Similarly famous in the history of oil painting for students today is Antoine Caron's (1521–99) *Dionysius the Areopagite Converting the Pagan Philosophers/Astronomers Studying an Eclipse* from 1571, on view in the J. Paul Getty Museum at the Getty Center in Los Angeles (Figure 11.18).[27] Depending on one's perspective, the iconography would have relayed either a miracle or a fraud. By his legend, Dionysius the Areopagite converted the pagan philosophers by relating the happening of a total solar eclipse to his visionary attendance at the Crucifixion. The intense mannerist distortion of some of the architectural elements in the background and its charged historical context have opened the painting's meaning and significance in the religiously complex environment of its time to further speculation. In the France of a Catholic Italian Queen, Catherine de' Medici, who was interested in astrology and opposed Protestantism, the work of her court painter Caron could comment on several of these issues at once.

With an increasingly realistic style and figurative investment in the history of European painting throughout the nineteenth century, shadows became a powerful iconographic tool in the different practices of oil painting. Between the poles of landscape realism and symbolism, on one end of the spectrum they could be employed to indicate directly the biblical darkness at the Crucifixion as, for instance, in Jean-Léon Gérôme's 1867 *'Consummatum est'/Jerusalem*, as a threatening spectacle. On the other, it could invest Christ with the symbolic foreshadowing of his death by his own shadow capturing the shape of his body on the cross in William Holman Hunt's 1873 *Shadow of Death*—seen through the eyes of his mother Mary.

Jean-Léon Gérôme's dramatic landscape in oil on canvas depicts the moment of the Crucifixion; it is titled *'Consummatum est'* and also *Jerusalem*, the shadow of the cross marking its own path of totality underneath the darkened sky (Figure 11.19). While almost all of the Romans are on their way back from Golgotha to the city, two of the soldiers have turned around and seem to react in shock and awe to what they see—an appearance that we can only intimate, the intense glory of light throwing the shadow of the three crosses onto the ground (Christ is in the middle, the two thieves executed with him to his left and right). Maybe the two figures that notice and acknowledge the supernatural happening are supposed to be understood as the converted Centurion Cornelius, named in the Acts of the Apostles as the first Gentile to convert to the Christian faith, and an unnamed companion of Cornelius with whom everyone shall feel invited to identify. While Gérôme often selected the moment of the aftermath instead of the dramatic height of an event (such as in his renditions of *The Execution of Marshal Ney*, in *The Duel After the Masquerade*, and in *The Death of Caesar*), one could argue that in this case the history painter does indeed show the exact moment of the death of Christ and the conversion of the two agitated bystanders, given their

---

[27] See the discussion of this painting in Olson and Pasachoff, *Cosmos*, pp. 62–4. See also Mary Kerr Reaves and Gibson Reaves, 'Antoine Caron's Painting "Astronomers Studying an Eclipse."' *Publications of the Astronomical Society of the Pacific* 77, 1965, p. 153, and Margaret Aston, 'The Fiery Trigon Conjunction: An Elizabethan Astrological Prediction.' *Isis* 61, 1970, pp. 158–87.

Figure 11.18 Antoine Caron, *Dionysius the Areopagite Converting the Pagan Philosophers / Astronomers Studying an Eclipse* (1571).
Source: UC Berkeley Historical Slide Library, Baxandall and Partridge Collection.

reaction together with the shadow of the lifeless body of Christ. As characteristic for the visualization of eclipses, it looks strangely removed and suspended. There remains the ambiguity of painting the biblically reported 'darkness' as an eclipse, or not: beyond the central scene out of our view, we witness a metaphorical darkness protruding from the top right, and we also see something that looks like a sickle Moon

Figure 11.19 Jean-Léon Gérôme, '*Consummatum est*' / *Jerusalem* (1867).
Source: UC Berkeley Historical Slide Library.

in the darkness over the horizon, while the overall effect of strong highlights and shadow over layers of the landscape and the sky do not read as a total solar eclipse. Looking at it with a scientific gaze, where Caron merged allegory with much more astute observational detail, Gérôme takes poetic and painterly licence in outdoing the drama of the total solar eclipse by adding the additional strong light source fighting the extreme darkness of the upper right corner.

On the symbolic–literal end of the spectrum, William Holman Hunt shows the viewer through the *Rückenfigur* of Mary the cross-shaped shadow of Christ, foreshadowing the image of the crucifix in his 1873 *The Shadow of Death* in oil on canvas, today in the Manchester Art Gallery (Figure 11.20). Mary looks up from the gifts of the Three Kings, startled. In the Christian tradition of light-and-shadow symbolism, Hunt had already drawn criticism for his explicit iconography on his earlier painting, *The Light of the World* (Chapel of Keble College, Oxford). In the mid-1850s, he had shown Christ with a lantern literally as saviour and light of the world. Hunt, however, doubled down on such direct iconographies of shadows in his portrayal of Christ with Mary in their humble home before entering his mission.

It is a short step from the nineteenth-century academic realism on one hand, and the Pre-Raphaelite pictorial reinvention of the past on the other, to European classical modernism's painting into abstraction, and to British surrealist Paul Nash (1889–1946) and his early twentieth-century rethinking of the eclipse as a natural phenomenon—as a sunflower (see Figure 11.21a and cover). His 1945 *Eclipse of the Sunflower* is only a part of an unfinished project, originally planned to include the *Solstice of the Sunflower* (today in the National Gallery of Canada, Ottawa) as well as two more pieces, *The Sunflower Rises* and *The Sunflower Sets*, that he could not finish before his death. The project was inspired by Dante as well as by William Blake's famous 'Ah! Sun-flower' from Blake's *Songs of Experience* (1794) and the

Figure 11.20 William Holman Hunt, *The Shadow of Death* (1873).
Source: UC Berkeley Historical Slide Library.

religious–mythical exploration of themes of renewal and the overcoming of death in James George Frazer's *The Golden Bough*.

Nash's Sunflower project remained unfinished, but already the subtle changes in his development of the oil painting from a preparatory watercolour (Figure 11.21b) show how much thought and procedural care went into the artist's increasing reconciliation between a cosmic look and the still life's detail realism of the sunflower's seeds

Figure 11.21a Paul Nash, *The Eclipse of the Sunflower*, oil painting (1945).
Source: British Council Collection, London.

Figure 11.21b Paul Nash, *The Eclipse of the Sunflower*, watercolour (1945).
Source: British Council Collection, London.

and more textured appearance.[28] In the version of his sunflower eclipse in oil, the hovering sunflower is much more blended into the waves, flames, and smoothness of the intuited cosmic environment of its apparition, uniting the contradictory elements that each and every recording and representation of total solar eclipses must tackle as a part of the experience for the observer of the event and the viewer of the artwork.

Conclusion: Time, Experience, and Representation (or, Welcome All Wonders)

Hannah: It's the wanting to know that makes us matter.

Tom Stoppard, *Arcadia*

Thomasina: Then we will dance. Is this a waltz?

Tom Stoppard, *Arcadia*

In the numinous happening of the total solar eclipse, the self-determined joy and self-assertive expansiveness of knowledge and experience in the sciences and arts merge with the wonder and humility of *not* knowing—of seeing attentively yet *not* grasping the entirety of the cosmic event. Neither history's ritual (e.g. the faux substitute king in Chapter 5), nor the scientific registration of solar structure through photography, have reduced or diminished the sense of awe. The specific recognition of the impossibility of phenomenological representation of total solar eclipses opens a new, even higher challenge either of accepting that some things cannot be recorded or communicated, or to find ever new media channels and creative methods—but regardless the outcome of any viewer's pondering of their options, the experience has left its imprint and changed everything forever, like synapses are changed by the act of reading and learning even if the material memorized seems to fade away. What remains, one would hope, would be then to be anew awake to wonder, like Tom Stoppard's Thomasina in *Arcadia* ultimately demanding to dance between the endless possibilities of seeing, knowing, and experiencing everything across time and space.

Monuments already pose challenges for artistic representation as well as for historical discourse, but an eclipse monument falls into the most counterintuitive category. Respective monuments in Uganda, such as the Biharwe Eclipse Monument in commemoration of the total eclipse of the Sun which took place on 17 April 1520 and that in Pakwach for the 3 November 2013 eclipse, resort to abstract architectural and sculptural forms, but the event literally escapes monumentalization even more than visualization in general, as we have seen. The location was only once meaningful for something that was happening above, and something so fleeting and escaping representation that the 2001 Pinehurst Civic Group's plaque for the 28 May 1900 eclipse

---

[28] See Jemima Montagu, *Paul Nash: Modern Artist, Ancient Landscape*. London, 2003. See also James King: *Interior Landscapes: a Life of Paul Nash*. London, 1987, p. 218. See also Tom Overton, 'Paul Nash: Eclipse of the Sunflower,' http://visualarts.britishcouncil.org/collection/artists/nash-paul-1889/object/eclipse-of-the-Sunflower-nash-1945-p114.

Figure 11.22 The impossibility of a monument: The 2001 Pinehurst Civic Group's plaque for the 28 May 1900 eclipse.

seems rather fitting in its laconic avoidance of representation, specificity—even style, ornament, or iconography (Figure 11.22).

Since the early noughties, digital photography and further imaging techniques have opened new worlds of visualization, so that spectacular sights such as Turkish amateur astronomer, photographer, and civil engineer Tunç Tezel's composite image of an annular eclipse (Figure 11.23a) can now show the timelapsed sequence over the silhouetted landscape of Monument Valley. Tezel's image of the annular eclipse also serves to illustrate the dramatic difference of the chromatic light effects in comparison to *total* solar eclipses in recent photography. Jay Pasachoff's photo from the aeroplane (Figure 11.23b) with the bending of the sunset/sunrise line over an orange marmalade horizon, given the speed at which it was taken contrasting the absolute stillness and calm of its skyscape, is a successful instance of paradoxical strategies for phenomenological representation when it comes to the illustration of what those experiencing a total solar eclipse perceive with their eyes, bodies, and minds.

Finally, Jay Pasachoff combined three particularly striking versions of visualizations for the 2020 eclipse in Patagonia/Argentina (Figures 11.24a–11.24d). These make themselves an important point that Olson and Pasachoff formulated in their

Figure 11.23a Tunç Tezel's timelapse sequence of the annular solar eclipse of 2012 over Monument Valley, USA.

Source: Tunç Tezel.

Figure 11.23b Jay Pasachoff's photo out the window of a charter flight due east from Punta Arenas, Chile, on 4 December 2021.

Figure 11.24a Compound image from 65 originals from Andreas Möller at Piedras del Aguila made by Roman Vaňúr/Jay Pasachoff, including a coronal mass ejection (CME) on the left.

Source: Jay M. Pasachoff, Vojtech Rušin, and Roman Vaňúr.

Figure 11.24b Prominences at the solar limb (1) by Verónica Espino, director of the Planetario Galileo Galilei, Buenos Aires, from El Condor near Las Grutas, Argentina.

cross-disciplinary book *Cosmos*: both astrophysics and art history are highly visual disciplines.[29] A focus on the visual can indeed be yet another method for developing interdisciplinary practices. In these cases, the possibilities of digital image generation and processing render visible aspects of the eclipse that would not ever have been seen otherwise.

[29] See *Cosmos*, p. 9.

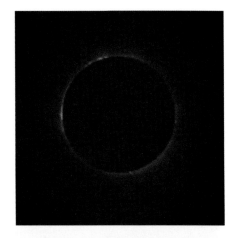

Figure 11.24c Prominences at the solar limb (2) by Verónica Espino, director of the Planetario Galileo Galilei, Buenos Aires, from El Condor near Las Grutas, Argentina.

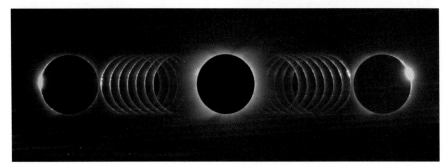

Figure 11.24d A composite of the chromosphere and both diamond rings, as seen from Valcheta by Guillermo Abramson, Física Estadística e Interdisciplinaria—Centro Atómico Bariloche / Instituto Balseiro.

When it comes to alternatives to modern digital or aerial photographic visuality, two other extremes of representation can then ultimately be identified as particularly apt for bypassing the ultimately lacking, unsatisfying figurative mode: On one hand, there are the earliest surviving moving images of an eclipse, from 1900—recorded on kinematograph film by the magician and inventor Nevil Maskelyne (1863–1924, not to be confused with Nevil Maskelyne FRS, 1732–1811, who was the fifth British Astronomer Royal, holding the office from 1765 to 1811) on an expedition by the British Astronomical Association to North Carolina. The film was forgotten, only recently digitally restored by the Royal Astronomical Society and the British Film Institute National Archive, and published on the BFI's YouTube channel, so that the trace of the eclipse from then became available for a global audience's eye after many decades of lying dormant in the archive. The Victorian film is thought to provide the first known and preserved moving images of the astronomical phenomenon, originally captured over a century ago. Reassembled and retimed frame by frame, through the disturbances of speckles and glimmering light, the fascinating segment invites a reflection on time and cosmos through the flimsy materiality of the early film—because its staticky, silent-movie aesthetics are so insufficient to show the event,

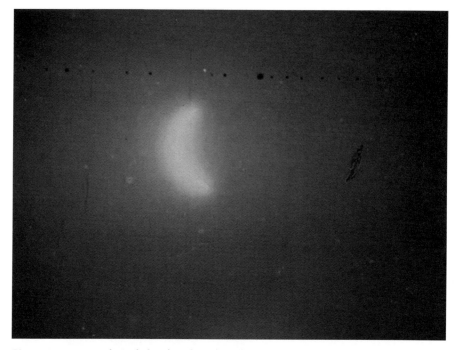

Figure 11.25a Nevil Maskelyne's *Eclipse* (1900), sample still before totality.

the self-representation of the Sun's light in 1900. However, knowing the historical distance and the wonder of recovery of this footage from the archive allows in its otherness a space for reflection on the cosmic phenomenon that is in the focus of each of the stills, and the film of one minute and eight seconds (Figures 11.25a–11.25c; see also the BFI archive on YouTube for the full footage online).[30]

On the other hand, there is still the upper half of Grandville's 'Conjugal Eclipse', with a classical Apollo Sun God embracing a Moon Diana/Venus in Victorian dress, both rising far above the progress-positivistic audience of the new optical machines as if mocking, with their celebration of both historicizing and mythologizing imagery, the humanized technology approach to the forces of life, desire, and humour. While the movie allows for reflection of time and the nature of technological progress—it is a good minute long in its retimed version—Grandville's projection of mythology and artistic styles into the technocratic heavens subverts the supposed sequence of progress and seems to insists on the necessity of telling a story (a love story) and investing cosmic phenomena with the meaning that human beings need to ascribe to them.

Finally, the paradoxicality of the total solar eclipse's nature connects to theological investments and interpretations, particularly within the Christian tradition. It is a tradition that uniquely requires a constant stretching of mind and soul to 'see' without

---

[30] See https://www.youtube.com/watch?v=q4jfPfMKBgU.

Figure 11.25b Nevil Maskelyne's *Eclipse* (1900), sample still during totality.

Figure 11.25c Nevil Maskelyne's *Eclipse* (1900), sample still after totality.

Figure 11.26 Grandville, 'A Conjugal Eclipse,' Sun–Apollo embracing
Moon–Diana/Venus (detail of Figure 11.11).

using one's eyes and grasp without using one's hands instances of invisible, seemingly paradoxical truth. Believers embrace unions of opposing principles such as a
Trinity–Unity, a virgin mother, and an infant king—the latter then effecting a reversal of powers and earthly logics as proclaimed in the Sermon on the Mount where
those that suffer are blessed and the powerless suddenly empowered.

Having witnessed (Chapter 9, 'Dante's Total Eclipses') a struggle with words and
speechlessness in facing the numinous with Dante Alighieri's symbolic cosmos of the
Divine Comedy in the fourteenth century, we shall close with a slightly more recent
poem. Richard Crashaw published it in 1646 with an emphasis on the witnesses of the
Nativity, the shepherds, and their humble tunes and simple language under the title
'In the Holy Nativity of our Lord: A Hymn Sung as by the Shepherds.' Anticipating
the most powerful proof of nature's inversions through paradox, the upside-down
logic of the poem declares in simple words the greatest wonder of God appearing in
human form. It thus stretches the mind in ways similar to how the Christian belief
requires a certain reciprocity between intellectual and emotional experience to allow
for the idea of 'seeing' without the eyes of the body, in a spiritual way, in the tradition
of Augustine's theory of vision.[31] With its multi-layered function of seeing, Christian

---

[31] See, for instance, Augustinus, *Über Schau und Gegenwart des unsichtbaren Gottes*, ed. Erich Naab,
Stuttgart-Bad Cannstatt 1998. For Latin and German versions of the texts see *De videndo Deo* and *De
praesentia Dei* in Augustinus, *Über Schau und Gegenwart des unsichtbaren Gottes*, pp. 118–91 and 214–
59, respectively. See also 'Inward Turn and Intellectual Vision.' In Phillip Cary, *Augustine's Invention of*

faith constituted in many ways an ideal nutrient medium for early modern science;[32] Crashaw, on the page of poetry, then unites the spiritual and the poetic for making the invisible seen. This union brings us back to Dante's eclipses as revelatory forces and sacred knowledge.[33] Through the eyes of the shepherds, the short text states such apparent contradictions for the occasion of Christmas. In tune with the profound union of Christmas as the beginning of *earthly* life in the middle of winter with the time of Passion and Easter shortly following to celebrate the beginning of *eternal* life in the spring—reaching from the star-of-Bethlehem comet at Christ's birth to the total solar eclipse at his death—the poem that begins with visual abundance ('all wonders in one sight') quickly eclipses its own vision of a comprehensive, universal, cosmic landscape imagination:

In the Holy Nativity of our Lord

Welcome, all wonders in one sight!
 Eternity shut in a span;
Summer in winter; day in night;
 Heaven in earth, and God in man.
Great little one, whose all-embracing birth
Lifts earth to heaven, stoops heav'n to earth.

---

the Inner Self: The Legacy of a Christian Platonist. Oxford, 2000, pp. 63–76. See also Cary on Augustine expressing, in *Confessions* 7:23, his thoughts on seeing intellectually 'invisible things, by means of the things that are made,' Cary, *Augustine's Invention of the Inner Self*, p. 65.

[32] See Peter Harrison, '"Science" and "Religion:" Constructing the Boundaries.' *The Journal of Religion* 86, 2006, pp. 81–106. See also Harrison, *The Territories of Science and Religion*. Chicago, IL, 2015.

[33] '[The] image of the Sun and Moon, the divine and the natural, yoked together, is an image of Christ as well as of man. In particular, the image fuses the Incarnation and the Crucifixion. A pervasive medieval tradition links the Incarnation and the Crucifixion by portraying Christ as both the rising sun (*sol oriens*) and the setting sun (*sol occidens*). This idea has never been more profoundly and perfectly expressed than in the dimensionless point or instant of *Paradiso* 29's exordium, a nexus-balance-knot-yoking of Creator and creation, which is simultaneously and undecidably a sunrise and a sunset. Incarnation and Crucifixion are one. What is eclipses itself by revealing itself as or through finite form; at the same time, it reveals itself by eclipsing itself as (sacrificing) finite form.' Christian Moevs, '"Il punto che mi vinse": Incarnation, Revelation, and Self-Knowledge in Dante's *Commedia*.' In Vittorio Montemaggi and Matthew Treherne (eds), *Dante's* Commedia: *Theology as Poetry*. Notre Dame, IN, 2010, pp. 267–85, at p. 278.

12

When Words Fail

Eclipse, Music, and Sound

Elaine Stratton Hild

Music, History of Music

We complete our selective survey of eclipses across the arts by listening for the seemingly impossible—for how could a phenomenon as utterly soundless as an eclipse generate responses in the inherently aural form of music? This genre-crossing chapter will lead from Handel to Pink Floyd through a broad variety of eclipse engagement in the history of Western music, from the classical style to Sir John Kenneth Tavener, to pop and rap music, including the famous eclipse tunes by Welsh popstar Bonnie Tyler and ecliptic imagery in the tunes of the American rapper YoungBoy Never Broke Again (Kentrell deSean Gaulden). The interaction of thematic questions and case studies invites readers/listeners to consider music in all dimensions—remembering that music has also the capacity of expressing the wordless and transcendent features of the total solar eclipse experience, the musical chasing of total solar eclipses becomes a more immanent endeavour.

aria An elaborate melody sung solo with accompaniment, usually in an opera or oratorio.

arpeggio The sounding of the notes of a chord in rapid succession.

crescendo A gradual, steady increase in loudness.

double stop Two or more notes bowed simultaneously on a string instrument such as a violin.

fortissimo Very loud.

Grim, Wayne San Francisco Bay Area-based American musician and sound artist who turns naturally occurring data into music (e.g. in his collaboration with the Kronos Quartet for the 2017 eclipse).

Handel, Georg Frideric (1685–1759) German-British Baroque composer (born Georg Friedrich Händel).

major key A key or mode based on a major scale.

Mingus, Charlie (1922–79) American jazz bassist, composer, and bandleader.

minor key A key or mode based on a minor scale.

pianissimo Very soft.

Pink Floyd English rock band, formed 1965.

Stifter, Adalbert (1805–68) Austrian writer, poet, painter, and pedagogue.

Tavener, John (Sir John Kenneth Tavener) (1944–2013) English composer.

Tyler, Bonnie (b. 1951) Welsh popular singer.
YoungBoy Never Broke Again (Kentrell deSean Gaulden) (b. 1999) American rapper.

A note on key structures: E-minor and A-major refer to the pitch and to the type of home-scale that the music is currently on; changes in key result in a shift in mood.

A note on translation: Different musical nomenclatures for rhythm have developed in America and Europe: The American 'measure' reads in Europe 'bar;' and the American 'quarter note' and 'eighth note' read in Europe 'quaver' and 'semi-quaver.'

A practical note: There is no substitute for listening to music while analysing it. Following the links in the footnotes throughout the chapter, readers can enjoy recordings of performances and music videos in their own real time.

> Because I can portray the matter pretty well on paper by drawing and calculation . . . I thought to be able to describe a total solar eclipse beforehand as faithfully as if I had already seen it. However . . . once I stood on a lookout point high above the entire town and regarded the phenomenon with my own eyes, there naturally occurred entirely other things, which I had never thought of either waking or dreaming, and about which no one thinks who has not seen this miracle.—Never, ever in my entire life was I so shaken, from terror and sublimity so shaken, as in these two minutes . . . I think there can be no heart upon which this phenomenon did not leave an indelible impression. You, however, who have experienced it in the greatest measure, be lenient with these poor words that tried to portray it and fell so short. Were I Beethoven, I would say it in music; I believe, I could do it better then.[1]
>
> Adalbert Stifter, 1842

> Even in the most rigorous of scientific accounts, one finds the admission that the experience of observing the eclipse is altogether different than anticipated . . . It is difficult to shake off less-than-rational feelings of dread and wonder—even in the face of a scientific phenomenon whose advent has been calculated to the tenth of a second.[2]
>
> Jocelyn Holland, 2015

The natural phenomenon that provokes an emotional response: Over 150 years apart, the author Adalbert Stifter and scholar Jocelyn Holland both suggest that even a scientific observer encounters the total solar eclipse in non-cognitive ways. A thorough knowledge of the eclipse's causes provides no protection from experiences of 'terror and sublimity.' The Sun concealed, we cannot deny that our lives depend

---

[1] Adalbert Stifter, 'The Solar Eclipse on July 8th, 1842.' Trans. Jocelyn Holland as 'Appendix.' *Configurations* 23, 2015, pp. 253–8; at 253 and 257.

[2] Jocelyn Holland, 'A Natural History of Disturbance: Time and the Solar Eclipse.' *Configurations* 23, 2015, pp. 215–33; at 217.

upon a small, sweet spot of habitational possibility that we neither create nor sustain. The sight of the shadow rushing across the Earth reveals our powerlessness over the mechanics of the universe. We are confronted with forces orders of magnitudes greater than our own. No wonder, then, that those who stand in the path of totality report feeling awe, dread, and insignificance. With the Sun hidden, our dependence is revealed.

For Stifter, his unexpected, full-bodied response to the eclipse prompted a reconsideration of his descriptive medium. Words no longer seemed sufficient. The experience so surpassed and overwhelmed the possibilities of language that he asked the reader to 'be lenient with these poor words that tried to portray it and fell so short.' The gifted writer looks wistfully towards a different mode of description and expression: 'Were I Beethoven, I would say it in music.'

Writing in 1842, it is not surprising that Stifter thought of music—particularly Beethoven's music—to portray an experience that broke the bounds of convention and expectation. Forty years earlier, Beethoven's third symphony ('Eroica') created new possibilities for musical composition by cracking open conventional forms to depict emotion. In the symphony's third movement, a funeral march, expectations established by the unfolding ABA form are thwarted when the composer develops the mournful, thematic melody into unexpected and intense expressions of lament. The melody fragments and dissolves as the movement ends, as if the music itself is exhausted with grief. The conventional musical form is shattered by an expression of anguish. This pain is so great, it will not conform to artificial boundaries, Beethoven's composition seems to say. It will not be confined within a neat, socially acceptable package. Stifter's reference to Beethoven's music suggests that the experience of the eclipse similarly defied his expectations and revealed the limited communicative abilities of human language. The two minutes of the eclipse exploded the boundaries of his previous, self-assured engagement with the phenomenon through the prediction of temporal events alone. Overwhelmed with emotions, Stifter descended from his point of observation 'high above the town' like Moses descending from Mount Horeb: 'confused and with a numb heart . . . shaken from terror and sublimity.'[3]

Although Stifter considered the Romantic music of his time to be an appropriate medium with which to depict an eclipse (and an observer's response to it), more than one hundred years elapsed after his writing before any composer made the attempt. It was not until the twentieth century that the natural phenomenon served as a stimulus for musical structures, forms, and compositions. However, as a *metaphor*, depicting overwhelming, emotional experiences and moments of great change, the solar eclipse

---

[3] Stifter, 'The Solar Eclipse,' 253. In 'Signs and Portents: Reflections on the History of Solar Eclipses,' David Bentley Hart describes his own contemporary, Stifter-like experience: observing the 2017 eclipse from a 'broad, open prospect atop a mountain outside Nashville,' and discovering he 'was not at all prepared . . . for the effect the event in its entirety would have on me. . . . I felt a tremor or two of existential vertigo as I stood there gazing into the abyss of space.' Mike Frost (in 'Testimony of an Eclipse-Chaser') details an umbraphile's 'passion . . . sometimes all-consuming' to repeatedly experience the phenomenon of the total solar eclipse. Chapters 8 (Bentley Hart) and 7 (Roos) document responses to total solar eclipses (and resulting political and social effects) in different historic contexts.

has been incorporated into musical compositions both before and after Stifter's lifetime.[4] More often than music being used to describe eclipses, as Stifter imagined, the image of an eclipse has been used in musical works to describe events of heightened importance. Musicians and composers have sought to create soundscapes adequate to experiences of heightened intensity, such as mystical encounter and religious conversion. Their compositions have taken on a wide variety of narratives as dramatic as the Moon's blocking of the Sun—being suddenly blinded, falling passionately in love, moving from extreme poverty to fame and fortune, to name just a few. The solar eclipse draws together musical compositions as disparate as Handel's oratorios, Pink Floyd's prog rock, Charles Mingus's 'pre-Bird' jazz, and Wayne Grim's recent compositional project using data collected by NASA.

This chapter presents a series of case studies to capture the wealth of musicians' responses to total solar eclipses. These case studies are organized according to the manner in which composers incorporated the eclipse into their musical works. The discussions begin with pieces that reference the natural phenomenon only in their verbal content—their sung words and titles. These include contemporary hits performed by Bonnie Tyler and YoungBoy Never Broke Again. Their songs use the eclipse as a metaphor depicting passionate obsession and dramatic transformation. The chapter continues with compositions that incorporate the eclipse more extensively: not only into a piece's verbal content, but also into its musical aspects. These more extended case studies explore both the verbal and musical elements of the works. Georg Frideric Handel and Pink Floyd create musical depictions of an eclipse's characterizing contrasts: darkness and light, routine and disruption. In pieces by Charles Mingus and John Tavener, the eclipse becomes a metaphor for relationships of love. The musical aspects of these compositions portray intimate encounters and moments in which one person willingly becomes subsumed into another. The chapter continues with an example of a musical composition focused exclusively on the natural phenomenon. Ola Gjeilo's composition for solo piano serves as an example of the type of music imagined by Stifter: a musical description of an eclipse. The chapter's final two case studies move into realms of composition and sound even more intertwined with the natural phenomenon. For his composition '233rd Day,' Wayne Grim used data collected by NASA during a solar eclipse as an element of a musical composition. The sonified data provided a source of sound for the performance, alongside traditional string instruments. The last case study allows for some final considerations of Stifter's idea of creating aural depictions of solar eclipses. The Eclipse Soundscapes Project relies exclusively on scientifically gathered data, using sound to provide an experience of an eclipse to those who cannot see it. Taken together, these case studies reveal how the media of music and sound depict the indescribable eclipse, and how

---

[4] In a related investigation with literary material, Alison Cornish considers the ways in which Dante utilizes the image of the eclipse in the *Divine Comedy* ('Dante's Total Eclipses'). Contributions by Henrike Lange and Roberta J. M. Olson and Jay M. Pasachoff gathered in this volume explore the appearances of the solar eclipse in visual art.

the eclipse itself serves as a reference point for human experiences of indescribable intensity.[5]

Verbal References: Eclipse as Metaphor

Listeners of contemporary music will recognize two enormously popular songs that incorporate the image of an eclipse: 'Total Eclipse of the Heart,' performed by pop star Bonnie Tyler, and 'Solar Eclipse,' performed by celebrity rapper YoungBoy Never Broke Again. These songs use the eclipse as a foundational metaphor depicting intense experiences: for Tyler, the experience of being overwhelmed with an illicit passion; for YoungBoy, the experience of moving from extreme poverty to extreme wealth. Because the metaphor of the eclipse remains within each song's verbal content (rather than shaping musical aspects), the following discussions focus primarily on the songs' lyrics.

Reader's discretion recommended: The following passage about the official music video of 'Total Eclipse of the Heart' alludes to paedophilia.

James Richard Steinman (songwriter and video producer), 'Total Eclipse of the Heart,' performed by Bonnie Tyler (1983), lyrics:

> Together we can take it to the end of the line
> Your love is like a shadow on me all of the time
> I don't know what to do and I'm always in the dark
> We're living in a powder keg and giving off sparks [. . .]
>
> Once upon a time I was falling in love
> But now I'm only falling apart
> There's nothing I can do
> A total eclipse of the heart

For many, Bonnie Tyler's pop hit from 1983, 'Total Eclipse of the Heart,' may be the first tune that comes to mind when thinking of eclipses and music. The song's popularity was both immediate—the number-one ranking in six countries and more than six million copies sold—and enduring—Tyler re-recorded it in 2009 and performed it in the path of totality during the 2017 eclipse. Spotify currently lists the song as having over 460 million streams, with user interest peaking around solar eclipses.[6] The song's distinctive melody and Tyler's passionate delivery have made it a classic

---

[5] Although it is worthy of its own study, this chapter does not include compositions that use the eclipse primarily as a marketing tool, such as the popular sheet music of the nineteenth century produced close to the times when solar eclipses were visible in North America. These pieces reference the eclipse exclusively in their titles and cover art, rather than in their musical aspects. Please see the Library of Congress's survey of such pieces by Cait Miller, 'Total Eclipse in the Music Division.' Library of Congress, In the Muse: Performing Arts Blog, 18 August 2017. Accessed 16 October 2021 at https://blogs.loc.gov/music/2017/08/total-eclipse-in-the-music-division/.

[6] BBC News, 'Tyler releases new Total Eclipse.' 2 September 2009, available at http://news.bbc.co.uk/2/hi/entertainment/8232149.stm (accessed 31 December 2021); Raisa Bruner, *TIME*, 'Bonnie Tyler will sing "Total Eclipse of the Heart" during the actual eclipse.' 16 August 2017, available at https://time.com/4902109/total-eclipse-of-the-heart-bonnie-tyler-cruise/ (accessed 31 December 2021). Spotify, 'Bonnie Tyler.' Available at https://open.spotify.com/artist/0SD4eZCN4Kr0wQk56hCdh2 (accessed 31 December 2021).

of the eighties. Quiet sections, with only solo piano and Tyler's pleading voice, alternate with dramatic crescendos driven by drums, keyboards, and a choir. The song is a tour de force of its genre, a highly successful commercial product, and a well-known connection point between eclipses and music.

Just as striking as the song's success is the negative interpretation it gives to the image of the eclipse. The natural phenomenon is invested with emotional and moral meaning as it describes a person overwhelmed with desire. The song's lyrics suggest an obsessive attraction so strong that the person experiencing it loses agency. She becomes an eclipsed object, powerless to change the situation: 'Nothing I can do—a total eclipse of the heart.' This passion overwhelms the emotions and ability to act, as completely as the Moon covers the Sun during an eclipse's totality. The song's lyrics imply a strongly negative aspect to the passion, by referencing it as a shadow that obscures the narrator's clarity of thought: 'Your love is like a shadow on me all of the time—I don't know what to do, I'm always in the dark.'

The official music video (produced by the song's writer, Steinman) suggests paedophilia as a specific example of such an overwhelming desire; Bonnie Tyler portrays a teacher attracted to her students and unable to diffuse the temptation. In the song's lyrics, and in the narrative of the music video, the shadow of the eclipse represents a powerful force that prevents clear thinking and right action. The eclipse's totality refers to the completeness of a pathological obsession. Desire darkens the moral capacity as the Moon darkens the Sun. At the end of the video, just as at the beginning, Tyler is seen in conflict with herself as she feels her attraction to the students. In Steinman and Tyler's hit song, the eclipse portrays a person struggling in a dark area beyond the limits of societal acceptability.

Aaron Lockhart, Joseph Steele, Kentrell DeSean Gaulden, 'Solar Eclipse,' performed by YoungBoy Never Broke Again (2017)

Rapper Kentrell DeSean Gaulden, performing as YoungBoy Never Broke Again, uses the image of the eclipse to describe a fundamentally different type of experience. While Tyler's song focused on the eclipse as a cause of darkness and shadow, in YoungBoy's 2017 song the eclipse reveals a delicate, unexpected light shining forth in the midst of darkness. The official music video begins with a static image of a solar eclipse at the moment of totality.[7] Although the Sun's primary light is completely blocked, the corona's filaments stream forth around the edges of the Moon. As the video's first image, the total eclipse frames and forecasts the account of YoungBoy's life that follows: a story of wealth arising in violent and impoverished circumstances; a story of light shining in dark places.

Using the precise, driving, rhythmic speech and repeating melodic gestures of rap performance, the song describes YoungBoy's life as a transformation, made possible through the success of his music career—a contemporary rags-to-riches story. The definitive change is summarized in the song's bridge, repeated three times:

> Back in 8th grade, I swear I ain't have a thing baby
> Since I got money I swear it ain't been the same baby
> Got money in my pocket, diamonds in my chain baby
> Won't ever hurt again I made it through the rain baby

---

[7] YoungBoy Never Broke Again, 'Solar Eclipse: Official Music Video.' Accessed 21 October 2021 at https://www.youtube.com/watch?v=xw0owsSgO3Y.

The video reinforces the narrative of transformation. Images show YoungBoy as a celebrity musician—performing for packed, dancing crowds—and living the life of the super-rich—burning cash and wearing jewel-encrusted necklaces, watches, and teeth. The images also show him with large responsibilities—his toddling son kisses him as they play together.

Yet this new life appears within a context of vulnerability, just as the Sun's corona is surrounded by blackness in the video's opening image. For YoungBoy, the possibility of loss is as real as his current success. Thoughts of violent death and poverty remain, even in the new life:

> As I ride on the city lights
> I wonder who gon' ride for me if it go down
> And I start to get this feeling like
> Who gon' be there for my sons the day that I'm not around?
> I hope you love me as much as I love you . . .
> If I die right now, it's so much that I would lose . . .
> If you go broke now they ain't gon' do shit for you

The image of the Sun's corona—a delicate light in the midst of darkness—echoes throughout the video. Small points of light encompassed in blackness appear repeatedly, becoming a consistent visual motif. YoungBoy drives in a car at night, while streetlights and headlights appear as blurred, bright spheres. YoungBoy performs in a darkened venue in front of a dancing audience; the crowd's cell-phone lights appear like stars in a dim sky. These images recall the Sun's corona at the moment of totality and relate the possibility of light existing in darkness. YoungBoy is portrayed as experiencing illumination in dark surroundings through his commercial success and newfound wealth.

In fact, the song and video might be interpreted as suggesting that darkness allows such light to be seen. As the Moon's position at the moment of totality blocks the primary light of the Sun, and thus reveals the corona, the darkness of YoungBoy's former poverty provides a background for the bright new success—tenuous as it seems. The Moon's darkness uncovers a fragile light.

Musical Depictions: Eclipse as Darkness and Disruption

Separated by more than two hundred years, compositions by Georg Frideric Handel and Pink Floyd's Roger Waters use the image of the solar eclipse similarly. In both pieces, the eclipse appears in the sung texts as a metaphor for experiences of profound disruption. Handel's work depicts Samson, of the Hebrew Bible narrative, newly captured and blinded by his enemies. Samson compares his loss of sight to the Moon's darkening of the Sun. Waters's composition recalls the harmonious rhythms of daily life and then abruptly interrupts them, just as the Moon disrupts the Sun's light during an eclipse. The eclipse forms the focal point of both pieces' texts; it also shapes their musical elements. Using the eclipse as a central image, these two works present musical depictions of darkness and light, routine and disruption.

Georg Frideric Handel (1685–1759), 'Total Eclipse!' and 'O first created beam' from the oratorio *Samson* (HWV 57, 1741–2); lyrics by Newburgh Hamilton[8]

Recitative (Samson)
O loss of sight, of thee I most complain!
Oh, worse than beggary, old age, or chains!
My very soul in real darkness dwells!

Aria (Samson)
Total eclipse! No sun, no moon!
All dark amidst the blaze of noon!
Oh, glorious light! No cheering ray
To glad my eyes with welcome day!
Why thus depriv'd Thy prime decree?
Sun, moon, and stars are dark to me!

Chorus
O first created beam,
and thou, great word,
Let there be light . . .

In the oratorio *Samson*, a dramatic shift from profound darkness to dazzling illumination resembles the progress of a total solar eclipse. Lyricist Newburgh Hamilton and composer Georg Frideric Handel depict the biblical story of Samson, a hero whose superhuman strength has finally been overcome by his enemies. Although blinded and chained, Samson prays that he can succeed in one last effort—to crush his enemies by pushing away the building's supporting columns. Using an orchestra composed of strings and woodwinds, along with brass and percussion, Handel depicts both the darkness of Samson's blindness as well as a subsequent transformation to brightness. The depiction of darkness begins with Samson's recitative and aria. 'Total Eclipse!' shows the hero lamenting the darkness of his world. Samson uses the solar eclipse as a primary image to describe his loss of sight. The first words announce the metaphor as an exclamation of grief: 'Total eclipse!' The following statements list the lost sources of light ('No sun, no moon!') and the resulting blackness ('All dark amidst the blaze of noon!').

Handel creates a musical depiction of darkness as definitive as the singer's words. The aria's opening musical gesture, played by the orchestra, leaves no doubt as to the piece's harmonic framework: the key of E minor. The instruments strongly articulate a direct, stepwise descent from B down to E—a decisive outlining of the music's key, using the prominent pitches of scale degrees five and one. This music displays absolute harmonic certainty, anticipating the definitive announcement of the singer's words, and evoking the fixed and inalterable nature of the celestial mechanics that bring forth an eclipse.

---

[8] A performance of the aria by Aaron Crouch is available at: Carnegie Hall, 'Handel's "Total Eclipse!" from Samson.' See https://www.youtube.com/watch?v=4KoDAbNYsrg (accessed 18 October 2021).

The composer goes even further with the musical depiction of certainty. Following this opening gesture, the music leaps upwards, twice, from B to E. Each landing on the first scale degree of E—the most central pitch of the piece—occurs with the emphasized beats of the measure. In this way, the pitch E is hammered home in the listener's ear, without ambiguity or nuance. The singer's first musical statement repeats these strong, definitive melodic gestures. The words 'Total eclipse!' are conveyed with a repetition of the orchestra's first gesture, the stepwise descent from B to E. The predictable, certain progression of the melodic line lends credence to the singer's declaration of finality. The musical content of the aria expresses the same irrefutability as its text, and the same irrefutability as the movements of the Moon and Sun during a total eclipse.

On this bedrock of musical certainty, an intensification occurs as Samson sings the words 'all dark' (measures 7–8). The melody in this position leaps *down* from B to E: a musical gesture of complete finality, underscoring Samson's total loss of light. The aria ends with similar, gut-wrenching melodic expressions. A G-sharp diminished chord conveying the word 'dark' (measure 29) begins a harmonic progression that unfolds to E minor. This progression concludes with a falling scale that leads, with a sense of inevitability, back to the pitch E. Samson's aria ends, then, with a leaden descent, a definitive expression of unalterable darkness.

Yet the darkness is altered. Handel and Hamilton guide the listener away from this place of dismay. The piece's next movement (a chorus following the aria) reveals a contrasting world of brightness. The composition introduces an interlude to the narrative of Samson that is an explication of light, drawing on the creation narrative from the first chapter of Genesis. The transformation from darkness to light is heard in the text that shifts from 'Sun, moon, and stars are dark to me' (in the aria) to 'O first created beam' (in the chorus); the music reflects this profound transition. Handel's musical conveyance of these words creates an eclipse-like change—a musical transformation that could be considered an audible counterpart to the dramatic shifts between darkness and light seen in a solar eclipse. While the chorus sings the words, 'O first created beam,' the pitch E, which had been the focal point of the aria's depiction of darkness, appears again in a prominent position. But now it is transformed by its context. In the chorus's first measure, the pitch E is the fifth scale degree of an A minor chord; by the end of the second measure, it is the first degree of an E major chord. Instead of being the centre of the dark, minor harmonies, as it was in Samson's aria, the pitch is now the centre of a bright, harmonic counterpart, the parallel key of E major. While the aria created a convincing description of darkness, Handel's music with the following chorus creates a contrasting sense of illumination as it moves from E minor to E major.

Handel uses additional means to create this musical transformation. The composer mimics the striking gesture heard in the aria in the chorus that follows. The gesture conveying the words 'all dark' in the aria (measures 7–8, a descending leap from scale degree five to one), appears in an altered form in the chorus, conveying the words 'be light.' The distinctive leap down to the first scale degree of a minor key is now mirrored with an upward leap to the first scale degree of a major key. This upward leap and the shift to a major key convey the aural equivalent of brightness as the singers articulate the word 'light.' The distinctive, profiled gesture from the aria remains recognizable, even as it is transformed to depict illumination.

With these compositional nuances, the darkness of Samson's aria is countered with an equally decisive portrayal of luminescent radiance. The rapid movement from 'total' darkness to the 'first created beam' of light mimics the startling shift in brightness experienced during a total solar eclipse immediately after totality. Where Hamilton's lyrics refer to an eclipse, Handel's musical conveyance resembles one, in its dramatic shift from darkness to light. Indeed, this striking musical conveyance suggests that the darkness of Samson's blindness is temporary. The aria and chorus present a transformation from a small world of despair to a broadened perspective of hope. The total solar eclipse provides a visual template for the musical depiction of Samson, a hero who refuses to succumb in his darkest moments.

Roger Waters, 'Eclipse,' performed by Pink Floyd (released on the album 'The Dark Side of the Moon,' 1973)

All that you touch
And all that you see
All that you taste
All you feel
And all that you love
And all that you hate
All you distrust
All you save
And all that you give
And all that you deal
And all that you buy
Beg, borrow or steal
And all you create
And all you destroy
And all that you do
And all that you say
And all that you eat
And everyone you meet
And all that you slight
And everyone you fight
And all that is now
And all that is gone
And all that's to come
And everything under the sun is in tune
But the sun is eclipsed by the moon

In 'Eclipse,' written by Roger Waters and performed by Pink Floyd as part of the *Dark Side of the Moon* album, the solar eclipse also depicts the potential for profound change. Even while drawing on a similar interpretation of the eclipse, Pink Floyd's composition reflects the musical aesthetic of progressive rock in the 1970s. Handel's acoustic instruments are replaced by electric guitars, and the singers' voices are processed electronically. The song itself is produced in a sound studio, allowing for nuanced mixing of each musician's contribution. The drum kit produces a forceful, steady beat that drives listeners through the song.

This insistent beat lays the groundwork for the depiction of the eclipse. In the lyrics of Waters' composition, the Moon's disruption of the Sun's light illustrates the fragility of our lives. The song portrays disruption by first creating patterns of predictability, in both its lyrics and music. In fact, repetition is a defining characteristic of the song. The steady pulse throbs throughout; an eight-bar chord progression repeats in the harmony; and the melody duplicates its rhythmic and pitch content every eight measures as well. Within each eight-measure iteration, the vocal melody consists primarily of a single, repeated pitch, varying only by one half-step above and below. The repeating musical material conveys lyrics that describe the actions and emotions of daily living—touching, seeing, tasting, feeling, eating, meeting, loving, hating—formulated in a generalizing, repetitive way: 'All that you touch, all that you see, all that you taste . . .'

These patterns of repetition provide the foundation for a depiction of disruption. The repeated material builds a sense of expectation and predictability that is broken in the song's ending. The composition creates and then fractures established patterns. The first hints of disruption occur after six repetitions of the same chord progression. At the end of the sixth iteration, the vocal line sustains into the beginning of the seventh, rather than resting (as it had with earlier iterations). This musical disruption occurs with the final words of the sentence, 'Everything under the sun is in tune.' Ironically, the predictability of the vocal line begins to break down when the lyrics claim that everything is in harmonious order.

At the first moments of disruption—when the repeated material begins to vary—the image of the solar eclipse makes its appearance. Listeners experience the disintegration of musical patterns as they hear the words 'the sun is eclipsed by the moon.' The metaphor of the eclipse gains meaning from this musical conveyance. The solar phenomenon becomes the signal of disruption.

Complete interruption occurs with the word 'moon,' the final word of the song and also the word that completes the textual reference to the solar eclipse. In this position, the repetitive chord progression changes: specific chords (A major and A dominant seventh) are omitted in the harmonic line for the first time. Additionally, the established, repeating rhythm of the vocal line stops entirely. The melodic movement grinds to a halt, as if the patterns of daily life, referred to in the lyrics, have ceased. As the harmony and melody sustain the song's final chord, and then stop, the sound of a heartbeat is heard. The bulk of the song had focused on the frenetic minutiae of daily living, but now the music is reduced to a stark depiction of life's most essential function. The song shifts abruptly from an energetic list of daily activities to a small, fragile pulse.

In Waters' composition, the eclipse creates an uncontrollable and transformative change that stops the predictable motions of daily life. Not only with its textual reference to the eclipse, but also with its musical conveyance of this text, the song depicts a disruption of established order. The solar eclipse becomes a metaphor for the fragility of life. As the Moon can eclipse the Sun, the song implies, so can human life be eclipsed and reduced to an essential heartbeat—or less.

Musical Depictions: Eclipse as Love

In works by jazz composer Charles Mingus and classical composer John Tavener, the eclipse serves as a metaphor for relationships of love. Mingus's piece, at both the textual and musical levels, depicts an intimate encounter between two people. The image of the solar eclipse references the differences between the two lovers. In the eclipse, as in the intimate encounter, two distinct bodies encounter each other. In Tavener's work, the relationship depicted is religious and mystical. The Christian saint Paul (also referred to as Saul) experiences a religious conversion as he encounters the resurrected Christ (related in the New Testament, Acts 9). Tavener's piece depicts both the saint's conversion and his death as a willing subsumption of Paul into Christ. The eclipse serves as an image of the complete incorporation of one person by another.

Charles Mingus, 'Eclipse,' recorded with Lorraine Cusson (vocalist) in 1960 and released with the album 'Pre-Bird'

> Eclipse, when the moon meets the sun,
> Eclipse, these bodies become as one
> People all around, eyes look up and frown, for it's a sight they seldom see.
> Some look through smoked glasses hiding their eyes, others think it's
> tragic—sneering as dark meets light.
> But the sun doesn't care and the moon knows no fear, for destiny's
> making her choice.
> Eclipse, the moon has met the sun,
> Eclipse, two bodies become as one

In his recording from 1960, Charles Mingus uses the eclipse as a metaphor for a topic that was highly controversial at the time: an amorous relationship between people with different skin colours. Mingus addresses this racist aspect of his society in a composition with a musically diverse ensemble. While the vocalists are featured, the upright bass and drum kit structure the performance with a languid pulse. A saxophone inserts fast figurations between the vocalists' statements, and brass and woodwind instruments generate underlying swells of sound. The pianist plays a cadenza-like interlude before the vocalists resume their position in the spotlight.

The song's lyrics describe an eclipse as an intimate encounter—a melding—between two different celestial bodies: 'Eclipse, when the moon meets the sun. Eclipse, these bodies become as one.' As the lyrics continue, the metaphorical reference becomes clearer. Written at a time when interracial marriage was illegal in the United States, the song's text depicts negative reactions from people observing the lovers together: 'Eyes look up and frown, for it's a sight they seldom see . . . Others think it's tragic, sneering as dark meets light.' In the song's narrative, the lovers dismiss these racist reactions. 'The sun doesn't care and the moon knows no fear.' Mingus uses the metaphor of the eclipse to associate each of the two lovers with a celestial body, and to depict an experience that was considered illicit in a society guided by discriminatory, bigoted norms.

The song's melodic material makes these veiled textual references more explicit. In the performance recorded in 1960, two voices encounter each other as they convey—and embody—the metaphor of the eclipse. The first vocalizations are heard from a male voice, singing the song's first line: 'Eclipse, when the moon meets the sun.' The female voice improvises a contrasting, embellishing melody around the male vocalization. The singers thus begin the performance with an interactive musical gesture—an encounter. This musical gesture is repeated at the song's ending. Although the female voice has proven to be primary and has sung the entirety of the song's lyrics, the male voice echoes the last line of text in the final moments of the performance. Again in this position, the female voice improvises a contrasting melody without words. These interactive gestures create an aural depiction of the text's subject matter—an encounter between two different people. The poetic, metaphorical expression of the encounter—the image of the eclipse—is underscored by the song's performance.

The understanding of eclipse as an encounter is reflected in the pitch content of the melodic line, as well. The pitches used to convey the first occurrence of the word 'eclipse' (E and D) are repeated with the words 'moon' and 'sun.' A similar melodic association appears with the second occurrence of the word 'eclipse.' The pitches are similar (only the E has been moved up a sensuous half-step to F) and the words 'bodies' and 'one' repeat the pitch content. This repetition brings the words referencing the lovers ('moon' and 'sun'; 'bodies' and 'one') into the same melodic gestures that convey the image of encounter ('eclipse'). This aspect of the melody emphasizes the portrayal of eclipse as an interaction. The bodies become more than individuals as they move in intimate relationship with each other.

The song's primary focus is on this sensuous encounter. Although the lyrics in the middle section momentarily change focus (describing how others react to the lovers), the song begins as it ends, with a metaphorical depiction of the encounter between the two. The final lines of the song again articulate the image of an eclipse, and repeating melodic content again conveys the words 'eclipse'; 'sun' and 'moon'; 'bodies' and 'one,' thus drawing these images together in performance. The interplay between the male and female voices is heard at the end of the song, as at the beginning. The vocal encounter provides a framework for Mingus's lyrical meditation on physical encounter.

With these compositional nuances, Mingus uses the image of an eclipse to depict a relationship that was problematic to speak about openly at the time the song was written. The metaphor provided a way to portray controversial subject matter. The image of a solar eclipse veiled the topic enough that it could be revealed in song. The jazz composition became a vehicle for expression and a bold rejection of racist restrictions.

John Tavener, _Total Eclipse_, for solo treble, countertenor, and tenor voices, solo saxophone, chorus, and orchestra (including baroque wind and percussion instruments and Tibetan temple bowl), published by Chester Music (London, 2000).[9]

The title of John Tavener's orchestral piece, *Total Eclipse*, refers to profound experiences in the life of Paul of the New Testament: his religious conversion and his martyrdom. Tavener engages the pivotal moment of Paul's life, when he changes from persecuting followers of Christ to becoming one of the most dedicated and passionate followers himself. This moment is depicted in several accounts in the Christian New Testament, each time as an encounter between Paul and the resurrected Christ. In Tavener's composition, a dialogue with Christ overwhelms Paul at his conversion; in the musical depiction of his death, Paul's final words are, 'Even so Lord Jesus, come.' Paul is shown being eclipsed by Christ.

A unique array of historically and culturally diverse instruments allows Tavener to create strong musical contrasts as he portrays Paul before, during, and after his conversion. The first section of the piece depicts Christ's crucifixion, with timpani 'like rolling thunder' (measure 2) and extreme dynamic changes in the orchestra. The listener senses earthquakes and cataclysms. While this is occurring, the unconverted Paul (here referred to as Saul) is represented with the soprano saxophone, 'screaming' sustained notes that convulse from pianissimo to fortissimo (beginning measure 3). In notes to performers prefacing the piece, the composer states, 'his saxophone screams abuse.' This is a musical portrayal of Saul standing by at Christ's execution, echoing the New Testament narrative of Saul at the stoning of Stephen (Acts 7–8). As the chorus repeats the Greek word for 'crucifixion' (stavroménos), the saxophone taunts in rolling, arpeggiated waves (beginning in measure 34). Saul is depicted with abrasive sounds, unharmonious and unsympathetic to the upheaval of Christ's crucifixion.

A profound change, beginning at measure 47, moves the voice of Saul towards coherence with the depicted voice of Christ. As Christ sings out 'Saul,' the performance indications for the saxophone part (representing the voice of Saul) change to 'becoming more pure.' Rather than playing its own, contrasting rhythms, the saxophone's part now begins to resemble the melodic gestures heard in Christ's voice, with crescendoing whole notes and leaps of perfect fourths. From the perspective of the music, Saul no longer stands in opposition, but instead takes on a more pliant posture, questioning and becoming progressively oriented towards Christ.

Once the voice of Christ calls out to Saul, the rhythmic impulses and orchestral textures are also transformed. Time feels suspended as the intensity of the timpani rolls and the arpeggiated scalar movements cease. No instrument or voice defines an audible pulse. The only rhythmic movement comes from languid eighth notes, echoing back and forth in the parts of Christ and Saul. In a section that feels ethereal in its cessation of rhythmic drive and melodic narrative, the depiction of Saul begins to resemble the depiction of Christ. The profound change, of Saul being eclipsed by Christ, has begun. As with the celestial phenomenon, Saul's eclipse progresses

---

[9] The composer's programme notes are available at Wise Music Classical, 'Total Eclipse: John Tavener.' See https://www.wisemusicclassical.com/work/11842/Total-Eclipse-John-Tavener/ (accessed 18 October 2021).

inexorably. The person listening to Tavener's piece and the person witnessing the solar eclipse might sense something similar: large-scale events in motion, and an unwillingness and inability to resist.

Saul's transformation continues while the chorus sings 'Metanoia,' followed by 'Divine love' (Agapi). Indications in the musical score encourage the performing musicians to create music of inexpressible serenity: 'Like a day of calm sunshine' (measure 155) and 'With a tenderness beyond human comprehension' (measure 211). The voice of Saul (now depicted by the countertenor), echoes that of Christ exactly. Where Christ says 'Beareth all things' (measure 171), the voice of Saul responds two beats later with the same words and the same melodic gesture. The orchestral texture thins, while retaining the languid, suspended quality. The listener focuses only on the voice of Christ and the voice of Saul (now referred to as Paul), calling to each other with whole-note gestures spanning a fourth: 'Paul.' 'Lord.' 'Paul.' 'Lord.' This is a depiction of Christ's claim on Paul and Paul's willing and complete subsumption into Christ: a caressing, hypnotizing love song.

The final section of music, marked 'Dying into Christ' (beginning measure 277), references Paul's martyrdom. Even in its depiction of a violent death, the music maintains its peaceful, hypnotic qualities. 'Even so Lord Jesus, come,' are the final words given to the voice of Paul. Death is portrayed as the final, willing submission of Paul to Christ. The music depicts Paul's martyrdom as a welcome process, as frictionless as the Moon gliding in front of the Sun.

The title of the piece frames Paul's martyrdom—a death that could be avoided, were it not for religious devotion—as a total eclipse. For Tavener, the eclipse provides a visual template for a spiritual awakening that can overtake an individual's self-oriented plans and make religious devotion a higher priority than self-preservation. For the sake of his fidelity and attraction to Christ, Paul willingly submits to being overtaken by death, as the Sun silently submits to being blocked by the Moon. Paul practises complete self-negation twice in this musical biography, and the serene nature of Tavener's music makes clear the willing, peaceful nature of his sacrifices.

Tavener uses the image of the eclipse to refer to Paul's devotional acts. But it is also possible that Tavener's music offers insight into our devotion to total eclipses, as well. Human observers are drawn towards a celestial phenomenon over which we have no control. As eclipse-chasers, we willingly subordinate a portion of our lives to an event that defies human agency. In Tavener's hands, the medium of music depicts passionate devotion in a way that extends beyond the languages available to scientific disciplines.

Eclipse as Compositional Impulse

Ola Gjeilo, 'Eclipse,' for solo piano (2019)

In contrast to these composers, who describe various human experiences with the eclipse as an illustrative reference, Ola Gjeilo uses the eclipse as a compositional starting point. In his three-minute piece for solo piano, the composer creates a soundscape depicting the natural phenomenon itself. As Stifter imagined, the medium of musical composition—with its broad sonic and emotional palette—creates an aural illustration of an eclipse. Gjeilo takes on the challenge articulated by Stifter of depicting an eclipse using the medium of music. But while Stifter's essay on the eclipse included both a description of the natural phenomenon and his own emotional responses, Gjeilo's composition focuses only on the celestial phenomenon. His work includes no references to the emotions provoked in a human observer.

Using extremely economical musical means, Gjeilo creates a luminous, ethereal soundscape. The piece is characterized by an austere simplicity within a slow-moving, unhurried pulse. The pianist's right hand plays a melody so unadorned that every note sounds with poignancy; the left hand creates triplets with rising pitch content, forming a consistently undulating surface rhythm and a hypnotic, wave-like pattern. The generally wide spacing of the triplets' pitch content creates a sense of spaciousness and breadth.

Gjeilo uses a dramatically restricted number of musical elements in his depiction of an eclipse. Only a single pitch sounds in the melody at one time. This unembellished line most often makes stepwise movements among the pitches belonging to the key of C minor. But the composer restricts his palette of pitches even further than the key would permit: the majority of the piece uses only five pitches, which appear and reappear in shifting combinations. Similarly, the melody's rhythm makes use of an extremely limited number of possibilities. Only half notes and quarter notes appear, aligning with the slowly pulsating accompaniment in the pianist's left hand. Patterns of repetition circulate these limited numbers of pitches and rhythms throughout the piece.

Yet the music creates an engaging, hypnotic drama with these recurring patterns. Moments of subtle novelty draw the listener into an intricately shifting soundscape. As one example, the melodic gesture heard in the piece's first measures recurs in measure 9, with different harmonies underlying it. The gesture appears again in measure 14, now with the pitches doubled at the octave. In another example, the triplets in the pianist's left hand move from open patterns to closely spaced pitches, at times at the end of a melodic gesture (measure 17), and at times underneath melodic movement (measure 11). The introduction of the key's fifth and sixth scale degrees in the final portion of the work creates a nuanced change in sound; in measure 24, the omission of a note within a triplet leaves the listener in a moment of vertigo and suspension. These finely crafted changes, occurring within the piece's circulating patterns, create a texture of quiet complexity. The listener's attention is mesmerized by gradually shifting forms and frictionless transformation. Gjeilo's 'Eclipse' glides across the sky with effortless movement.

In this piece, we encounter no large-scale key changes, no pitch (apart from B natural) that appears outside of the home key of C minor, no dotted rhythms, no changes in dynamics—in short, we find none of the typical elements that composers use to create emotional narrative, suspense, and engagement. The piece eschews the hooks that normally draw a listener in. Gjeilo's composition depicts beauty so remote that it seems to open a gulf between itself and the listener.

The eclipse, as depicted by Gjeilo, partakes of neither human narrative nor human emotion. This piece does not detail an observer's experience of an eclipse or her reaction to it; instead, it provides an artistic depiction of the phenomenon itself. In this, the composer has perhaps echoed and articulated our limited possibilities when encountering the celestial phenomenon. Rather than participating, we remain observers and admirers only. In his musical depiction of an eclipse, Gjeilo reveals the immensity of the solar system and the corresponding insignificance of the human observer. The composer has perhaps successfully portrayed the aspect of the eclipse that prompted Stifter's terror—its vast remove from any human affairs or concerns.

Eclipse as Compositional Sound

Composer Wayne Grim follows Stifter's idea of combining scientific observation and musical composition in a way that could not have been imagined by the nineteenth-century writer. For his composition from 2017, Grim incorporated data gathered with scientific instruments as one source of sound alongside traditional musical instruments. In his use of scientifically generated data, Grim contributed to a growing body of collaborations among musicians, composers, and scientists.[10]

Wayne Grim, '233rd day: Before and After Totality; Total Solar Eclipse Sonification,' performed by Kronos Quartet, 21 August 2017

Grim's composition differs from the pieces examined earlier in this chapter, which were created as fixed, repeatable entities, with their details stabilized in musical notation.[11] In contrast, '233rd day' was created as an experience and reflection of a specific, singular event—the total solar eclipse of 21 August 2017. The performance of the piece can be viewed in a video that was recorded live, but because '233rd day' incorporated elements of improvisation and sonified data from the solar eclipse itself, it cannot be repeated in the same way that a pianist could, for example, perform Gjeilo's piece in multiple recitals. '233rd day' was conceived and performed as a real-time accompaniment to the total solar eclipse, an accompaniment that facilitated a shared experience for listeners and observers.

Since the composer structured the musical performance but did not fully determine its details, '233rd day' is an example of composition as collaboration. This collaboration included the eclipse itself. Through a process of sonification, the eclipse became a source of sound during the performance. NASA provided a live feed from its telescopes positioned in Casper, Wyoming (in the path of totality); Grim created software that filtered this data and converted it to sound. These feeds occasionally include the distorted voices of NASA employees working during the eclipse. Even the more traditional element of the performance—the music composed by Grim and

[10] Elie Dolgin, 'The Sounds of Science: Biochemistry and the Cosmos Inspire New Music,' *Nature* 569, 2019, pp. 190–1, also available at https://www.nature.com/articles/d41586-019-01422-0. Yuval Avital, 'Unfolding Space.' 2012, accessible at https://www.youtube.com/watch?v=2dX7ZKKP23I.

[11] Images of the performance, score, and, software are available at the composer's website: Wayne Grim, 'Musician, Composer, Sound Artist,' accessed 22 October 2021 at https://waynegrim.com/. Advance press releases described the musical event: Tom Huizenga, NPR News, 'The Sun, The Moon, and a String Quartet: Kronos Quartet Plays Live to the Solar Eclipse.' 18 August 2017, accessed 22 October 2021 at https://www.npr.org/sections/deceptivecadence/2017/08/18/544417454/kronos-quartet-plays-a-duet-with-t-the-sun-the-moon-and-a-string-quartet-kronos.

played by the Kronos Quartet—incorporates aspects of improvisation and choice. According to Grim's notes and the score (Figure 12.1), the composition was structured as a series of discrete 'cells.' The composer provided guidance for the sounds created by the string players within each cell, as well as the approximate progression of the cells. But many of the specifics were left to the performers. In addition to this improvisation, the Kronos Quartet maintained responsibility for coordinating the musical composition with the progress of the eclipse. A clock positioned in the centre of the performance space allowed them to control the timing precisely. The sounds heard by listeners during the performance are a combination of the sonification of the data feed from NASA's instruments, previously recorded sections of music played by the Kronos Quartet and manipulated electronically by the composer, and the Kronos Quartet's live performance. As the ensemble relied on the eclipse as a source of sound and a reference point for the piece's progression, the eclipse itself became a vital part of the musical collaboration.

Grim's composition was a collaborative effort, not only between performers and composer and electronic sound and acoustic instruments, but also between visual and aural elements. The progress of the eclipse was visible for participants as images from NASA's telescopes were projected onto large screens behind the string quartet. As people gathered at the site of the Kronos Quartet's performance and live-streamed the event from other locations, Grim's composition guided their viewing and listening

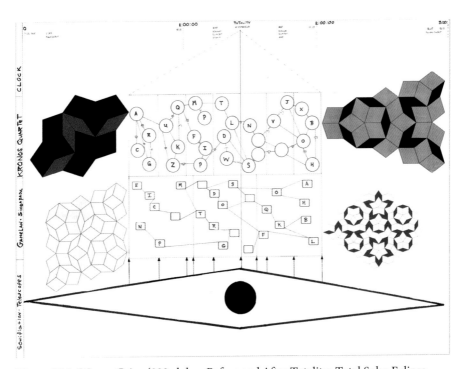

Figure 12.1 Wayne Grim, '233rd day: Before and After Totality; Total Solar Eclipse Sonification.'

experience. '233rd day' offered a way of experiencing the eclipse with an audible accompaniment. Conversely, the visual images of the eclipse provided a focal point and gave coherence to the auditory component created by Grim. Interestingly, the composer was reluctant to mirror the visual aspects of the eclipse too closely at the auditory level: 'I'm trying for something that follows an arc other than light–dark–light,' the composer stated.[12] Perhaps, with real-time images of the eclipse so prominent, Grim felt that the event's aural component did not need to duplicate the visual drama. The musical sounds added a layer of complexity to the visual experience, not a repetition or a reflection of it.

Even without duplicating the 'light-dark-light' progression of the eclipse, Grim's musical composition creates a corresponding structure with its interplay between electronic sound and acoustic instruments. The piece follows a large-scale form of three parts, with the first and third sections consisting only of electronically processed sound and the middle section (approximately 30 minutes before, during, and after totality) incorporating a live performance by the Kronos string quartet on their acoustic instruments.

The piece's three sections are also differentiated in their musical content. The first and third consist of slower rhythmic pulses and slower transformations between timbres and pitches. The middle section is marked not only by acoustic instrumental sound, but also by faster transformations. The Kronos Quartet produces soundscapes of greater contrast, and the shifts between these contrasting soundscapes occur more quickly. The third section returns to electronic sound, with less urgency of rhythmic and pitch transformations.

This musical structure creates a sense of remoteness with the first and third sections (corresponding to the beginning and ending of the eclipse) given that human involvement is present only through distorted, electronically processed sounds. In contrast, the middle section of the piece (corresponding to the time period immediately prior to, during, and after totality) contains visible human involvement, with the Kronos Quartet in front of the eclipse images and producing acoustic sound. Even while avoiding the eclipse's progression of light–dark–light, Grim shaped a three-part structure with visible human actors only in the middle section. The presence of human performers and acoustic sound creates a sense of shared experience at the eclipse's moments of greatest intensity.

Within the piece's middle section, Grim's composition accompanies the pivotal moments of the eclipse's totality with musical contrasts. Immediately prior to totality, a sense of increasing anticipation is created with rising glissandi. Listeners then hear the quartet play with energy and urgency, with double stops, pizzicati, and strong articulations of pitches' beginnings. After the first moments of totality, the urgency gives way to a sense of suspension. Listeners hear electronically manipulated pitches and a percussive bell, followed by less urgent rhythms from the acoustic string instruments. The activity and energy return a minute into totality, but the

[12] Liz Ball, Exploratorium, 'Turning Light Into Sound with Wayne Grim and the Kronos Quartet.' 13 July 2017, accessed at https://www.exploratorium.edu/blogs/eclipse/turning-light-sound-wayne-grim-kronos-quartet on 22 October 2021.

anticipated first moments of the Moon completely eclipsing the Sun are accompanied by a striking quietness.

Grim's use of music to accompany the experience of a total solar eclipse differs from the descriptive potential articulated by Stifter. Grim does not provide an aural equivalent to the visual; he does not attempt to replicate the natural phenomenon with sound. Nor does he musically depict a human observer's emotional response. In fact, Grim's music was not intended to stand alone as an independent expression. '233rd day' accompanied and incorporated a visual experience of the solar event. The prominent position given to the images of the eclipse might have allowed Grim to conceive of the aural component more freely. Since the visual images focused and engaged the audience's attention, the details of music and sound could be less controlled. Grim offers an example of music freed from any descriptive function.

Indeed, Grim's work reaches far beyond any function of music that Stifter imagined. The nineteenth-century writer could not have predicted our ability to make visual images broadly accessible. Our technology releases us from a reliance on others to describe the phenomenon of the eclipse. Rather than using music as a descriptive tool, Grim exploits its potential to draw people together into a listening, experiencing community. The natural phenomenon offered the composer sonic material to incorporate into a musical work; the musical work, in turn, provided viewers and listeners with a shared experience of the solar drama.[13]

Coda: Eclipse as Sound

Eclipse Soundscapes Project, accessible at https://eclipsesoundscapes.org/.[14]

Like Stifter, the creators of the Eclipse Soundscapes Project strive to accurately and fully describe a total solar eclipse to those who cannot see it. This ongoing project moves away from musical composition and instead relies entirely upon scientifically generated and sonified data as a descriptive medium. The project offers an additional, aural possibility for eclipse-chasers and all who are interested in a multi-sensory experience of an eclipse.

For the solar eclipse of 21 August 2017, the project developed innovative resources. A downloadable app—still available on the project's website—includes a tactile 'rumble' map. As the user moves her finger across the surface, the map creates vibrations and pitches that correspond to the brightness level of the area the finger touches. Higher pitches and more intense vibration indicate higher levels of light. The eclipse's progress can be felt and heard by tracing the high-contrast images.[15] The app also includes audio files containing verbal descriptions of an eclipse's specific features,

---

[13] The composer plans a subsequent, collaborative sonification of the eclipses in the Valley of the Gods (Utah) on 14 October 2023, and in Junction (Texas) on 8 April 2024.

[14] The project's website names as current partnership organizations: NASA's Space Science Education Consortium, the National Center for Accessible Media, the National Federation of the Blind, and Science Friday.

[15] An introduction to the app for the visually impaired is available at Blind Abilities: 'Feel, Hear and Interact! The Eclipse Soundscapes Project App Overview and Demonstration.' Accessed 27 September 2021 at https://blindabilities.com/?p=2786.

such as helmet streamers and Baily's beads. For the image of totality, as one example, the narrator uses poetic language: 'As if the moon were the center of a celestial sunflower, the corona appears as soft white petals.'

The project also gathered sounds from the animal world during the eclipse of 2017.[16] Brigham Young University (Idaho) and the National Parks Service helped to facilitate recordings in natural areas, documenting the audible changes in animal behaviour—such as birds beginning their dawn chorus as the eclipse ended.[17] Processes of sonification provide additional possibilities for creating audible descriptions of an eclipse. As described by Kelsey Perrett of the Eclipse Soundscapes Project: 'Scientists use instruments to collect radio waves, microwaves, infrared rays, optical rays, ultraviolet rays, X-rays, and gamma-rays, then convert them into audible sound waves.'[18] In developing these possibilities, the Eclipse Soundscapes Project reveals new prospects for completing Stifter's challenge—how to describe a solar eclipse to those who have not seen it themselves; and how to depict a phenomenon so extraordinary that it calls forth a multi-faceted response. Not musical composition (as Stifter imagined), but a new use of technology meets Stifter's challenge.

Conclusion: Chasing Stifter's Insight

Stifter imagined sound—in the form of composed music—as a medium that could describe the total solar eclipse and encompass the full spectrum of human responses to it. Composers of the late twentieth and early twenty-first centuries, such as Augusta Read Thomas, Lance Eccles, and Ola Gjeilo, have indeed used the celestial phenomenon as an impulse for musical creations.[19] These compositions reveal the insight Stifter offered when he looked to the medium of music for a depiction of an eclipse. Using no mathematical formulas or scientific measurements, Gjeilo's musical composition reveals the vastness of the solar system and the insignificance of the human observer. The piece invites listeners to feel the chill, perhaps even the terror, that Stifter experienced as he viewed the solar eclipse.

But more often than depicting the eclipse, composers and musicians have drawn upon it to depict human experiences with highly charged emotional content. When words fail to describe an event of life-changing import, the image of an eclipse offers a possibility. The shocking sight of the Sun being blocked by the Moon can portray the events of our lives that overwhelm our minds and tongues, rendering us speechless. In Bonnie Tyler's hit song, the sudden darkness of the eclipse conveys the despair of a person attracted to acts so illicit that they would place the perpetrator outside of the boundaries of human society. For Pink Floyd, the shocking disruption of an

---

[16] Chapter 13 (by Steven J. Portugal) documents scientific studies of animal behaviour during eclipses; Chapter 14 (Giles Harrison) considers the changing atmospheric conditions associated with eclipses.

[17] Kelsey Perrett, 'Will Wildlife Sound Off during Eclipse 2017?' Accessed 27 September 2021 at Eclipse Soundscapes Project, https://eclipsesoundscapes.org/will-wildlife-sound-off/.

[18] Kelsey Perrett, 'A Brief History of Sound in Space.' Accessed 27 September 2021 at Eclipse Soundscapes Project, https://eclipsesoundscapes.org/a-brief-history-of-sound-in-space/.

[19] Lance Eccles, *Solar Eclipse*, for recorder quintet; Augusta Read Thomas, *Eclipse Musings*, for flute, guitar, and chamber ensemble.

eclipse reveals the hidden fragility of our self-assured, daily activities. Only an event as dramatic as the Moon blocking the Sun could describe the transformation of an impoverished child into a celebrity like YoungBoy, so wealthy he can burn money. Just as the corona streams around the Moon's impenetrable darkness, a teenager rises from a violent gang community to fortune and stardom. In Tavener's work, the totality of an eclipse offers the visual counterpart for a saint's total religious devotion. These pieces show the potential for music to reveal the hidden connections between the celestial event and the experience of being human.

At the same time, sound offers a way to describe the eclipse for those who do not have access to the arresting images. Humans primarily approach the phenomenon of the solar eclipse through visual means, yet the sight calls forth a desire to share the experience. We write, we compose, we perform, we develop technologies and applications, in order to create common avenues of access to those without sight and to those who observe alongside us. The medium of sound depicts the indescribable eclipse, and the indescribable experience of viewing an eclipse brings together communities of observers. When words seem insufficient, sound—collected by scientists and organized by composers and musicians—embodies experience. As Stifter knew, the solar eclipse exceeds any single medium of description; it inspires the rich and multifaceted possibilities of human intellect and expression.

IV
ANIMALS, WEATHER, ENVIRONMENT

13

Animal Behaviour and Eclipse

Steven J. Portugal

Animal Physiology and Behaviour, Comparative Ecophysiology (including Physiology, Sensory Ecology, and Behaviour of Vertebrates)

Humans are not the only species trying to make sense of eclipses. Though animals have not developed the same cultural–artistic responses to the stunning cosmic event, many of them show a fine-tuned sensitivity to the sudden and unusual change in the Sun. Indeed, their circadian rhythm, migration patterns, and mating behaviours are set to the regularity of sunlight, twilight, day, and night in their predictable seasonal and diurnal patterns. Animals' behavioural reactions to eclipses have attracted great interest – surely in part because they confirm something about total solar eclipses that runs deeper than the human mind's painstaking calculations, representations, and interpretations. The field of animal behaviour studies is rapidly changing with the exponential growth of progress in digital technology, especially when it comes to tracking devices that, only in recent years, have opened an entirely new spectrum of knowledge about individual animals' responses to total solar eclipses: they help create an outdoor lab that did not exist before. Looking for an imaginative understanding of the vast ranges of animal experiences of the total solar eclipse, the following chapter takes us into various ecospheres and to many creatures around the world— from fireflies and spiders to elephants, from indigo buntings to zooplankton. The richly illustrated chapter is also a reminder of the variety of species around the world, their specificity, and the all-encompassing reach of the total solar eclipse as a phenomenon that touches all of them.

biologging The use of small, animal-mounted devices to measure aspects of physiology and behaviour in free-ranging animals.

circadian rhythm Biological processes recurring naturally on a 24-hour cycle.

circannual Characterized by or occurring in approximately yearly periods or cycles.

dawn chorus The sound of a large number of birds singing before dawn each day (particularly during breeding season).

Emlen funnel An inverted cone with an ink pad placed at the bottom, and wire across the top. A form of bird cage, an Emlen funnel is used to study migratory bird behaviour.

endogenous A behaviour having an internal cause or origin, rather than being driven by external factors.

lux A unit of illumination. In SI units, one lux is the amount of illumination provided when one lumen is evenly distributed over an area of one square metre.

melatonin A hormone released at night by the pineal gland in the brain. Melatonin is associated with synchronizing both circadian rhythms and seasonal events, such as moulting, hibernation, migration, and breeding.

passerine birds Birds of a large order distinguished by feet that are adapted for perching, including all songbirds.

roosting Settling for the night on a perch.

torpor A state of physical or mental lethargy, typically accompanied by reductions in metabolic rate and body temperature.

zeitgeber A rhythmically occurring natural phenomenon that can provide cues to organisms to assist in the regulation of the body's circadian rhythms.

Zugunruhe A period of migratory restlessness, typically observed at night, observed in migratory birds. An internal cue to coincide with the natural migratory period.

Animal Behaviour and Eclipses

Humans have been fascinated by how animals respond to solar eclipses for centuries (Figure 13.1). A first-known formal report comes from the Italian monk, early encyclopaedist, and scientist Ristoro d'Arezzo, who, on 3 June of the year 1239, wrote '*all the animals and the birds became frightened, and the wild beasts could be captured with ease.*'[1] From spiders deconstructing their webs to nocturnal mammals leaving their sleeping quarters, the unusual behaviour animals exhibit during eclipses is met with intrigue by human observers. Historically, part of this fascination was associated with the poor understanding of what was happening during the event. Thus, humans looked to animals in wonderment, at the strange and unusual behaviours that they started to exhibit during eclipse events. Historic accounts describe 'animals behaving as they would at nightfall.' This is despite the period of darkness associated with the absolute totality of an eclipse often only lasting between 2 and 3 minutes, and the decrease in temperature being noticeable for only around 10–15 minutes. One continuing challenge to questions of animal behaviour during eclipses is the paucity of data available; most of our relevant knowledge has been gathered in a largely opportunistic and descriptive fashion, typically lacking actual empirical data or evidence. At times it has been difficult to draw strong conclusions about the impact that solar eclipses have on animals and their behaviour. Nevertheless, there are many

---

[1] 'Convinced that "it is a dreadful thing for the inhabitant of a house not to know how it is made" (Book I, ch. 1), Ristoro collected in the *Composizione del mondo* all available knowledge of cosmology. [. . .] There are also noteworthy personal observations, including the description of the total solar eclipse that he observed in Arezzo in 1239. "The sky was clear and without clouds, when the air began to turn yellow, and I saw the body of the sun being covered little by little until it became obscured and it was night; and I saw Mercury near the sun; and all the stars could be seen . . . . All the animals and the birds became frightened, and the wild beasts could be captured with ease . . . . And I saw the sun remain covered the length of time a man can walk 250 paces; and the air and the earth began to grow cold"' (Book I, ch. 15), quoted after the most reliable text (Biblioteca Riccardiana, Florence, Codex 2164) and the edition of the first book: *Il primo libro della Composizione del mondo di Ristoro d'Arezzo, a cura di G. Amalfi*. Naples, 1888. See Francesco Rodolico: 'Ristoro (Or Restoro) Arezzo.' *Complete Dictionary of Scientific Biography*, https://www.encyclopedia.com/science/dictionaries-thesauruses-pictures-and-press-releases/ristoro-or-restoro-darezzo.

Figure 13.1 A flock of birds flies in front of the Sun during totality, seen from La Serena, Chile in 2019.
Source: Leo Caldas.

fascinating anecdotal reports spanning a significant period of time that do indeed truly suggest that, for many species of animals, eclipses have a significant and dramatic impact on their behaviour. More recently, controlled experiments have taken place, in an effort to gather more robust and systematic data. Thankfully, eclipses are predictable, meaning that research scientists can now prepare for solar eclipse events and design experiments to garner empirical data. Until recently, these sorts of experiments were beyond the realms of possibility, both in terms of technology and methodology, and with respect to access to the animals themselves. Only in the last 15 to 20 years or so have scientists really begun to plan and undertake experiments specifically to try and determine how animals are responding to eclipse events. This combination of new technologies and increased accessibility to locations around the world is now synergistically creating new possibilities for the expansion of our knowledge regarding how animals behaviourally respond to solar eclipses.

Typically, for many aspects of animal behaviour research, experiments take place within a laboratory—essentially, an artificial setting. Undertaking such experiments in a controlled environment permits control of numerous environmental aspects, for example temperature, light levels, noise, and access to food. However, it is not possible to truly recreate the scenario of a solar eclipse within a laboratory setting. This is because a solar eclipse is not simply a change in light levels. Other parameters are changing in the atmosphere that the animals will be sensing and detecting—some we are aware of, such as drops in temperature during totality, and other parameters we are likely entirely unaware of. Thus, in this instance working with captive animals under laboratory conditions is not the solution. We must therefore continue to work

Figure 13.2 A Belgrade zookeeper holds a young chimpanzee Olgica wearing solar viewing glasses while they look up into the sky to watch a solar eclipse as it passes over the Serbian capital.
Source: Reuters.

with wild animals (Figure 13.2), and study how they alter their behaviour in response to solar eclipses. While this limits the number of studies that can be undertaken, it will ensure that such studies remain ecologically relevant.

Why Should Animals Respond to Eclipses?

The longstanding reason why it is thought animals would and/or should show responses to solar eclipses is due to the role that light plays in structuring the typical daily cycle for most animals. Not only do animals notice total solar eclipses, but they also react to them. The main reason for both their awareness and their reaction has long been considered to be the central importance of light in structuring the typical daily cycle for most animals. Most (but not all) species exhibit a strong circadian rhythm (from *circa* meaning 'approximately,' and *dion*, day). Circadian rhythms include behavioural and physiological changes that follow a 24-hour rhythmic cycle. Animals use the largely predictable light–dark cycle throughout the day, and in turn throughout the year, to ensure that specific events are timed to match the animal's requirements. These events can include daily occurrences, such as foraging and sleeping, and annual events, such as breeding, moulting, and migrations. Day-light and the associated daylength are critical for most species and the biological processes taking place within them. The regularity of daylength and the alternation between night and day has led to the evolution of these internal circadian clocks that allow animals to make anticipatory adjustments in behaviour and physiology, in a predictable fashion. Circadian rhythms are controlled by endogenous molecular mechanisms (discussed later) that are then fine-tuned by the light–dark cycle. For many species, the rhythmic release of the hormone melatonin when it is dark is a key factor of their circadian rhythm. An eclipse presents an interesting scenario for an animal, as the internal molecular circadian clock is suddenly in conflict with the anticipated light–dark cycle. We know how important the light–dark cycle is for many animals, due to their strong responses to dawn and dusk.

For example, the dawn chorus, observed in many passerine birds, is typically asso-ciated with the first signs of light at dawn. The reason that songbirds choose dawn as the time to start singing is thought to encompass many possible explanations. Sound travels furthest at this time of day due to the atmospheric conditions and ambient pressure. Thus male songbirds, who are looking to both defend their territory and attract potential mates, use this time of day to ensure that their song carries over as long a distance as possible. The timing of the dawn chorus is also thought to act as a signal to females as an indicator of male quality. Males that can wake up and immedi-ately begin singing before engaging in any foraging activity are likely to be in a better physiological condition than those males who must forage first, prior to commencing their territorial singing. Thus, in this instance, light is an important signal to the birds to start singing. In contrast, nocturnal mammals respond to the decrease in light lev-els towards the end of the day as their signal to begin waking up and commencing their nightly activities. This reliance on the light–dark cycle for most species is why it has long been thought that the temporary, yet extreme, changes in the light environ-ment experienced during a solar eclipse should elicit dramatic behavioural changes in animals.

What Are the Anecdotal Reports of Animal Responses to Eclipses?

As described, early written accounts of how animals respond to solar eclipses were opportunistic and anecdotal. Nevertheless, they offer a fascinating insight into how animals are perceiving and responding to the dramatic event that an eclipse presents. Observations of behavioural responses come from all corners of the animal king-dom, from spiders, bees, and butterflies to fish, amphibians, birds, and mammals, suggesting that the impacts of eclipses are felt across an extremely broad range of species.

An eclipse across Europe on 11 August 1999 provided the opportunity to establish how dairy cows responded to the sudden onset of darkness. During totality, light lev-els dropped to less than 10 lux, essentially mimicking levels experienced at night-time. Focusing on grazing behaviour, the authors[2] determined that the eclipse had little, if

---

[2] S. M. Rutter, V. Tainton, R. A. Champion, and P. Grice, 'The effect of a total solar eclipse on the grazing behaviour of dairy cattle.' *Applied Animal Behaviour Science* 79, 2002, pp. 273–83.

any, impact at all on the cattle. Overall grazing behaviour was similar before and after the eclipse, suggesting the rapid onset of darkness did not 'trick' the cows into thinking it was night-time, nor did the changes in light levels and temperature unsettle them greatly. During the same eclipse, however, other mammalian species did respond to the period of totality, with bats seen leaving roosts and beginning foraging,[3] and diurnal squirrels reducing activity levels. Covering a wider range of species, it was noted that, during an eclipse in Zimbabwe in 2001,[4] warthogs, crocodiles, zebras, elands, and lions showed no notable changes in behaviour in response to the eclipse at all. So perhaps animals are not so sensitive to rapid short-term changes in light levels, and simply wait out the eclipse for the light to return? This would make sense if all species responded to eclipses in the same fashion, yet, they do not.

Orang-utans (Figure 13.3),[5] chimpanzees,[6] and hamadryas baboons (Figure 13.4)[7] in rehabilitation centres around the world showed a similar response to the diurnal squirrels, increasing resting time while decreasing feeding, movement, and social

Figure 13.3 Orang-utans.

Source: Photograph by Steven J. Portugal.

[3] J. L. Kavanau and C. E. Rischer, 'Ground squirrel behaviour during a partial solar eclipse.' *Journal of Zoology* 40, 1973, pp. 217–21.

[4] P. Murdin, 'Effects of the 2001 total solar eclipse on African wildlife.' *Astronomy & Geophysics* 42, 2001, p. 4.

[5] D. Prasteyo, S. S. Utami-Atmoko, D. Kurniawan, and E. R. Vogel, 'The response of orangutans to a total eclipse event.' *Journal of Tropical Biology* 1, 2021, pp. 74–83.

[6] J. E. Branch and D. A. Gust, 'Effect of solar eclipse on the behaviour of a captive group of chimpanzees.' *American Journal of Primatology* 11, 1986, pp. 367–73.

[7] C. Gil-Burmann and M. Beltrami, 'Effect of solar eclipse on the behaviour of a captive group of Hamadryas baboons.' *Zoo Biology* 22, 2003, pp. 299–303.

Figure 13.4 A hamadryas baboon, Tierpark Hellabrunn, Munich, Germany.
Source: Image by user Rufus46, Wikimedia Commons.

activity. However, a comparison study on wild primates showed quite different results. Wild orang-utans did not change their behaviour during the same total solar eclipse, while wild proboscis monkeys actually increased the amount of time spent moving and feeding. This species-level contradiction in behavioural responses to an eclipse led the authors to conclude that the predictability of feeding and comparative safety in captivity was driving the lack of, or different, responses to the eclipse in those individuals. Moreover, these observations of differences between captive and wild individuals suggests that a species-level approach to unexpected changes in light levels is not necessarily what is needed, and responses can be individual, context-dependent, and dependent on location.

Diurnal birds typically reduce activity during the totality part of eclipses,[8] but evidence of avian responses to eclipses are not entirely ubiquitous. During the total solar eclipse of Nova Scotia on 10 July 1971, warbler species (small passerine songbirds) ceased all singing activity, but sparrow vocalizations were largely unaffected. General observations during the 1974 eclipse over Western Australia[9] noted that birds stopped calling entirely and began moving towards their roosting sites. A large-scale study in Ethiopia during the eclipse on 21 June 2020 found similar results, with flying and calling behaviour significantly reduced in avian species, and roosting behaviour observed.[10] This effect was witnessed at the onset of the general eclipse, and was greatest during the period of totality.

Flying insects seem particularly susceptible to the effects of solar eclipses. Numerous accounts, albeit with small sample sizes, have documented 'odd' behaviours in insects during the totality of an eclipse.[11] The behaviour of dragonflies was described as 'abnormal', with activity ceasing immediately during totality, while butterflies reduce their movements and seek shelter. Mosquitos respond in the same manner as when they are artificially transferred to a dark area, and crickets and cicadas begin calling, as they typically would at dusk. Similarly, flies seek shelter, with one study demonstrating the number of flies on a carcass decreased by 25% during the totality of an eclipse. Spiders are particularly fascinating in how they respond to eclipses. Once light levels reach a certain low, the spiders begin to deconstruct their webs, as many species traditionally do at dusk.[12] However, they instantaneously reverse their behaviour as soon as the light levels begin to increase, and recommence web building. Or rather, they reconstruct the areas of the web they had begun to destroy when the light levels dropped. Web reconstruction in spiders in this form is relatively rare, as typically spiders make a new web each day, rather than repair existing webs. The spiders likely thought it was night-time during eclipse totality, but the return of the light just a few minutes later triggered the web-construction-sequence behaviour associated more typically with the dawn. An eclipse essentially causes an overlap of two behaviours—web construction and web deconstruction—that would never be

[8] P. Veerman, 'A record of avian and other responses to the total solar eclipse—23rd October 1976.' *Australian Bird Watcher* 9, 1982, pp. 179–209.

[9] N. Lomb, 'Australian eclipses: the western Australian eclipse of 1974 and the east coast eclipse of 1976.' *Journal of Astronomical History and Heritage* 24, 2021, pp. 475–97.

[10] S. Mekonen, 'Bird behaviour during June 21, 2020 solar eclipse in Debre Berhan Town, Central Ethopia.' *Egpytian Academic Journal of Biological Sciences (B. Zoology)* 13, 2021, pp. 103–15.

[11] C. Galen, Z. Miller, A. Lynn, M. Axe, S. Holden, L. Storks, E. Ramirez, E. Asante, D. Heise, S. Kephart, and J. Kephart, 'Pollination on the dark side: acoustic monitoring reveals impacts of a total solar eclipse on flight behaviour and activity schedule of foraging bees.' *Annals of the Entomological Society of* America, 2018, pp. 1–7.

[12] G. W. Uetz, C. S. Hieber, E. M. Jakob, R. S. Wilcox, D. Kroeger, A. McCrate, and A. M. Mostrom, 'Behaviour of colonial orb-weaving spiders during a solar eclipse.' *Ethology* 96, 1994, pp. 24–32.

temporally so close together under normal circumstances. Due to the closer-than-usual time between the two events, coupled with the incomplete nature of the web deconstruction, spiders rebuild portions of the web, rather than either continuing to deconstruct their webs or building an entirely new one. Such a rapid and dramatic response to changing light levels makes one wonder if the spiders' neurology has essentially been tricked into experiencing an entire day, somewhat rapidly.

Bees respond in a particularly dramatic fashion to solar eclipses (Figure 13.5).[13] Honeybees start returning to the hive, while relatively few individuals attempt to leave. Interestingly, those bees returning to the hive just before eclipse totality do so at a much faster speed than those individuals returning before and after totality. This accelerated return to the hive just before the total eclipse is more pronounced in drones (the stingless male worker bees), compared to the workers (non-reproductive females). This is not due to naturally differing flight speeds between the two, as workers can fly faster than drones, but rather because drones are unlikely to survive a night outside the hive, unlike workers, which can do so.

Thus, a rapid onset of darkness prompts the drones to return to the hive rapidly, for safety reasons. During the totality of the eclipse, both worker and drone bees were

Figure 13.5 Bees returning to the hive.
Source: Creative Commons.

[13] P. Waiker, S. Baral, A. Kennedy, S. Bhatia, A. Rueppell, K. Le, E. Amiri, J. Tsuruda, and O. Rueppell, 'Foraging and homing behaviour of honeybees during a total solar eclipse.' *The Science of Nature* 106, 2019, p. 4.

significantly slower in returning to the hive, with the drones being slower than the workers. This demonstrates how sensitive bees are to subtle changes in light levels, while also showing that flight during low light levels, and being out at night, are considered hazardous. Indeed, under normal circumstances bees avoid flying in the dark.

Conversely, the flight activity of beetles seems unaffected by solar eclipses. During the eclipse of 21 August 2017, researchers in South Carolina measured the number of beetles (of 16 species) attracted to a UV light during the daytime darkness of the eclipse and compared this to sampling sessions run the following day at the same time.[14] They noted there was no significant difference in flight activity in the beetles between sampling sessions, leading the authors to conclude that solar eclipses did not affect beetle flight activity. During the same eclipse, however, fireflies (another group of beetle species) in North Carolina began flashing during the eclipse totality; a behaviour that normally only occurs at night (Figures 13.6 and 13.7).

Figure 13.6 A firefly.
Source: Creative Commons.

---

[14] M. L. Ferro, 'Beetles collected in light-traps during the North American total solar eclipse of 21 August 2017 with a review of arthropod behaviour during solar eclipses.' *Amateur Entomologists' Society* 129, 2020, pp. 348–59.

Figure 13.7 Fireflies in the woods.
Source: Creative Commons.

Similarly, during the total eclipse of 1954, on 30 June, European June beetles in Poland emerged from the ground, where they typically remain buried during daylight hours, and began flying around the treetops. Such behaviours were not witnessed during a partial eclipse, suggesting that the low light levels experienced during a full eclipse are required for the beetles to mistake the eclipse for dusk—partial eclipses are not sufficient to elicit a change in behaviour.

The total solar eclipse of 21 August 2017 was studied in particular detail concerning animal behaviour, due to the time of day and time of year the eclipse took place. Although the period of totality only lasted a maximum of 2 minutes 41 seconds, the eclipse occurred in late summer and near midday, potentially resulting in maximum impact, as air temperature and solar illuminance are at both their daily and seasonal maxima at this point.

In the aquatic environment, Rohu fish significantly reduced their feeding rate during totality, thought to be in response to the decrease in water temperature. Perhaps some of the most rapid and distinct responses to solar eclipses can be found in reef-dwelling fish (Figure 13.8).[15]

On 26 February 1998, there was a solar eclipse over the Galapagos Islands, with a period of totality lasting three and a half minutes. The magnitude of the darkness matched that recorded at night, to the extent that planets and stars could be seen. During this period, diurnal fish species significantly decreased activity levels. For

[15] S. Jennings, R. R. Bustamante, K. Collins, and J. Mallinson, 'Reef fish behaviour during a total solar eclipse at Pinta Island, Galapagos.' *Journal of Fish Biology* 53, 1998, pp. 683–6.

Figure 13.8 A school of reef-dwelling *Lutjanus kasmira* (common bluestripe snapper).
Source: Creative Commons.

these species, night-time brings a considerable increase in potential predators and the chance of being eaten. It is perhaps, then, unsurprising that diurnal reef fish immediately seek shelter in rocks and crevices when the light levels drop during an eclipse. Marine zooplankton (Figure 13.9) showed a similar dramatic response to the eclipsing Sun. The plankton responded to the period from the midpoint of the eclipse to its end in the same manner as they would normally behave at dawn.[16]

Zooplankton are vertical migrators, being found at greater depths during daylight hours, away from potential predators, and moving up towards the surface at night to feed. The response of the plankton to the incoming darkness was rapid, with quick decreases in depth noted. Such an event will involve billions of these tiny, microscopic animals moving, and is widely considered the largest migration, over any timescale, to take place globally.

As discussed, in an attempt to take investigations of eclipses from anecdotes into the realms of empirical data, in the last 20 years scientists have tried to plan experiments to coincide with periods immediately before, during, and after a solar eclipse, to fully determine behavioural responses. These studies aim to have some element of experimental control by having before, during, and after comparison groups. Designing specific experiments to study how animals respond to eclipse events is pivotal,

[16] K. Sherman and K. A. Honey, 'Vertical movements of zooplankton during a solar eclipse.' *Nature* 227, 1970, p. 12.

Figure 13.9 A vastly magnified view of zooplankton.
Source: Creative Commons.

because prior observations had been heavily biased towards species which had been, or were, showing dramatic or extreme responses to the sudden change in light levels. This is of course because these species would be acting differently, in obvious ways. Different individuals or different species not responding to such events would largely go unnoticed or be considered not interesting enough to warrant people document- ing their behavioural responses to eclipse events. Therefore, this may have actually offered an overly dramatic view of how animals respond to solar eclipses, and indeed early observers often wrote in such a way to assume that these casual observations would be commonplace across most species around the world. However, what these controlled experiments actually began to confirm was that there is a species-level difference in responses to solar eclipses; different species respond differently to unex- pected darkness. Interestingly, this suggests that for some reason, different species are perceiving the changes in light levels differently, and thus there is not one singular response to an eclipse event, observed across all animal species.

Another study undertaken during the 21 August 2017 total eclipse used a continental-scale radar network to track flying animals over the United States.[17] Com- bining data from 143 weather radar stations, the authors wanted to establish if the low light levels associated with the eclipse would prompt insects to take flight, as is

[17] C. Nilsson, K. G. Horton, A. M. Dokter, B. M. van Doren, and A. Farnsworth, 'Aeroecology of a solar eclipse.' *Biology Letters* 14, 2018, 20180485.

typically witnessed at dusk. The study found that overall activity levels dropped in the period leading to totality, but a small peak in activity during totality was observed. Such a pattern in flight behaviour suggests, interestingly, that while the changing light levels are sufficient to suppress the activity levels of diurnal species, they are not sufficient to initiate nocturnal activity.

Why Is It Important for Animals to Respond to Changing Light Levels?

Animals structure their day to maximize their foraging and reproductive opportunities while looking to minimize their chances of being predated. Mistiming activity levels, for example, may mean insufficient foraging occurs during daylight hours. For diurnal species, the onset of dusk is the cue to return to safe roosting and sleeping sites and to ensure travel at a time when sight is still possible.

Throughout the year, animals must ensure breeding coincides with the greatest abundance of food, and must avoid extreme weather. Behavioural reactions, as opposed to those which are physiological or neurological, have been the most commonly documented responses in animals to solar eclipses, due to it being the most obvious aspect of animal biology to record. However, what we see in terms of changes in behaviour is akin to the tip of the iceberg concerning what will be occurring inside the animal. Outward responses to changes in light levels are accompanied by several physiological and biochemical responses within the animal.

Diurnal birds and mammals display a distinct increase in heart rate and body temperature at dawn when the animal wakes up, and subsequently a gradual decrease in these parameters as the animal prepares for sleep. This decrease in heart rate and body temperature observed overnight in many diurnal mammals and birds is important for the animal in terms of energy conservation. This is particularly vital during the winter months when ambient night-time temperatures can drop to very low levels throughout much of the temperate world. This decrease in body temperature and heart rate observed in animals means they are not using as much energy to maintain themselves to a specific body temperature as they do during the daylight hours when they are awake. For some species, particularly small birds and small mammals, this reduction in body temperature can be very extreme, and miscalculations are costly, at times fatal. Extreme drops in heart rate and body temperature are referred to as torpor and this is a vital energy-saving mechanism that allows small animals to survive cold nights and save energy for the next morning. Both field-based and laboratory studies have shown that changes in light levels during the night have significant impacts on the physiology of diurnal species, which at that point are in the natural resting phase of their daily cycle.[18] For example, diurnal mammals that are sleeping during the night show a less dramatic reduction in heart rate and body temperature in the presence of artificial anthropogenic lighting. Over a long period, this is likely to have significant implications for the energy budgets of these animals; they are not getting

---

[18] T. Le Tallec, M. Perret, and M. Thery, 'Light pollution modifies the expression of daily rhythms and behaviour patterns in a nocturnal primate.' *PLoS One* 8, 2013, e7925.

the required reduction in heart rate and body temperature overnight which enables them to conserve energy. In laboratory-based studies, this night-time reduction in heart rate and body temperature in response to the dark has been recorded in multiple species. What research has not yet produced are physiological measurements taken from wild animals during a solar eclipse. It is largely unknown how the physiology of animals responds to unexpected darkness, and whether their bodies instantaneously prepare them for sleep.

Over the last three decades, there has been one crucial technological development that now helps bridge the worlds of laboratory-based settings and fieldwork for recording these physiological parameters in wild animals: the advent of biologging devices has dramatically changed the possibilities of the discipline and increased our knowledge of animal behaviour and physiology. These devices are either attached to the animal or implanted within its body, and can subsequently record for long periods, sometimes for many years. Biologging revolutionized how we understand animal movement, and how animals respond to specific events within their annual cycle and changes in their environment.

Biologging technology can incorporate elements which measure the global positioning of animals (through GPS technology), may record depth and altitude to determine how deep an animal is diving or how high it is flying, and can contain accelerometers to record every movement an animal undertakes, up to 300 times a second. Biologging has helped us understand how animals respond to both natural light cycles and artificial anthropogenic light, and also how they respond to natural yet unusual events in the annual cycle. One such example utilized biologging technology to study wild barnacle geese (*Branta leucopsis*), a population of which breed in Svalbard, and winter in south-west Scotland.[19] In this study, biologgers were measuring the heart rate and body temperature in wild birds for around 12 months. The authors noted an odd yet recurring phenomenon throughout the winter months when the geese were experiencing around 15 hours of darkness per day. At rhythmic points throughout the winter, the barnacle geese showed a rapid increase in heart rate and body temperature between 20:00 h and 04:00 h. Further investigation uncovered the likely cause behind these unusual night-time winter peaks in these important physiological parameters; they coincided with a specific phase of the Moon. Throughout the cycle of the Moon, as it orbits Earth, there are two distinct points which are referred to as apogee and perigee. Apogee is the point at which the Moon is furthest from Earth (*c.*253,000 miles), while perigee is the point at which the Moon is closest (*c.*226,000 miles). The nocturnal peaks in heart rate and body temperature in the geese coincided with perigee, when the Moon looks considerably bigger in the sky and emits more light. This effect was exacerbated when a full Moon coincided with perigee, yet was absent on nights where cloud cover meant the Moon would not have been visible in the sky. These data demonstrate that naturally occurring phenomena of light level do interrupt the typical circadian rhythms of animals, and thus that it is likely that eclipses would be driving changes in the physiology of diurnal species.

---

[19] S. J. Portugal, C. R. White, P. B. Frappell, J. A. Green, and P. J. Butler, 'Impacts of "supermoon" events on the physiology of a wild bird.' *Ecology and Evolution* 9, 2019, pp. 7974–84.

Why Do Eclipses Particularly Confuse Birds?

Many reports on how animals respond to eclipses have focused on birds.[20] This may be due to their conspicuousness, being typically easier to observe than certain cryptically behaving mammals, or could be a result of their sensitivity to changes in light. The avian circadian pacemaking system—that is, how birds structure their time and day—is complex, consisting of three components that contribute to the regulation of general behaviour and physiology, and maintain its rhythmicity:[21] (1) the pineal gland, by rhythmically releasing the hormone melatonin; (2) the hypothalamic oscillator, possibly acting through neural output pathways; and (3) the retinae of the eyes. Several lines of evidence suggest that these components interact with each other to produce a stable organismic circadian rhythmicity. Most birds have pineal glands that are well developed and rhythmically release melatonin. The pineal gland is often colloquially referred to as the third eye, and studies which have covered the pineal gland demonstrated that the birds were no longer able to sense light–dark cycles to the extent they previously could.

Even when birds are under constant conditions (e.g. continuous darkness), rhythmic melatonin can be found in the blood, demonstrating firstly that melatonin production is regulated by endogenous circadian oscillators, but also how important melatonin is in structuring a bird's daily cycle. With one exception (Japanese quail, *Coturnix japonica*), all avian pineal glands studied contain autonomous circadian oscillators that rhythmically regulate melatonin synthesis. Pineal cell cultures from the chicken express circadian oscillations of melatonin release for several cycles in constant darkness with a period close to 24 hours. Changes in the light–dark cycle are mirrored by a corresponding shift in the melatonin rhythm and the phase shifts persist after transfer to darkness, showing that the underlying circadian oscillator was entrained. Circadian rhythms in melatonin release were also found in cultured pineal glands of house sparrows, where the amplitude of *in vitro* melatonin release was high for a week or more.[22] If the pacemaking system of birds can continue under conditions of total darkness, for at least a short period, it raises the question of why some bird species are sensitive to solar eclipses.

In house sparrows (Figure 13.10), a pinealectomy abolishes the circadian rhythm of locomotor activity, body temperature, and feeding. There are several lines of evidence that these effects are primarily due to the lack of a rhythmic melatonin signal after the pineal gland had been removed. Rhythmicity could be restored when the pineal gland of a sparrow was implanted into the anterior chamber of the eye of an arrhythmic pinealectomized host. The emerging rhythm had the phase of the rhythm of the donor bird, indicating that circadian clock properties were transplanted with the pineal gland.

---

[20] S. Wiantoro, R. Pramesa-Narakusumo, E. Sulistyadi, A. Hamidy, and F. Fahri, 'Effects of the total solar eclipse of March 9, 2016, on the animal behavior.' *Journal of Tropical Biology and Conservation* 16, 2019, pp. 141–53.
[21] E. Gwinner and R. Brandstätter, 'Complex bird clocks.' *Philosophical Transactions of the Royal Society London B* 356, 2001, pp. 1801–10.
[22] R. Brandstätter, V. Kumar, U. Abraham, and E. Gwinner, 'Photoperiodic information acquired *in vivo* is retained *in vitro* by a circadian oscillator, the avian pineal gland.' *Proceedings of the National Academy of Sciences of the USA* 97, 2000, pp. 12324–8.

Figure 13.10 A male house sparrow in Prospect Park, New York.
Source: Creative Commons.

The pineal gland is also involved in regulating complex circadian performance. Studies have indicated that photoperiodic information acquired by house sparrows and reflected in the pattern of melatonin release *in vivo* are retained for some time in isolated pineal glands cultured *in vitro*, thus suggesting that the pineal gland can store and retain biologically meaningful information about time. The temporarily stored information might be used by birds to determine whether day length increases or could allow birds to buffer the effects of weather-dependent short-term variations in photoperiod which might impair precise photoperiodic time measurement.

Is it likely that a system composed of several interacting components with distinct physiological properties—such as the retina, the pineal gland, and the hypothalamic oscillator—allows a higher degree of adaptation to changing environmental situations than a highly centralized singular system. To be able to regulate behaviour and to counteract environmental disturbances (or *zeitgeber*; 'cues'), the system that interprets the ambient conditions needs to have a high level of self-sustainability and stability, as well as remaining flexible, and it would seem that the special feature of the pineal melatonin rhythm dictates this. This feature is particularly important for birds that are exposed to different photic environments, for enabling high-latitude birds to remain synchronized with the low-amplitude *zeitgeber* conditions in midsummer, or for allowing birds to adjust their behavioural rhythms to rapidly changing environmental conditions such as those before, during, and after a solar eclipse. It is this mechanism that may explain why there is between-species variation in how birds respond to eclipses. In theory, the temporarily stored information in the pineal gland should cushion birds from responding to unexpected and comparatively brief periods of darkness, yet birds have been shown to respond to eclipses. What is needed is an extensive cross-species comparison to understand the links between those species which are seemingly 'tricked' by the sudden light change during an eclipse and those species which are not.

Why Could Total Solar Eclipses Cause Issues for Birds?

For many species, migration is a key feature of their annual cycle, and in birds the time to migrate is triggered primarily by daylength. Animals migrate for a multitude of reasons, including predator avoidance, to escape particularly harsh weather, and to take advantage of seasonal resources. Migration is particularly evident in birds. Most bird species—but not all—typically migrate twice a year; in the spring, heading towards their breeding grounds, and in the autumn, returning to their wintering quarters. It is estimated that over 50 billion birds around the world migrate each year, often in flocks large enough to be detected by radar. Some mammals undertake seasonal migrations over smaller distances, albeit in equally impressive numbers. Long-distance migrations, however, are particularly evident in birds, due to the greater areas they can cover with flight. One of the most famous examples of amazing migrations is that of Arctic terns. Arctic terns summer and breed in the Arctic, but winter in Antarctica. The birds fly approximately 35,000 kilometres during this journey, meaning they fly 90,000 kilometres a year. Over the course of its 25-year lifespan, an individual Arctic tern would have flown to the Moon and back three times. It is the mechanism by which birds know when and where to migrate that means they can be potentially susceptible to eclipses. Thousands of bird species migrate by night; they are nocturnal migrants. Migrating at night-time can reduce your exposure to predators, help you keep cool, and enable you to use the stars to help you navigate. It is this propensity to travel at night that may explain why birds can be confused, albeit temporarily, by eclipses.

Bird migration is a complex phenomenon that requires several behavioural and physiological adaptations, and these adaptations are particularly complex in species that are normally active during the day but perform these migratory flights mainly or exclusively at night. In these species, migration is associated with a major switch in the circadian pattern of activity. Many migratory passerine birds display nocturnal activity, or 'restlessness,' during the migratory seasons when kept captive in cages. Often referred to as 'Zugunruhe,' this fascinating behaviour is striking for someone to witness for the first time. During the middle of the night, the bird will wake up, and begin flapping its wings at the same intensity and frequency as if it was mid-flight during migration. Migratory restlessness in willow warblers is most intense in September, the time when free-living conspecifics (members of the same species) cross the Sahara Desert at a high speed, while captive birds initiated spring migratory restlessness during February and March, the time of departure of their free-living conspecifics.

Subsequently, migratory restlessness slowly fades away just as the actual migratory performance of free-living birds declines. Fascinatingly, the degree of migratory restlessness varies between populations of the same species. Blackcaps, *Sylvia atricapilla*, from Germany (Figure 13.11) develop much more migratory restlessness than do conspecifics from the Canary Islands, corresponding to the fact that German birds migrate over relatively long distances to the Mediterranean area and north Africa, while Blackcaps from the Canary Islands are essentially non-migratory.[23] When eclipses occur during the spring and autumn, they can act as a temporary, but misleading, cue to migratory birds to depart on a migratory flight.

More conspicuous is a relationship between migratory distance and the overall amount of nocturnal activity; the longer the distance a species normally travels, the more nocturnal activity tends to develop. Even populations of the same species travelling different distances may show corresponding differences in the amount of migratory restlessness. But these birds do not only have an innate desire to be actively engaging in nocturnal flight, they angle their flight to the same direction that they would be travelling in if they were free-flying in the wild. Early studies that established the night-time migratory restlessness of birds used a simple but highly effective

Figure 13.11 A Eurasian blackcap, photographed from Lullington Heath, East Sussex, by Ron Knight.
Source: Wikimedia Commons.

[23] F. Pulido, P. Berthold, and A. J. van Noordwijk, 'Frequency of migrants and migratory activity are genetically correlated in a bird population: evolutionary implications.' *Proceedings of the Nationa. Academy of Sciences of the USA* 93, 1996, pp. 14642–7.

device known as an Emlen funnel. This is an inverted cone, with chalk or an ink pad placed on the bottom. When a bird exhibits nocturnal migratory restlessness and attempts flight, its feet leave a trace on the walls of the funnel. Emlen funnels were used in a fascinating study by Stephen Emlen in 1967, looking at how indigo buntings (Figure 13.12) used the night sky to navigate.[24] Using a planetarium, Emlen was able to manipulate the night-time star alignment that baby buntings were seeing as they grew up. He raised indigo buntings from hatching in a planetarium. Instead of circling the Pole Star, Emlen's artificial night sky rotated about Betelgeuse, a star in Orion.

That made the stars move slowest and arc most tightly around Betelgeuse, in essentially an unnatural fashion. When his baby indigo buntings grew up, and the time for migration approached, they felt impelled to fly away from Betelgeuse. They behaved as if they had formed a concept of 'south,' as the direction opposite the place that Betelgeuse occupied in the sky. Emlen surmised that outdoors in nature, baby indigo buntings in their nests spend their nights studying the starry sky. They notice that the stars close to the Pole Star move the slowest, tracing the path of a small circle around

Figure 13.12 Indigo bunting. Image by Dan Pancamo.
Source: Wikimedia Commons.

[24] S. T. Emlen, 'Migratory orientation in the indigo bunting. Part II: mechanism of celestial orientation.' *The Auk* 84, 1967, pp. 463–89.

that star. From then on, the night sky is a map and compass for them. And it will guide them on their migrations.

Do Animals in the Arctic and Antarctic Respond in the Same Way?

Sadly, data is lacking about how animals in polar regions respond to eclipse events. This is due, in part, to the lack of human observers present to make the required observations, but also because polar regions make up a comparatively small proportion of the Earth's land area. Zoologists are very interested in determining how polar animals respond to eclipses, because of the unusual light–dark cycles these species are experiencing. During the summer, tens of millions of seabirds enjoy the constant daylight that the Arctic provides, while male emperor penguins must endure constant darkness while they incubate eggs during the Antarctic winter. Given most species use light cues to structure their daily cycle, how polar species manage without these cues is a matter of interest.

Polar regions are challenging environments for animals and their circadian systems because they lack light–dark changes twice each year. In the high Arctic, the light intensity remains below civil twilight for more than two months in winter, and in late spring and summer, there is continuous light for five months, although daily changes in light intensity do occur. Laboratory studies have demonstrated that exposure to continuous periods of light or dark can suppress or disrupt circadian function in many animals, essentially meaning they don't know how to structure their day. In addition, rhythmic hormonal releases associated with the light–dark cycle are dampened or absent under the continuous light of the polar summer in Lapland buntings, Svalbard ptarmigans (Figure 13.13), seals (Figure 13.14), and penguins (Figure 13.15).[25]

Even minor changes in light intensity during the winter months can inhibit the production of melatonin in polar environments. During the Arctic winter, even the period of twilight at midday is sufficient to suppress melatonin production in Svalbard ptarmigans. This demonstrates that some species are very responsive to subtle changes in light levels, suggesting that even a short period of light intensity change, such as that experienced during an eclipse, could be sufficient to change an animal's behaviour and physiology quite dramatically. Indeed, polar species may show the most extreme response to eclipses, due to the dramatic nature of the change in light levels, particularly if they occur during the summer when constant daylight would be the standard condition. An interesting study[26] on Svalbard ptarmigans demonstrated that for these polar species, alternate mechanisms might aid them in structuring their time, which in turn may suggest some polar species at least may be potentially unresponsive to solar eclipses.

[25] K. A. Stokkan, A. Mortenson, and A. S. Blix, 'Food intake, feeding rhythm, and body mass regulation in Svalbard rock ptarmigan.' *Americal Journal of Physiology* 251, 1986, R264–7.
[26] E. Reierth and K. A. Stokkan, 'Dual entrainment by light and food in the Svalbard ptarmigan (*Lagopus mutus hyperboreus*).' *Journal of Biological Rhythms* 13, 1998, pp. 393–402.

Figure 13.13 Svalbard rock ptarmigan. Image by Per Harald Olsen.
Source: Wikimedia Commons.

Figure 13.14 A seal looking into the light.
Source: Creative Commons.

Figure 13.15 Penguins frolicking in sunshine in Antarctica.
Source: Creative Commons.

In 1998, Reierth and Stokkan proposed that birds' feeding themselves could potentially act as a *zeitgeber*. With animals that have continuous access to food in their natural environment (e.g. the herbivorous Svalbard ptarmigans), it would be advantageous to have *zeitgebers* that were not light only, particularly during the autumn when fat deposition is imperative to ensure sufficient energy stores for the winter. Using food availability as a *zeitgeber* may allow the birds to enjoy a longer feeding period each day than if the rapidly declining daylength was the only cue. Svalbard ptarmigans feed continuously throughout the summer months. In late winter, spring and autumn, when there is a daily light–dark cycle, Svalbard ptarmigans are mainly active during the light part of the day, with peaks of activity in the morning and evening. The photoperiod and the activity period do not change in parallel, however, causing the phase relationships between the morning and evening peaks of activity and the light–dark cycle to change progressively. In captivity, ptarmigans tended to begin and end their daily visits to their food box at about the same time as they did the previous day, suggesting feeding itself acts as a *zeitgeber*. During *ad libitum* feeding and light–dark cycles, Svalbard ptarmigans displayed a bimodal, diurnal activity pattern that is typical for diurnal birds. Through controlling their access to food, the researchers were able to shift the morning and evening peaks of activity in the ptarmigans, indicating that the timed access to food acted as a *zeitgeber*. When the birds had established a stable phase relationship with the food access interval, a marked anticipatory activity preceded the feeding interval by approximately 1 hour, demonstrating that species can act independently of the light–dark cycle.

Conclusion

Most species of animals at least partially use the light–dark cycle to structure parts of their daily and annual cycle. The rhythmic predictability of daylength provides a reliable environmental cue to organize time-critical events such as breeding and migration. Animals have evolved mechanisms to ensure their bodies are not 'tricked' by short-term rapid changes in the environment, such as bad weather or cloud cover, yet despite this, many species show remarkable changes in behaviour in response to solar eclipses. It is possible that the infrequency of solar eclipses results in a lack of general adaptation in response to them. However, a topic that is surely worthy of further investigation is the variation in behavioural responses, both within and between species. If responding to the instantaneous light-level changes associated with an eclipse was generally always advantageous, then all individuals of a species should demonstrate the same response. That this is not the case suggests that perhaps the impacts of eclipses are minimal.

Recent technological advances, particularly in biologging technology and remote monitoring, will allow us to study behavioural and physiological responses to eclipses in wild animals at a level of detail not previously possible. These advancements will be harder to implement in remote regions or inaccessible places, where a limited amount of study has taken place previously. These environments, such as the Polar regions, are particularly at risk from global climate change and disturbance. Understanding how the species which live there respond, generally, to changes in their environment, no matter how instantaneous, will provide valuable data for their capacity to adapt to more general environmental change. Environmental light pollution, a relatively recent phenomenon, has been repeatedly demonstrated to have significant deleterious effects on multiple facets of a species' biology, including energy expenditure, reproductive success, and longevity. The fact that many, but not all, species respond strongly to solar eclipses suggests that finding mitigations and solutions to artificial light pollution may indeed be problematic.

14

Weather and the Solar Eclipse

Nature's Meteorological Experiment

Giles Harrison

Meteorology, Atmospheric Physics, Eclipse Meteorology

This chapter brings our attention back to the larger context of human experience of the environment: the weather, commanding the unpredictable visibility or invisibility of the total solar eclipse. Fascinatingly, despite all human calculations of total solar eclipses across millennia, it is neither knowledge nor logistics, but rather the volatile weather, which has the final say about the eclipse-chaser's day. Meteorology is generally known as the branch of science concerned with the processes and phenomena of the atmosphere, especially as a means of forecasting the weather. It is also involved with the study of the climate and weather of a region. The exciting dynamics of weather and total solar eclipses not only determine the experience of the viewer from Earth (as we have seen in the former chapters of human and animal responses to eclipses), they also have their own interactions, which Giles Harrison comes to consider as a big meteorological experiment set up and run by Nature herself.

citizen science Scientific research and/or data collection that involves the collaborative efforts of non-scientists.

compositing (or **superposed epoch)** Statistical technique in which changes following a defined stimulus are combined to extract the generic response to that stimulus.

ECMWF European Centre for Medium Range Weather Forecasting.

ensemble forecasting Making multiple slightly different weather forecasts to assess the spread inherent in their predictions.

Met Office Popular name for the the UK's national weather service, formerly known as the Meteorological Office.

Newton's laws The three laws of motion and law of universal gravitation propounded by Sir Isaac Newton.

physical law Simplified description of conclusions arrived at from scientific investigations.

radiosonde Instrument package carried beneath a weather balloon, returning its observations to a ground station by radio.

solar radiation The energy emitted by the Sun.

thermodynamics The science concerned with the relations between heat and mechanical energy, or work, and the conversion of one into the other.

tropical storm A rapidly rotating storm system originating over tropical seas with sustained and strong surface winds.

The astronomical predictability of a solar eclipse contrasts sharply with the variability always present in Earth's weather, which is the window through which the human experience views the eclipse—or not. Weather conditions for viewing an eclipse can be uncertain, for example in important details such as cloud amount, right up until the moments of totality, except at locations where the climate is reliably stable. It is, however, incomplete to regard the weather as merely providing a passive but essential accompaniment to the astronomical spectacle, as removal of sunlight also has direct consequences for the atmosphere, such as near-surface cooling and changes in the wind. A solar eclipse can therefore be characterized as providing a well-defined 'cause,' with the weather's less predictable response the resulting 'effect.' This allows for interpretation of an eclipse as a natural meteorological experiment, within which long-reported anecdotal eclipse-generated phenomena can be investigated and understood. Practical applications which follow include testing and improving weather forecasting models, and the management of renewable energy supplies during an eclipse.

Introduction

Some natural phenomena have existed throughout human history, such as lightning and solar eclipses. Solar eclipses can provoke profound responses of awe and wonder, evident in the range of ways in which the inspiration they can provide has been expressed.[1] Whilst solar eclipses are usually regarded as rare and special events, they could also be considered as mundane consequences of the known motion and geometry of the solar system.[2] To do this would be to neglect the human experience of an eclipse which results from both the spectacle generated and, for some viewing locations, heightened anticipation due to uncertain weather. Although the laws of physics allow accurate mathematical predictions of the timing and extent of the solar eclipse as a celestial event, the same laws are unable to provide an exact determination of the accompanying meteorological conditions.

In this chapter, the differences between weather forecasting and eclipse forecasting will be explored and contrasted, and the complex relationship between weather and eclipses discussed. An eclipse can be understood as generating a natural experiment with the weather, which in turn may lead to improved weather forecasts, not least for providing better predictions of eclipse viewing circumstances and effective management of solar and wind energy generation during eclipses. The effect of weather—or of no weather—on viewing an eclipse is a central part of the human eclipse experience; almost no other event puts such demands on weather forecasting, essentially seeking to know the behaviour of clouds—and in particular their absence, during a few special minutes—many days or even weeks in advance.

---

[1] See Ian Blatchford: 'Symbolism and Discovery: Eclipses in Art.' *Philosophical Transactions of the Royal Society A: Mathematical, Physical and Engineering Sciences* 374 (2016), pp. 1–26. See also Olson (Chapter 10), Lange (Chapter 11), and Stratton Hild (Chapter 12).

[2] See also McLeish and Frost, Chapter 1.

Predictability: Eclipses and Weather

As the historical chapters of this volume illustrate, the desire to reliably predict the future is surely an ancient human aspiration, which to some extent weather forecasting attempts to address. There are, however, fundamental differences between predictions based on scientific methods and those that might be described in general terms as mystical methods, as perhaps exemplified by the soothsayer in Shakespeare's *Julius Caesar*. Physical science advances by a combination of theory and experiment, leading to the ability to make predictions about the natural world. Successful predictions, verified by experiment, are in themselves a test of theory, and unexpected experimental findings can indicate the limitations of existing understanding.

Prediction seems to have a complex status in human societies, and can also be exploited in various ways.[3] Christopher Columbus apparently used knowledge of the 1504 lunar eclipse to provoke the continued supply of provisions for his landing party in Jamaica.[4] However, the nineteenth-century founder of the UK's Meteorological Office, Robert Fitzroy,[5] was ridiculed for his weather predictions, despite his strong motivation to generate social benefit.[6] A difficulty then, as now, is that individual forecast failures can be held as examples of a widespread generalized failure, rather than taking a wider view that recognizes the presence of unpredictable fluctuations and weighs up whether, on balance, the forecasts are providing a net benefit. Even with greater societal recognition of the value of weather forecasts, and increasingly those of climate change, the modern Met Office still has critics.[7]

Disinclination towards scientific predictions in general, weather prediction in particular, or merely undesirable weather are all possible reasons for negative societal perceptions. A more light-hearted early twentieth-century view was that weather is described by Ambrose Bierce as "a permanent topic of conversation among persons whom it does not interest." Some early weather predictions used mechanical technologies, 'prognosticators' such as Fitzroy's storm glass, which was essentially a thermometer.[8] Fitzroy's Met Office also exploited the spread of telegraph networks to obtain barometer readings from coastal and other stations, to produce charts from which likely weather changes could be suggested. In contrast, a method underpinned by physical laws and mathematical calculations seems more likely to provide well-defined expectations of an event or occurrence. Whilst deterministic

[3] See Roos, Chapter 7.
[4] See Michael Hoskin: *The Cambridge Illustrated History of Astronomy*. Cambridge, 1996.
[5] Robert Fitzroy (1805–65), Naval officer and Captain of HMS *Beagle*. See J. Burton: 'Robert FitzRoy and the early history of the Meteorological Office.' *The British Journal for the History of Science* 19 (1986). https://doi.org/10.1017/S0007087400022949.
[6] M. Walker: *History of the Meteorological Office*. Cambridge, 2011, https://doi.org/10.1017/CBO9781139020831.
[7] See 'The Conversation: What's the point of the Met Office,' https://theconversation.com/whats-the-point-of-the-met-office-easy-to-miss-when-you-ignore-the-facts-45794.
[8] Tanaka, Y., Hagano, K., Kuno, T., & Nagashima, K.: 'Pattern formation of crystals in storm glass.' *Journal of Crystal Growth* 310 (2008), https://doi.org/10.1016/j.jcrysgro.2008.01.037.

scientific calculations were put to good use in Victorian science, it may be that the weather was considered too complicated for this approach.[9] For example, the trajectory of a projectile can be accurately described using such physical laws, with good expectations for where the projectile will land if all the relevant factors are either also known accurately or negligible. The projectile's landing point can be determined accurately, and the effectiveness of the calculation tested. Statistics can be used to test comparisons of outcomes which occur by some process, against those that would be expected to occur anyway by chance.[10] Mid-nineteenth century science was aware of the physical laws describing the atmosphere, but it was not ready to apply them to the complex system of the atmosphere for weather prediction.

In contrast with weather, solar eclipses are events for which all the relevant parameters, especially the orbital parameters of the Earth and Moon, are well known, and to such accuracy that good predictions of their future progression can be made. Furthermore, any slight changes in the orbit which could influence them are small.[11] The theory needed to predict eclipses is derived from Newtonian mechanics, which has been extensively evaluated and its limitations are now well understood. In contrast, although the processes of the atmosphere generating weather also obey physical laws—not least Newton's laws of motion and those of thermodynamics describing heat and moisture transport—the system is far more complex and less completely characterized, and very sensitive to small changes. This leads to different levels of predictability for different weather situations.

Weather Forecasting

As an astronomical event, a solar eclipse can be predicted highly accurately, but the effectiveness of prediction of the accompanying weather will be different in every case. For some solar eclipses, the region of the Earth affected may have so little variability in its local weather that it is possible to say with considerable certainty what the viewing conditions will be: eclipses in such places will probably encourage eclipse enthusiasts to book their travel long in advance (as Mike Frost testifies through personal experience in Chapter 4). The other scenario, of central interest here, is for an eclipse situation where the weather is variable, and hence the circumstances may or may not be favourable, and by when it will be known what the prospects for a weather forecast are. This additional consideration of how far ahead the weather associated with the eclipse will be predictable is rarely addressed, as episodes of weather can be more settled or more variable: fog is a settled period of weather conditions, showers

---

[9] This quest for a physical forecasting device is apparent from the work of Lord Kelvin, who invented a machine for tide prediction. He had also considered an instrument for atmospheric predictions, suggesting in 1872 that 'the electric indications, when sufficiently studied, will be found important additions to our means for prognosticating the weather; and [...] the atmospheric electrometer generally adopted as a useful and convenient weather-glass.' W. Thomson: *Reprint of Papers on Electrostatics and Magnetism*, 2011. https://doi.org/10.1017/CBO9780511997259.

[10] Weather forecasts are also evaluated against what would have occurred if conditions were just assumed to continue unchanged, due to the persistency in the system.

[11] Both systems can be considered in principle chaotic, but with different responses to small changes.

are not. Such variation leads to the need to consider not only what a weather forecast is indicating, but also how reliable that forecast for the likely weather situation will be around the time of the eclipse.

Early weather forecasting operated by forming judgements about how weather systems would evolve, by analogy with similar patterns that had occurred previously.[12] The modern revolution in weather forecasting has occurred through numerical calculations based on known physical laws. This approach was proposed by Lewis Fry Richardson,[13] in his influential book *Weather Prediction by Numerical Processes*, first published in 1922.[14] To undertake a numerical weather forecast, Richardson imagined a system of interlinked computation offices—sometimes characterized as a 'forecast factory'[15]—where weather information was passed around in a regularized way and advanced to a later time, ultimately providing a prediction. In Richardson's concept, the computation would have been carried out by hand, by an army of human computers. Subsequently, digital computers developed rapidly, allowing Richardson's approach to be implemented practically.

The processes involved in making a numerical weather prediction are summarized in Figure 14.1. First, the state of the atmosphere is observed at a fixed time, essentially capturing the motion, temperature, and moisture content throughout. This is achieved through a combination of surface measurements, sensors carried within the atmosphere (e.g. from balloons and aircraft), and sensitive detectors looking down on the atmosphere from satellites. Obtaining these values in a coordinated way allows a snapshot of the global atmosphere to be obtained. This provides input data for computer representations (models) of the atmosphere, which are constructed around the relevant laws of physics, concerned with motion and heat transfer. The computer models advance the data contained in the initial snapshot to a future time, in a way consistent with the relevant physical laws. These new data values represent a weather prediction, which can be communicated to end users and the public.

The effectiveness of a numerical weather forecast depends both on how accurately the initial state of the atmosphere can be captured and on the detail with which atmospheric processes are described in the computer representation. However, some situations are inherently more predictable than others, allowing a better forecast to be made further ahead. One method of estimating the predictability is to slightly perturb the initial set of observations made to capture the initial state, and to investigate whether the forecast is affected by these changes. If there is little sensitivity in the forecast to the small variations imposed in the initial state, the situation is likely to be more predictable than if large changes result.

---

[12] Artificial Intelligence (AI) seeks to do something similar, through training a machine system to provide the statistical recognition of patterns.

[13] Lewis Fry Richardson (1881–1953) was a mathematician and a Quaker whose pacifism led him to spend the First World War in the Friends Ambulance Unit in northern France. During this period of non-military service he developed methods to calculate, by hand, the first numerical weather forecast. Although a fundamental methodological advance, his actual forecast, made retrospectively for 20 May 1910, was unsuccessful.

[14] See Lewis Fry Richardson, *Weather Prediction by Numerical Processes*, Cambridge, 2007.

[15] See also the painting by Stephen Conlin, https://www.emetsoc.org/resources/rff/.

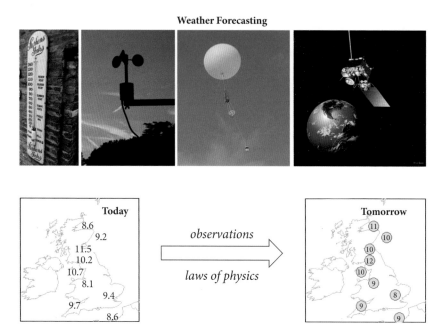

Figure 14.1 Aspects of numerical weather prediction. Upper panels: The state of the atmosphere at a fixed point in time is obtained by measuring systems and instruments, at locations from the surface to space. Lower panel: The process of weather forecasting takes the observed state of the atmosphere and uses known physical laws to step this forward to a future time. The new state of the atmosphere is determined throughout the atmosphere vertically, although a typical weather forecast will present information for the surface.

Figure 14.2 shows how meteorological predictions can be evaluated using this approach, through generating a set (or *ensemble*) of weather forecasts which evolve from slightly different initial circumstances. These examples from ensemble forecasting illustrate the spread of outcomes, for, in each case, a tropical storm approaching land. In Figure 14.2a, the expected progression of Cyclone Sidr in the Bay of Bengal to land is shown, from 12 November 2007. This can be regarded as essentially a predictable situation, allowing preparations for the storm's destruction to be made in Bangladesh. Figure 14.2b shows a different case, for the expected progression of Hurricane Katrina from 25 August 2005. The plume of possible trajectories is broader than in the Sidr case, demonstrating less predictability. In Figure 14.2c, calculations for the trajectory of Hurricane Nadine, from 2 September 2012 are shown. The wide range of possible trajectories—including movement in opposite directions—shows that this was a highly unpredictable situation.

These examples show that some weather systems are more predictable than others. In the case of forecasting the weather for a solar eclipse, the degree of predictability of the associated weather is as important as the most likely prediction. The

(a)   (b)   (c)

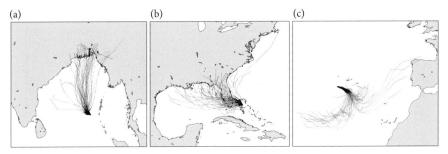

Figure 14.2 Predictions of possible trajectories (shown as a plume of lines) for the next 120 hours of three tropical storms potentially moving towards land, from the ECMWF Ensemble Prediction System Data. (a) Cyclone Sidr, (b) Hurricane Katrina, and (c) Hurricane Nadine.

Source: © OUP. Reproduced with permission from T.Palmer, 2022. *The Primacy of Doubt: From climate change to quantum physics, how the science of uncertainty can help predict and understand our chaotic world.*

range of predictability corresponds to the eclipse weather being reliably known perhaps a fortnight beforehand, or, less usefully, rather poorly known only a few days prior.

Beyond the fundamental problem of predictability, the detail required in a weather forecast for an eclipse is itself likely to be extremely demanding, essentially amounting, for interested observers at least, to whether optically dense cloud will occur within a narrow interval of a few minutes. Thin cloud, of only a few hundred metres' thickness, is usually sufficient to obscure the Sun, and it is notoriously difficult to forecast. Figure 14.3 demonstrates just how critical these conditions can be. Figure 14.3a shows a series of measurements of low cloud over the Reading University Atmospheric Observatory during the partial solar eclipse of 20 March 2015. In (b), the orange trace shows the predicted solar radiation at the top of the atmosphere, which shows the dip associated with the eclipse occurring between 0824 UTC and 1040 UTC; the red trace is the measured surface solar radiation at the site. A weather balloon was released through the cloud layer at 0847 UTC, partly to assess the prospects of the cloud dissipating (Figure 14.3c). At that time the cloud was found to be 432 m thick.[16]

Despite some thinning, sufficient cloud persisted at the site until 1220 UTC[17], to entirely obscure the eclipse. Nevertheless, on closer examination (Figure 14.3d), the surface measurements of solar radiation did show changes around the time of

---

[16] See Harrison, R. G., Marlton, G. J., Aplin, K. L., & Nicoll, K. A.: 'Shear-induced electrical changes in the base of thin layer-cloud.' *Quarterly Journal of the Royal Meteorological Society* 145 (2019), https://doi.org/10.1002/qj.3648.

[17] See Burt, S.: 'Meteorological responses in the atmospheric boundary layer over southern England to the deep partial eclipse of 20 March 2015.' *Philosophical Transactions of the Royal Society A: Mathematical, Physical and Engineering Sciences* 374 (2016), https://doi.org/10.1098/rsta.2015.0214.

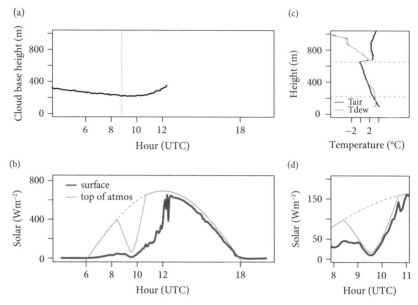

Figure 14.3 Measurements from Reading Observatory for the partial eclipse of 20 March 2015, which began at 0824 UTC and ended at 1040 UTC, with a magnitude of 0.88. (a) Height of the cloud base measured using a laser ceilometer, with the launch time of a weather balloon marked using the vertical dotted line. (b) Surface solar radiation on a horizontal surface (red trace), and calculated top of atmosphere solar radiation (orange trace, relative scale) for the same day. (c) Shows the lower atmosphere temperature profile obtained from a weather balloon released at 0847 UTC, with dashed horizontal lines to mark the cloud base and top identified from the sounding. (d) Increased resolution of solar radiation measurements around the eclipse time.

maximum eclipse, as effects are still discernible beneath cloud[18]. After the cloud has dissipated, the solar radiation trace of Figure 14.3b becomes typical of the second half of a cloudless day.

Eclipse Meteorology

A response in the atmosphere to reduced incoming solar energy driving circulation and weather is of course entirely expected from everyday experience of sunset, and eclipse effects on the atmosphere have long been implicitly acknowledged through descriptions of the associated cooling. An early modern example is by Edmond Halley.[19] However, the opportunities an eclipse provides for detailed investigations

---

[18] See Aplin, K. L., & Harrison, R. G.: 'Meteorological effects of the eclipse of 11 August 1999 in cloudy and clear conditions'. *Proceedings of the Royal Society A: Mathematical, Physical and Engineering Sciences* 459 (2003), https://doi.org/10.1098/rspa.2002.1042.

[19] Edmond Halley (1656–1741), second Astronomer Royal. See Edmond Halley: 'Observations of the late total eclipse of the sun on the 22d of April last past, made before the Royal Society at their house in Crane Court in Fleet-street, London. By Dr. Edmund Halley, Reg. Soc. Secr. With an account of what has

to advance understanding of atmospheric processes have only received attention relatively recently, following widely observed eclipses in the 1980s.[20] A pioneering early study was that of H. Helm Clayton, who brought together meteorological measurements from several sites during the May 28 1900 US eclipse, and suggested how the observed changes might be accounted for in terms of atmospheric structures.[21] The expression 'eclipse meteorology' began to be used some decades later,[22] and it was three centuries after Halley's account, in *Philosophical Transactions*, that the solar eclipse was highlighted in the same journal for its usefulness as a natural meteorological experiment.[23]

Although the concept of an experiment is a central part of the scientific method, it is more commonly associated with laboratory science than environmental science. This is due to cause and effect not being easily separated in the natural environment, as the atmosphere presents changing circumstances in an entirely uncontrolled way. In a controlled laboratory experiment, a stimulus is typically varied so that changes can be observed in one or two variables; but events occurring in the natural world are, as mentioned, uncontrolled, and frequently accompanied by many simultaneous responses in a large range of different parameters, and hence it may not be possible to determine a primary cause.[24] A solar eclipse offers an atmospheric change of prescribed intensity and duration, to which the responses can be studied. Even so, there is still the problem of working out what would have happened in the absence of the eclipse, to establish which effects can be definitively attributed to it. Determining eclipse-related effects in weather can be difficult as they may only generate small changes, although in clear skies the temperature changes are usually unambiguous. A few general approaches, shown here, have been used to extract eclipse-induced changes.

(1) **Extrapolation**. Non-eclipse-related changes, such as those of temperature already underway before the eclipse, are assumed to continue. Observations made during the eclipse are differenced from the pre-eclipse changes extrapolated forwards into the eclipse time.

been communicated from aboard concerning the same.' *Philosophical Transactions of the Royal Society of London* 29 (1714), https://doi.org/10.1098/rstl.1714.0025.

[20] See Aplin, K. L., Scott, C. J., & Gray, S. L.: 'Atmospheric changes from solar eclipses.' *Philosophical Transactions of the Royal Society A: Mathematical, Physical and Engineering Sciences* 374 (2016), https://doi.org/10.1098/rsta.2015.0217.

[21] See Clayton, H. H: 'The eclipse cyclone, the diurnal cyclones, and the cyclones and anticyclones of temperate latitudes.' *Quarterly Journal of the Royal Meteorological Society* 27 (1901), https://doi.org/10.1002/qj.49702712004.

[22] See Fergusson, S. P., Brooks, C. F., & Haurwitz, B.: 'Eclipse-meteorology, with special reference to the total solar eclipse of 1932.' *Eos, Transactions American Geophysical Union* 17 (1936), https://doi.org/10.1029/TR017i001p00129.

[23] See Harrison, R. G., & Hanna, E.: 'The solar eclipse: a natural meteorological experiment.' *Philosophical Transactions of the Royal Society A: Mathematical, Physical and Engineering Sciences* 374 (2016), https://doi.org/10.1098/rsta.2015.0225.

[24] 'A statistical approach to this problem is through averaging together similar events—*compositing* or *superposed epoch* analysis—to find the typical observed responses to a single clearly identifiable stimulus' See C. Chree: 'Some phenomena of sunspots and of terrestrial magnetism at Kew Observatory.' Philosophical Transactions of the Royal Society of London. Series A, Containing Papers of a Mathematical or Physical Character 212 (1913), https://doi.org/10.1098/rsta.1913.0003. Alternatively, a numerical model can be used to conduct a numerical experiment, by varying a parameter in a simulation and studying the response.

(2) **Similarity**. Non-eclipse days with similar weather conditions, perhaps averaged together, are compared with conditions on the day of the eclipse, and the differences obtained. Combining multiple similar days also allows the range of typical variations to be quantified, to provide an estimate of the natural variability. Often the day before or after can be used for this, or the same day in different years.

(3) **Modelling**. Calculations from a numerical model of the atmosphere which does not include the solar eclipse are compared with measurements made during the eclipse. This process essentially generates a weather forecast ignorant of the eclipse, for comparison with observations made during the eclipse.

(4) **Full simulation**. Two numerical models of the atmosphere are used, which are identical except that one includes internal representation of solar radiation changes without the eclipse, and the other with the eclipse. The difference between the two models' outputs is regarded as providing the eclipse-related changes. (Calculations from the eclipse-aware model may also be validated in some way against observations made during the eclipse day.)

There are many meteorological variables which can be monitored and analysed during an eclipse, from the top of the atmosphere down to the surface. Effects on some of the key near-surface quantities, specifically solar radiation, temperature, and wind, are now discussed further.

Solar radiation

The fundamental effect of a solar eclipse is briefly to reduce the atmosphere's incoming solar radiation, in a prescribed way. This will only be completely observable in a clear sky, and as Figure 14.3 shows, even a small amount of cloud can profoundly change what is experienced at the surface, compared with the expected variation from calculations. A comparison of the theoretical variation with measurements is therefore unlikely to be possible at sites where clear skies are not guaranteed. One option in such locations is to arrange to make measurements above the typical levels at which cloud may occur, for example using aircraft platforms. A simpler alternative is to use instrumented weather balloons, specially modified to carry solar radiation sensors and report the data back by radio. This was attempted for the 20 March 2015 eclipse, coordinated across three sites in Reading, Shetland, and Iceland.[25] The weather balloons were launched about an hour before the maximum eclipse at each site, to allow them to ascend sufficiently to be above the cloud (Figure 14.4).

The results from this coordinated atmospheric experiment across the well-separated sites are summarized in Figure 14.5. This shows that the expected solar

---

[25] Harrison, R. G., Marlton, G. J., Williams, P. D., & Nicoll, K. A.: 'Coordinated weather balloon solar radiation measurements during a solar eclipse.' *Philosophical Transactions of the Royal Society A: Mathematical, Physical and Engineering Sciences* 374 (2016), https://royalsocietypublishing.org/doi/10.1098/rsta.2015.0221.

Figure 14.4 Weather balloon launched from Lerwick Observatory, Shetland at 0850 on 20 March 2015, carrying a specially instrumented radiosonde package for solar radiation measurements.

radiation variations during the eclipse (left-hand column: a, b, c) agree with the measurements obtained from the weather balloons (right-hand column: d, e, f).

Temperature

By far the most common weather parameter remarked on or recorded is the near-surface air temperature, with the first systematic recordings probably occurring for eclipses in the early 1830s.[26] Discussion about temperature changes had occurred before then, including a reported 1°F decrease in air temperature during the eclipse of 1 April 1764,[27] and were considered further by the pioneer nephologist Luke Howard.[28]

---

[26] Aplin, K. L., Scott, C. J., & Gray, S. L.: 'Atmospheric changes from solar eclipses.' *Philosophical Transactions of the Royal Society A: Mathematical, Physical and Engineering Sciences* 374 (2016), https://doi.org/10.1098/rsta.2015.0217.

[27] J. Bevis: '"Observations of the eclipse of the sun, April 1, 1764:" In a letter from Dr. John Bevis, to Joseph Salvador, Esq; F. R. S.' *Philosophical Transactions of the Royal Society of London* 54 (1764), https://doi.org/10.1098/rstl.1764.0019.

[28] Luke Howard (1772–1864), originator of the cloud-classification system. See Luke Howard: *The Climate of London: Deduced from Meteorological Observations, Made at Different Places in the Neighbourhood of the Metropolis*, Vol. 1, London, 1833 (available in IAUC edition, https://urban-climate.org/documents/LukeHoward_Climate-of-London-V1.pdf).

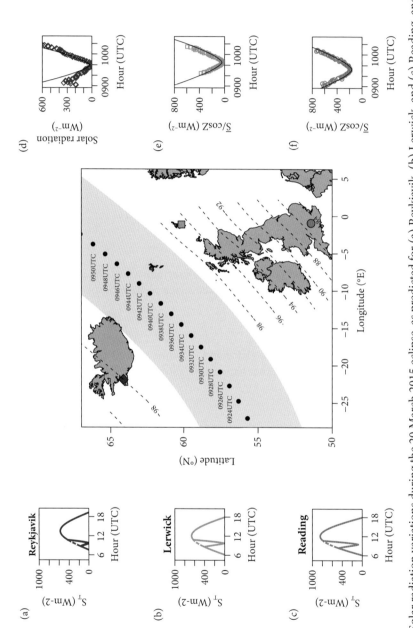

Figure 14.5 Solar radiation variations during the 20 March 2015 eclipse, as predicted for (a) Reykjavik, (b) Lerwick, and (c) Reading, and measured above the same three sites (d), (e), (f) respectively using instrumented weather balloons. The central panel shows the region of totality (grey band) with timings and the partial eclipse fractions. Reykjavik (diamond), Lerwick (square), and Reading (circle) are marked.
Source: R.G. Harrison et al, 2022. Coordinated weather balloon solar radiation measurements during a solar eclipse, Phil Trans Roy Soc A 374, 20150221).

Figure 14.6 demonstrates a range of air temperature and solar radiation measurements during eclipses. The values shown in Figure 14.6a are notable as being probably the earliest systematic UK temperature measurements during an eclipse, made by the Rev. Leonard Jenyns at Swaffham Bulbeck, Cambridgeshire, under the shade of a tree.[29] Small clouds were observed to dissipate at Greenwich during this eclipse,[30] probably associated with a reduction in convection.[31] Important conventions have since been developed in observational meteorology to control the exposure of thermometers, which will not register the air temperature correctly in direct sunlight or when the wind is calm.

Although temperature measurements are increasingly obtained at astronomical sites, the measurement circumstances can be less than ideal. Even if a conventional, naturally ventilated thermometer shelter is used, adequate wind is still required for accurate temperature measurements. At the solar radiation minimum, when the wind speed drops, the ventilation conditions may not be fulfilled, with the response time lengthened and the depth of the temperature minimum underestimated.[32]

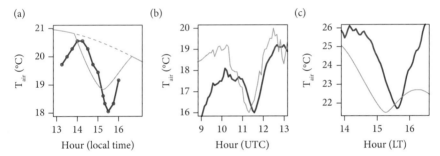

Figure 14.6 Air temperature variations observed (thick green lines) during solar eclipses. The measurements are from (a) Swaffham Bulbeck, Cambridgeshire on 15 May 1836, (b) Reading University Atmospheric Observatory, 11 August 1999, and (c) Mana Pools, northern Zimbabwe, 21 June 2001. Thin orange lines show (uncalibrated) solar radiation variations either calculated (in (a)) or measured (in (b) and (c)).

[29] Jenyns, L.: *Observations in meteorology: relating to temperature, the winds, atmospheric pressure, the aqueous phenomena of the atmosphere, weather-changes, etc.* London, 1858. https://doi.org/10.5962/bhl. title.133874.

[30] Birt, W. R.: 'Mr W R Birt's Meteorological observations made during the solar eclipse of May 15, 1836 at Greenwich.' *The London and Edinburgh Philosophical Magazine and Journal of Science* 9 (1836), pp. 393–394.

[31] Aplin, K. L., Scott, C. J., & Gray, S. L.: 'Atmospheric changes from solar eclipses.' *Philosophical Transactions of the Royal Society A: Mathematical, Physical and Engineering Sciences* 374 (2016), https://doi.org/ 10.1098/rsta.2015.0217.

[32] See R. Giles Harrison and Stephen D. Burt: 'Quantifying uncertainties in climate data: measurement limitations of naturally ventilated thermometer screens.' *Environmental Research Communications* 3 (2021), DOI 10.1088/2515-7620/ac0d0b.

Wind

The reduction in solar energy causes a temperature drop near the surface, and, due to changes in the atmospheric temperature structure, it has long been supposed that there are associated changes in the surface winds. Eclipse wind structures are of meteorological interest themselves, and they may be linked to processes associated with local cloud dissipation which are relevant to eclipse viewing circumstances. Eclipse-induced wind changes can occur in both strength and direction, and are relatively weak effects, requiring much more effort to detect than the associated temperature changes. Generally, averaging over a region is required to be sure that wind changes observed are not natural fluctuations unrelated to the eclipse.

Figure 14.7 show an analysis of temperature and wind changes for the 11 August 1999 solar eclipse, which was total in the south-west of the UK. The modelling approach (described as method 3 above) was used to extract the changes associated with the eclipse, by using a weather forecast model with a highly detailed calculating grid.[33] In (a), the temperatures predicted by the eclipse-unaware weather forecast model were subtracted from eclipse-influenced measurements made at 121 different sites, after having adjusted the measurements and model output onto a consistent set of positions. The darker colours identify the colder regions, which are in the central southern inland area, and a (blue, dashed-line) box has been drawn to emphasize this

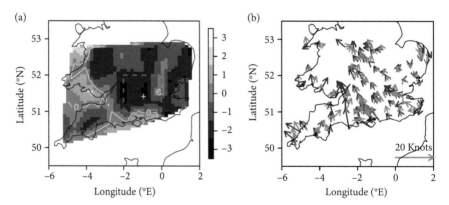

Figure 14.7 Temperature and (b) wind observed midway through the 11 August 1999 solar eclipse, at 11 UTC, showing the difference between measurements and modelled expectations of temperature and wind speed without the effect of the eclipse. In (a), the differences in temperature are shown, with the line of no difference marked in yellow. In (b), the modelled wind vectors are shown in grey and the measurements in blue. (The dashed blue box in (a) identifies the region from which a further analysis is made.)

Source: S.L. Gray and R.G. Harrison, 2012. Diagnosing eclipse-induced wind changes. Proceedings of the Royal Society A, 468(2143).

[33] See Gray, S. L., & Harrison, R. G.: 'Diagnosing eclipse-induced wind changes.' *Proceedings of the Royal Society A: Mathematical, Physical and Engineering Sciences* 468 (2012), https://doi.org/10.1098/rspa.2012.0007.

region; (b) shows effects on the wind speed and direction, in this case comparing the modelled (no eclipse) values with the values measured during the eclipse. In the inner spatial region identified for (a), the wind speeds are small, but there is some evidence of a change in wind direction. (Coastal regions can show very local effects, so are not considered further. Note that the cooling inland where the eclipse was only partial was greater than in the region of totality, due to the absence of cloud for the inland region.)

The changes in the wind vectors of Figure 14.7b in the identified inland region are considered further by combining the variations at individual sites. This averaging approach is commonly used when variations are small and difficult to observe individually because of the presence of other large changes, as in this case—the eclipse has a small effect in comparison with the bulk weather, but the small effect is nevertheless expected to be consistently present at all the sites. Figure 14.8 shows wind observations from various sites as speeds (a) and directions (b). For comparison, the expected variations obtained by modelling without the solar eclipse are given in (c) and (d). By averaging the changes at all the sites together, the effects of variability are reduced

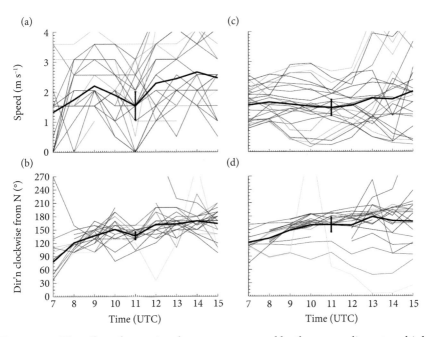

Figure 14.8 The effect of averaging the responses caused by the same eclipse at multiple sites. Variations in measured wind speeds (a) and directions (b) for sites in the region identified in Figure 14.7, at 10 m above the surface. Equivalent modelled values, without considering the eclipse, are given in (c) and (d). The thin lines refer to different sites, with the thick black line in each case the mean value, showing the 95% confidence range with the vertical bar.

Source: S.L. Gray and R.G. Harrison, 2012. Diagnosing eclipse-induced wind changes. *Proceedings of the Royal Society A*, 468(2143).

allowing any consistent pattern to emerge. In the eclipse day observations (a and b), there is a noticeable deviation around the maximum eclipse at 11 UTC, which is not present in the eclipse-free modelling (c and d). The wind speed was reduced, and the wind direction shifted in an anticlockwise direction, known as 'backing'. A similar analysis with more observation sites was used for the partial solar eclipse of 20 March 2015. The wind speed decrease accompanied by backing previously observed was repeated.[34] confirming eclipse-induced wind speed and direction changes within the southern UK.

Observations and Measurement Networks

The increase in interest in eclipse meteorology could well be associated with the steady and widespread increase in automatic weather recording stations, and the ready availability of data storage allowing rapid sampling.[35] These items of equipment are especially well suited to temperature and wind speed measurements. However, they are not suited to providing cloud information, the variation in which is of particular interest in defining the viewing conditions. Satellite cloud data is not especially suitable for this either, as both the time and spatial resolution is insufficient.

Solar eclipses are of such widespread interest that they provide an opportunity for science outreach, which brings a possible 'citizen science' route to obtaining more information. Seeking this type of data to help analyse eclipses is not a new idea, as it was used by Edmond Halley: 'I have added the following synopsis of such observations as have hitherto come to my hands, acknowledging the favour of those who have been willing to promote our endeavours to perfect the doctrine of eclipses.'[36] This approach was attempted in the UK for 20 March 2015 solar eclipse to provide cloud information, by harnessing the efforts of citizen-science observers.[37] A potential scientific advantage of this approach was that small changes in cloud could become detectable, by averaging across many observations. A carefully trialled webform was constructed to capture simplified weather information, including cloud. After national promotion through a variety of organizations, about 3,000 individuals participated in what was called the National Eclipse Weather Experiment (NEWEx). Analysis of the temperature data obtained showed that the observations

---

[34] See Gray, S. L., & Harrison, R. G.: 'Eclipse-induced wind changes over the British Isles on the 20 March 2015.' *Philosophical Transactions of the Royal Society A: Mathematical, Physical and Engineering Sciences* 374 (2016), https://doi.org/10.1098/rsta.2015.0224.

[35] See Hanna, E., Penman, J., Jonsson, T., Bigg, G. R., Bjornsson, H., Sjuroarson, S., Hansen, M. A., Cappelen, J., & Bryant, R. G.: 'Meteorological effects of the solar eclipse of 20 March 2015: analysis of UK Met Office automatic weather station data and comparison with automatic weather station data from the Faroes and Iceland.' *Philosophical Transactions of the Royal Society A: Mathematical, Physical and Engineering Sciences* 371 (2016), https://doi.org/10.1098/rsta.2015.0212.

[36] Edmond Halley: 'Observations of the late total eclipse of the sun on the 22d of April last past, made before the Royal Society at their house in Crane Court in Fleet-street, London. By Dr. Edmund Halley, Reg. Soc. Secr. With an account of what has been communicated from aboard concerning the same.' *Philosophical Transactions of the Royal Society of London* 29 (1714), https://doi.org/10.1098/rstl.1714.0025.

[37] Portas, A. M., Barnard, L., Scott, C., & Harrison, R. G.: 'The National Eclipse Weather Experiment: use and evaluation of a citizen science tool for schools outreach.' *Philosophical Transactions of the Royal Society A: Mathematical, Physical and Engineering Sciences* 374 (2016), https://doi.org/10.1098/rsta.2015.0223.

were comparable with those obtained from more sophisticated measurement networks.[38] For cloud, a simplified set of cloud descriptions were used. Figure 14.9 summarizes the NEWEx dataset. The reduction and recovery of the air temperature is quite evident, but an eclipse variation in cloud is not conclusively apparent.

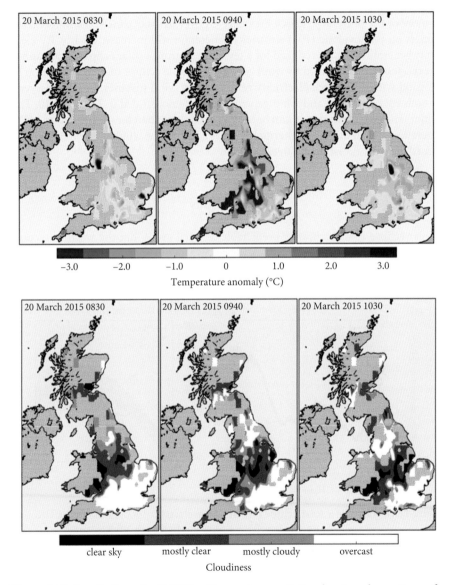

Figure 14.9 Results from the NEWEx citizen-science activity, showing observations of (upper row) temperature and (lower row) cloud, during the eclipse of 20 March 2015 (see also Figure 14.5)

Source: L. Barnard et al., 2016. The National Eclipse Weather Experiment: an assessment of citizen scientist weather observations. Phil Trans Roy Soc A 374(2077).

[38] See ibid.

Renewable Energy Generation

Although solar eclipses are brief and infrequent, they will impact renewable energy generation. Electricity distribution systems are carefully managed to balance the supply and demand, hence, as renewables increase their contribution to global energy needs, some consideration is needed in how to plan for the brief drop in generation. As more of the planetary surface area is given over to solar panels, the influence of eclipses on photovoltaic (PV) electricity generation will increase. Clearly, PV electricity generation will be reduced for a few hours during a solar eclipse. This cannot be expected to be mitigated by wind generation, as the eclipse-induced reduction in the wind will also reduce electricity generation by wind turbines. Effects observed on both PV generation and wind generation are demonstrated in Figure 14.10. The minimum in wind generation can be seen to occur later than for the minimum in PV generation, due to the slower thermal changes affecting the winds compared with the immediate solar radiation changes.

Conclusions

A solar eclipse is an accurately predictable celestial event, but the weather accompanying it, whilst also predictable, is not known in advance to remotely the same level of accuracy. The same laws of physics are used for the prediction of both an

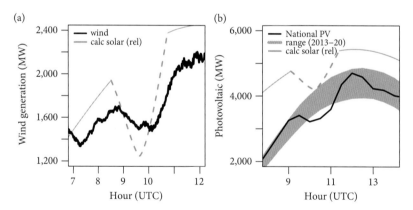

Figure 14.10 Effect of a solar eclipse on renewable electricity generation in the UK. (a) National wind generation of electricity during the 20 March 2015 solar eclipse. (b) Reduction of national photovoltaic (PV) electricity generation during the 10 June 2021 partial eclipse. Black lines show the power generation and orange lines the calculated top-of-atmosphere solar radiation on a horizontal surface in central England, on a relative scale, not shown. The grey band in (b) shows a range of one standard deviation on the PV generation in June from 2013–20.

Source: (a) modified from R.G. Harrison and E. Hanna, 2016. The solar eclipse: A natural meteorological experiment. Phil Trans Roy Soc A74(2077). (b) modified from E. Hanna et al., 2022. Meteorological effects and impacts of the solar eclipse of 10 June 2021 over the British Isles, Iceland and Greenland, Weather 78, pp. 124–135.

eclipse and the weather, but one system is much more complex than the other. In addition, the more complex system—the atmosphere—exhibits chaotic behaviour, causing weather situations to vary in their predictability: their sensitivity to small changes can be very large.

The predictability of solar eclipses offers an expected and characterized impulse to which the atmosphere responds, allowing this response to be studied and investigated. This can lead to the refinement and improvement of the computer models used for those used in weather forecasting. Although weather forecasts can always be compared with what ultimately occurs, the special circumstances of a total solar eclipse offer an unusually well-defined atmospheric experiment, allowing a range of different tests to those used in weather forecast verification. The predictable occurrence in one physical system can therefore lead to the improvement in the description of another.

Improving the predictions of eclipse weather requires a good understanding of weather in general and an eclipse's effect on the specific local weather especially. Eclipse meteorology has progressed in recent years due to the abundance of automatic weather stations. Through such networks, it has been possible to demonstrate that previous anecdotal accounts of eclipse winds do have some validity, although the effects are weak. The need to improve forecasting of cloud to predict eclipse viewing conditions is exceptionally demanding, and may only be possible in a narrow range of circumstances; it almost certainly requires much more detailed cloud measurements in space and time than are currently available. Existing understanding is, however, already sufficient to aid planning of renewable energy supplies under short-term changes in supply and demand.

Combining the understanding of physical science with mathematical calculations provides what is almost certainly the best method ever developed of predicting future weather. Forecasting of weather and solar eclipses demonstrates the wide applicability and effectiveness of these techniques for systems of hugely different complexity, and both approaches illustrate real progress in the human desire to predict.

Conclusion

The Moon and the Sun in the Afternoon

Henrike Christiane Lange and Tom McLeish

> So when we see the phases of the moon, we conceive not of a little penny that grows from a sliver into a round and back to a sliver again during the month, but of a sphere, illuminated by the sun in fully three dimensions. In other words, we do not merely record perceptions; we recreate the entire universe in model form.
>
> Tom McLeish, 'Science Is a Long Story'[1]

Interdisciplinary Fieldwork

This book is the result of a different kind of eclipse-chasing: We set out to consider approaches to total solar eclipses by a vast variety of scholarly, scientific, and creative disciplines—to hear what they might have to say about the phenomenon and its interpretation, both historical and present. We inquired how a spread of methods and analytical practices can illuminate specialists' understanding not only in each of those siloed categories separately, but for specialists in other disciplines mutually, in communication with each other, and, finally, in a presentation that speaks to a wide audience—to anyone interested in the phenomenon itself. We would then open books, archives, museum collections, box record players, labs, and each discipline's fieldwork luggage to give a glimpse into their methods and approaches. But with all this work of the eye, the ear, and the mind, the fundamental act of the total solar eclipse experience is one of heart and soul—of being supremely aware and in awe, of living a scientific–humanistic–artistic practice of attention and focus, of perception and processing, of excitement and enthusiasm, of curiosity and questioning, and last but never least, of sharing and mutual appreciation.

---

[1] 'My favorite debate as to how one tells whether mind is reducible to its matter or components in terms of thoughts—call it soul, if you wish—goes back to Gregory of Nyssa, one of the great Cappadocian theologians, and his elder sister, Macrina, who's also one of the great Cappadocian theologians, in her deathbed discussion, which is often translated *On the Soul and the Resurrection*. It should really be *On the Mind and the Resurrection*. And they go hammer and tongs about this. Interestingly, Gregory takes the view that everything is reducible to atoms, material properties—just for fun. He doesn't believe that, but he takes that view for the sake of debate. And Macrina takes the view that the mind is a causal agent: we make real decisions, we have free will, and so forth. And she ends up by pointing out that we know this is true because mind has the ability effectively to do science, to reconstruct nature. So when we see the phases of the moon, we conceive not of a little penny that grows from a sliver into a round and back to a sliver again during the month, but of a sphere, illuminated by the sun in fully three dimensions. In other words, we do not merely record perceptions; we recreate the entire universe in model form.' Full quote from Tom McLeish: 'Science is a Long Story,' see https://themarginaliareview.com/science-is-a-long-story/.

Attention and Awe

Attention is easy to pay once one realizes that the stars are always there, just at times more, at times less visible in the quotidian appearance of the cosmos on any unremarkable day and night. This thought alone alerts one to a numinous quality of the cosmos independent from something as rare as the total solar eclipse. Then attention can result in awe on any unspectacular day of any weather. The goal therefore, for the long periods of time between eclipses, seems to be the development of a fine-tuned perception for moments of subtle breakthrough, much like the 'Moon in the Afternoon' ('*Luna del Pomeriggio*') when it appears as almost transparent in a memorable passage from Italo Calvino's *Palomar*:

> Nobody looks at the moon in the afternoon, and this is the moment when it should most require our attention, since its existence is still in doubt. It is a whitish shadow that surfaces from the intense blue of the sky, charged with solar light; who can assure us that, once again, it will succeed in assuming a form and glow? It is so fragile and pale and slender, only on one side does it begin to assume a distinct outline like the arc of a sickle while the rest is steeped in azure. It is like a transparent wafer, or half-dissolved pastille, only here the white circle is not dissolving but condensing, collecting itself at the price of gray-bluish patches and shadows that might belong to the moon's geography or might be spillings of the sky that still soak the satellite, porous as a sponge.[2]

These observations lead us from keen perception to contemplation. While a more positivistic character in Tom Stoppard's play *Arcadia* stated: 'It's the wanting to know that makes us matter,' the Moon is a good example for the other end of knowledge– a more open and emotionally charged celestial vision can inspire the chase of knowledge but can also stand for itself as an experience without a teleological purpose in the material world.

[2] See Italo Calvino, *Mr. Palomar*. Trans. William Weaver. San Diego, New York, and London, 1985. In the section 'Mr. Palomar looks at the Sky,' see 'Moon in the afternoon' pp. 34–6, and 'The contemplation of the stars' pp. 43–8. For Calvino in the context of this volume, see especially the rich interdisciplinary approach in Gianni Cimador, 'La scienza del possibile: Italo Calvino e il superamento delle "due culture".' *Italianistica* 44, 2015, pp. 155–78. Calvino himself also explored the cosmos with a sense of humour to add a lighter mood to the big questions, as in his 1965 *Cosmicomiche* (*Cosmicomics*, translation from Italian by William Weaver, first published in 1968). In 'The Distance of the Moon,' Calvino builds the idea on the promise from the nineteenth century that millions of years ago, the Moon was much closer to the Earth and was gradually pushed away by the tides, letting one of his senior characters (old Qfwfq) proclaim, 'the rest of you can't remember, but I can. We had her on top of us all the time, that enormous Moon: when she was full—nights as bright as day, but with a butter-coloured light—it looked as if she were going to crush us; when she was new, she rolled around the sky like a black umbrella blown by the wind; and when she was waxing, she came forward with her horns so low she seemed about to stick into the peak of a promontory and get caught there. But the whole business of the Moon's phases worked in a different way then: because the distances from the Sun were different, and the orbits, and the angle of something or other, I forget what; as for eclipses, with Earth and Moon stuck together the way they were, why, we had eclipses every minute: naturally, those two big monsters managed to put each other in the shade constantly, first one, then the other.' For some of the rich interdisciplinary engagement of Calvino's *Palomar* and its discussion of perception, knowledge, and interpretation, see also Elio Attilio Baldi: 'Art and Science in Calvino's Palomar: Techniques of Observation and Their History.' *Italian Studies* 74, 2019, pp. 71–86.

Among cosmic bodies, the Moon in particular has always invited a sense of nostalgia, longing, and connection. It has kept this appeal for lovers and poets over the millennia, from Sappho to Leopardi to Dvořák's *Rusalka* to today's stargazers.[3] We might then broaden out what we have learned from the topic of total solar eclipses to the entirety of experiences within the contemplation of sky and heaven, striving like Calvino's character Palomar for the right balance between ego and surrender.[4] The famous evening song 'Der Mond ist aufgegangen' ('The Moon is risen') from 1779 by Matthias Claudius (1740–1815) combines the contemplative attention to, and enjoyment of, the serene beauty of this world with the rejection of hubris. The song celebrates not only the cosmos as place of encounter for human beings and the natural world, and not only the Christian aspirations of care, empathy, and love, but is also at its core about knowledge and the loss of perspective on the essential, if giving in to an overconfidence in one's own scientific and scholarly capacities. Thus 'Der Mond ist aufgegangen' stands as a musical and poetic example for the unity of experience, knowledge, and empathy in the service of our relationship to the cosmos, to others, and to the self. The wanting to know might often be a search for meaning, and to be comforted, if not by meaning found, then at least by the possibility of meaning—as we say, *when it rains, look for rainbows, when it's dark, look for stars.*[5] Striving for the interdisciplinary unity of knowledge is reaching for the

---

[3] See Sappho's fragment, 'Δέδυκε μεν ἀ σελάννα / καὶ Πληΐαδεσ, μέσαι δὲ / νύκτεσ πάρα δ᾽ ἔρχετ᾽ ὤρα, / ἔγω δὲ μόνα κατεύδω' (Prose version by H. T. Wharton: 'The moon has set, and the Pleiades; it is midnight, the time is going by, and I sleep alone.'). See Rusalka, 'Song to the Moon,' 'Svetlo tve daleko vidi, / Po svete bloudis sirokem, / Divas se v pribytky lidi. / Mesicku, postuj chvili / reckni mi, kde je muj mily / Rekni mu, stribmy mesicku, / me ze jej objima rame, / aby si alespon chvilicku / vzpomenul ve sneni na mne. / Zasvet mu do daleka, / rekni mu, rekni m kdo tu nan ceka! / O mneli duse lidska sni, / at'se tou vzpominkou vzbudi! / Mesicku, nezhasni, nezhasni!' ('Moon, high and deep in the sky / Your light sees far, / You travel around the wide world, / and see into people's homes. /Moon, stand still a while / and tell me where is my dear. / Tell him, silvery moon, / that I am embracing him. / For at least momentarily / let him recall of dreaming of me. / Illuminate him far away, / and tell him, tell him who is waiting for him! / If his human soul is in fact dreaming of me, / may the memory awaken him! / Moonlight, don't disappear, disappear!'). See also a performance of Rusalka's 'Song to the Moon' at https://www.youtube.com/watch?v=UwVYFpY3VL4.

[4] Palomar develops this most clearly in his contemplation of the 'Spada del sole,' the sun's reflection on the surface of the ocean that seems to be pointing always at him, the swimmer in the great sea of being. See 'The sword of the sun,' pp. 13–18, at 14, '"This is a special homage the sun pays to me personally," Mr. Palomar is tempted to think, or, rather, the egocentric, megalomaniac ego that dwells in him is tempted to think. But the depressive and self-wounding ego, who dwells with the other in the same container, rebuts: "Everyone with eyes sees the reflection with him; illusion of the senes and of the mind holds us all prisoners, always." A third tenant, a more even-handed ego, speaks up: "This means that, no matter what, I belong to the feeling and thinking subjects, capable of establishing a relationship with the sun's rays, and of interpreting and evaluating perceptions and illusions."' See also p. 18, 'Mr. Palomar thinks of the world without him: that endless world before his birth, and that far more obscure world after his death; he tries to imagine the world before eyes, any eyes; and a world that tomorrow, through catastrophe or slow corrosion, will be left blind. What happens (happened, will happen) in that world? Promptly an arrow of light sets out from the sun, is reflected in the calm sea, sparkles in the tremolo of the water; and then matter becomes receptive to light, is differentiated into living tissues, and all of a sudden an eye, a multitude of eyes, burgeons, or reburgeons . . .'. See also Calvino: *Mr. Palomar*, 'The contemplation of the stars,' pp. 43–8. On the necessity of wanting to see, at 43: 'When it is a beautiful starry night, Mr. Palomar says, "I must go and look at the stars."' See also, on the connection of others to existence, at 48: 'Some silent shadows are moving over the sand: a pair of lovers rises from the dune, a night fisherman, a customs man, a boatman.'

[5] 'Der Mond ist aufgegangen / Die goldnen Sternlein prangen / Am Himmel hell und klar: / Der Wald steht schwarz und schweiget, / Und aus den Wiesen steiget / Der weiße Nebel wunderbar. // Wie ist die Welt so stille, / Und in der Dämmrung Hülle / So traulich und so hold! / Als eine stille Kammer, / Wo ihr

union of experience, knowledge, and meaning in our relationships to the cosmos, within the cosmos. Pausing to look at the sky is a *modus operandi* and way of living for the editors, and the shared perspective that underlies this kind of unified visual–sensual–rational–creative–interpretative appreciation—the unity of knowledge with an emphasis on visual evidence—was a main motor of this collaboration.

The Interdisciplinarity Challenge

Over the span of these seven years extending themselves between the great American total solar eclipse of 2017 and that of 2024, we have deepened our understanding of such interlocking perspectives. We as a *family* (Tom McLeish), we as a *constellation* of scholars (as Roberta J. M. Olson put it so well in the final team conference of January 2023), found our lives so merged with our research, scholarship, observations, experiences and writing, constantly building and sustaining community and connection across the appearance of pandemic isolation and other experiences of hardship. The project itself sustained us with its unbroken connection to the life-affirming energy of the Sun and the serene force of the Moon—we remarked more than once that, for *this* book project, the timescale had no flexibility, but was set by the very phenomenon we were approaching in more than one sense. This would afford the editors an opportunity to chase authors for time-sensitive submissions with a humorous note on the unforgiving and non-negotiable timeframe set by the Sun, the cosmos itself. But truly, we kept looking up from our tables-turned-desks in our sunny garden and nightly back yard at the vast extension of the sky above us. As Giles Sparrow put it in *The Stargazer's Handbook*, '[the] shifting motions of objects in the sky are largely a reflection of our planet's own movement through space. With a basic understanding of them, the heavens become a far less intimidating challenge, and we can begin to

---

des Tages Jammer / Verschlafen und vergessen sollt. // Seht ihr den Mond dort stehen? / Er ist nur halb zu sehen, / Und ist doch rund und schön. / So sind wohl manche Sachen, / Die wir getrost belachen, / Weil unsre Augen sie nicht sehn. // Wir stolze Menschenkinder / Sind eitel arme Sünder, / Und wissen gar nicht viel; / Wir spinnen Luftgespinste, / Und suchen viele Künste, / Und kommen weiter von dem Ziel. // Gott, laß uns dein Heil schauen, / Auf nichts vergänglichs trauen, / Nicht Eitelkeit uns freun! / Laß uns einfältig werden, / Und vor dir hier auf Erden / Wie Kinder fromm und fröhlich sein! // Wollst endlich sonder Grämen / Aus dieser Welt uns nehmen / Durch einen sanften Tod, / Und wenn du uns genommen, / Laß uns in Himmel kommen, / Du lieber treuer frommer Gott! // So legt euch denn, ihr Brüder, / In Gottes Namen nieder! / Kalt ist der Abendhauch. / Verschon' uns Gott mit Strafen, / Und laß uns ruhig schlafen, / Und unsern kranken Nachbar auch!' ('The Moon is risen, beaming, / The golden stars are gleaming / So brightly in the skies; / The hushed, black woods are dreaming, / The mists, like phantoms seeming, / From meadows magically rise. // How still the world reposes, / While twilight round it closes, / So peaceful and so fair! / A quiet room for sleeping, / Into oblivion steeping / The day's distress and sober care. // Look at the Moon so lonely! / One half is shining only, / Yet she is round and bright; / Thus oft we laugh unknowing / At things that are not showing, / That still are hidden from our sight. // We, with our proud endeavour, / Are poor vain sinners ever, / There's little that we know. / Frail cobwebs we are spinning, / Our goal we are not winning, / But straying farther as we go. // God, make us see Thy glory, / Distrust things transitory, / Delight in nothing vain! / Lord, here on Earth stand by us, / To make us glad and pious, / And artless children once again! // Grant that, without much grieving, / This world we may be leaving / In gentle death at last. / And then do not forsake us, / But into heaven take us, / Lord God, oh, hold us fast! // Lie down, my friends, reposing, / Your eyes in God's name closing. / How cold the night-wind blew! / Oh God, Thine anger keeping, / Now grant us peaceful sleeping, / And our sick neighbour too.')

find our way around their wonders.'[6] Here and there, we noticed the subtly changing views of Sun and Moon, contemplated the time that it took to shift light, clouds, colours, and time's own embedded urgency, relative to the questionable meaning of our work of the eyes and the mind in the face of the numinous. As scholars, scientists, historians, thinkers, the production of our work is so deeply and existentially important for us to stay connected, and—strangely, a likewise comforting thought— the expansion of our work into the cosmos and the projection of our investments in it onto its vastness only reminds us that it is also so *little* in the face of a cosmos that we find most powerfully described in the Lord's answer to Job, a text at the centre of McLeish's *Faith and Wisdom in Science*.[7]

There is a way to think of interdisciplinarity as a superstructure over disparate islands of knowledge. However, in our experience, we must dive deep rather than build lofty structures on the surface—indeed, rather than building superficial bridges, it is possible to dig deep down to a common foundational core. Our conversations around the original panel at the Notre Dame Institute for Advanced Study and every exchange since helped facilitate such connecting to the core not only of academic understanding but, more importantly, the very excitement and motivation of the work itself. There we discovered connections in our scholarship, studies, methods, and knowledge that unite rather than divide. The quest for the unity of knowledge depends on rather deep and long labour: for thorough interdisciplinary work, one needs to be expert in at least one relevant discipline first, and as in language acquisition, with each new aquired language, it gets easier to add to one's linguistic competence. Mutual respect and empathy have also been key to the desire of understanding another discipline's questions, positions, methods and instruments, data collection, and points of view. The years of working collaboratively on the theme of total solar eclipses across disciplines and fields, across chronological and geographical areas, were particularly fruitful; their boundaries were set by the natural limitation of having started from a small cohort of NDIAS Fellows from 2017–18 and expanded into our specific and already interdisciplinary professional networks. This condition helped us avoid overwhelming the book with any unrealistic expectations of completeness or full coverage: We knew from the beginning with our first panel discussion that we could not possibly press a complete global survey of total solar eclipses from all ages, places, cultures, and disciplines into one book. Being aware of this limitation, we would rather keep the discourse oscillating between broad-sweeping approaches and very precise and narrow case studies, giving each author complete freedom in

---

[6] Giles Sparrow, *The Stargazer's Handbook: An Atlas of the Night Sky*. London, 2010, p. 9.

[7] 'Then the Lord spoke to Job out of the storm. He said: 'Who is this that obscures my plans with words without knowledge? Brace yourself like a man; I will question you, and you shall answer me. [. . .] 'Where were you when I laid the earth's foundation? Tell me, if you understand. Who marked off its dimensions? Surely you know! Who stretched a measuring line across it? On what were its footings set, or who laid its cornerstone—while the morning stars sang together and all the angels shouted for joy? Who shut up the sea behind door when it burst forth from the womb, when I made the clouds its garment and wrapped it in thick darkness, when I fixed limits for it and set its doors and bars in place, when I said, "This far you may come and no farther; here is where your proud waves halt"? [. . .] "Can you bind the chains of the Pleiades? Can you loosen Orion's belt? Can you bring forth the constellations in their seasons or lead out the Bear with its cubs? Do you know the laws of the heavens? Can you set up God's dominion over the earth?"' (Job 38.1–11, 16–21, 28–33). See also McLeish, *Faith and Wisdom*.

their approach and presentation of the issue from their individual perspectives. They would know best what is relevant in the context of their fields, and how to present it to interested general audiences. The coexistence of these perspectives with the scaled approaches is part of the method—to learn from zooming in and zooming out on the topic, and to consider both the wide panorama of history, science and scientific method, theology, religion, and ritual, and the unique personal narratives such as those offered by historically disconnected voices (e.g. Dante Alighieri, Adalbert Stifter, and Pink Floyd).

This long collaboration was inspired and governed by real-life phenomena and experience—challenging the usual dividing principles of the academy—and set up on the grounds of our former experiences in our own multidisciplinary backgrounds on the sides of sciences and theology on the one hand, and arts, history, and humanities on the other (e.g. natural philosophy spanning soft matter physics, astrophysics, and theology of science; histories of art, architecture, literature, and languages; all of the above phenomena engaged in their enduring presence throughout individual and collective time). Our experience of writing together, of conceptualizing and of editing the book over so many years, confirmed again and again that most of our best ideas and insights can only be sparked in dialogue with a complementary other. The deeper we dig, the closer we come together, and the closer that we can get to the foundational connections of a unity of knowledge, the more can be learned from other disciplines and scholarly practices. The learning happens along the way, in all kinds of ways: in the initial approach to the phenomenon, in the evaluation of methods and adjustment of instruments, in the comparison of data, in philosophical or intellectual frameworks, in the theory and practice involved with the production of scholarship, in the different ways in which evidence can be gathered, analysed, and presented, and in which evidence reveals its own prismatic shimmer.

A total eclipse in 2017 inspired our first collaboration in science, arts, humanities, social studies, and theology; now, in 2023, we are looking back at a journey that brought unexpected wonders and disasters, especially, on a global scale, the pandemic that prevented our in-person gatherings—from our long-anticipated workshop in Berkeley for the Medieval Academy of America conference in March 2020 ('Visionaries of Vision: A Modern Scientist's Reading of Grosseteste and Giotto') to the long-awaited workshops at the Royal Society in London as well as in Cambridge, Oxford, Durham, York, and Madrid that we had prepared yet could not host quite as planned. By now, the San Francisco Bay Area-based half of the editorial team has lived through heretofore inconceivable environmental catastrophes—the massive Californian wildfire smoke of 2020 occluded the Sun over the Bay Area for weeks. While teaching out of Berkeley via video call, the virtual lecture hall that her desk had become gathered students scattered all over the American continent, Canada, Europe, India, China, and Australia. During the fires, the Sun, the Moon, and the stars were all occluded over this corner of the world—over such an extended period of time that we feared they would never reappear and we would never breathe clean air again. We could not tell the difference between day and night. We left the curtains closed, the blinds down because the artificial light in our darkened rooms was preferable to the disturbing orange glow of that seemingly post-apocalyptic atmosphere outside.

Teaching Dante's *Divine Comedy* under such extreme and existentially threatening conditions was a lasting transformative experience—and writing about Dante's way from Hell to hope while simultaneously surviving for an uncertain future moment of, we hoped, once more seeing the stars again. Then, one evening, the frightening thick smoke finally had been transported away by more benign winds. We opened the door to the garden: The night was clear. The sky once more visible. We sat down on the little backdoor stoop in awe. The revelation of the stars, their silent comfort—*Ursa Major* a reversed eclipse, no shadow, no occlusion—was marvelous.[8] The sudden vision of the stars after the catastrophe was like seeing them for the first time, with no less an emphasis than the Italian word for stars, 'stelle,' ends each instance within the triple experience of Dante emerging from Hell, moving beyond Purgatory, and transcending from Paradise (see Cornish, Chapter 9, p. 159): 'E quindi uscimmo a riveder le stelle.' (*Inferno* 34, 139)... 'puro e disposto a salire a le stelle.' (*Purgatorio* 33, 145)... 'l'amor che move il sole e l'altre stelle.' (*Paradiso* 33, 145).

Everything has only become more real for us these years, despite the disorienting, unreal atmosphere of the virtual means of communication on which we depended, in the end, for our team meetings and authors' workshops, joining in one virtual space from around the globe. In some ways, we have gathered more experience than we would have wished for, considering the devastating personal hardship of this time individually and collectively. Yet the suffering has rather increased our dedication for the furthering of the ideals of this optimistic and idealistic project, and led to the completion of a fascinating sequence of perspectives. Even suffering is a perspective to include, and an important one—unavoidable in human experience and therefore indispensable in empathic, aesthetic, and philosophical approaches to life.[9] Maybe the depth and truth of this experience is similar to the fundamental difference between a partial and a total solar eclipse—a less dramatic interim might also have been less illuminating; we had to work harder, and endure more, but we also received more precious insight into what we need and what keeps us connected and alive as human beings in this cosmic clockwork. In that sense, the theme of the eclipse 'hiding to reveal' likewise has become more urgent and more illuminating.

The Natural Experiment

Another lesson came from the interdisciplinary conversations between scientists and scholars in the arts and humanities. We came together from the sides of the sciences and of history, arts, and humanitites to confirm a point on theoretic universality,

---

[8] *Ursa Major* (Latin: "Greater Bear") is a constellation of the northern sky at about 10 hours 40 minutes right ascension and 56° north declination, referred to in Job 9:9 and 38:32. The Big Dipper, Plough, or *Sternenwagen* asterism forms part of Ursa Major.
[9] See Elizabeth V. Spelman: *Fruits of Sorrow: Framing Our Attention to Suffering*. Boston 1997. With warm thanks to Steven Botterill for our discussions of the book, especially the passages on suffering and the economy of attention, compassion, virtue and feeling, empathy, the 'aesthetic usability of suffering,' and suffering as the human condition.

formulated brilliantly by Philip Ball in a review of Tom McLeish's *Soft Matter: A Very Short Introduction*,[10] that 'there is a universality to scientific theory that transcends the immediate subject matter or even disciplines':

> Find the right analogy—it is in truth more than that—and you can import the conceptual framework from one area of science to another. The reasons for this universality are not fully understood, although they seem clearly to be saying that deep and generic principles such as symmetry and topology govern the way the universe works, more than do the fine details.[11]

Gödel, Escher, Bach comes to mind.[12] This truth applies to the disciplines of science, arts, humanities, social and historical studies, and theology alike—the multiple interconnections are so manifold and varied that they usually go unrecognized, but a strong focus on a universal phenomenon such as that of our 'eclipse and revelation' theme in total solar eclipses renders this principle visible.

Most excitingly, the project has brought out a rich web of similarities between disciplines as well as their differences—and has illustrated how each discipline in turn gains from the understanding of differences, friction, and productive dialogue with the conditions of other areas of research, knowledge-formation, and communication. One framing in common is the sense of a total solar eclipse as a 'natural experiment'— 'what happens if the Sun is extinguished for a few minutes on a localized area of the Earth?' This question can be asked of human communities and civilizations (see Steele, Gasper, Roos, Hart) as well as it can be of animal populations and behaviour (Portugal), or of the atmosphere (Harrison), or of artistic, representative, and creative traditions (Cornish, Olson, Lange, Hild). For such a categorization to be valid, conditions must be highly rare, unusual and powerful, yet predictable—all three conditions are met by the total solar eclipse. In all these cases the 'hidden' Sun then 'reveals' much that would otherwise be unseen. This generalization of the meaning of 'experiment' into the humanities and arts is not unprecedented, for the same has been said of the early novel (e.g. Pat Waugh) in regard to artificial experiment—the eclipse arranges a natural version of special conditions and subsequent response.[13] In both novel and experiment much of the complexity of the world has to be hidden in order that the sought structures, characters, and behaviours be revealed: from eclipse emerges apocalypse. There are strong resonances here with the theological foundations of early

---

[10] See McLeish, *Soft Matter*. Oxford, 2020.

[11] Full quote: 'The third lesson is, for many scientists, the most profound and indeed the most beautiful: there is a universality to scientific theory that transcends the immediate subject matter or even disciplines. Both de Gennes and Edwards were able to solve some problems in soft matter by drawing on their knowledge of other topics, such as quantum mechanics and superconductivity (the ability of some metals and other solids to conduct electricity without any resistance). Find the right analogy—it is in truth more than that—and you can import the conceptual framework from one area of science to another. The reasons for this universality are not fully understood, although they seem clearly to be saying that deep and generic principles such as symmetry and topology govern the way the universe works, more than do the fine details.' Philip Ball, 'Philip Ball on Tom McLeish.' *The Marginalia Review*, December 4, 2020, available at: https://themarginaliareview.com/what-is-soft-matter/.

[12] See Douglas Hofstadter: *Gödel, Escher, Bach: An Eternal Golden Braid*. New York, 1979.

[13] See McLeish, *Poetry and Music of Science*, Oxford, 2019, ch. 4 on the novel.

modern science. As Peter Harrison and others have pointed out, the efficacy of experiment in revealing truth about the natural world is by no means a given, and in many ways counterintuitive—how could a system so simplified, artificial, and isolated teach anything about the complex and multiply connected real world?[14] This argument was, for the great early-modern natural philosopher Margaret Cavendish, a devastating blow to the new experimental philosophy.[15] That experiments function was, for near-contemporary Francis Bacon, a perception of grace at work.[16] To witness the cosmos performing an experiment on itself is, from this perspective, a shocking (and doubly graceful) experience of gift.

Like all the best experiments, the conditions of their construction are well controlled, though their outcomes are often unpredictable. Contemplation on this aspect of the eclipse reminded us of the polyvalency of science, and of its mathematical representation. The 'cosmic clockwork' of prediction of the when and where is now a high-precision application of Newtonian mechanics—the historical emergence of this greater precision, and the concomitant mathematical sophistication, has been fascinating to chart from ancient times to modern.[17] Yet the consequences of the natural experiments in human, animal, and even meteorological domains remain uncertain, being of a quite different order of complexity.

Perhaps the greatest gift of the eclipse to the framing of science is its possession of the numinous for even the most hardened of scientists who experience the glow of the corona against a black daytime sky. The visceral sense of connectedness between the celestial and the terrestrial that the experience of an eclipse engenders is arguably unsurpassed. As David Bentley Hart has written, the phenomenon presents itself as a corrective to a purely materialistic and dualistic late-modern framing of science into which public discussion so easily and misguidedly slips. Maybe the rarity of the event helps its effectiveness: On an observational level, one aspect of note that has arisen across the entire disciplinary palette of this survey is the vast qualitative difference between a total solar eclipse and a partial eclipse, no matter how nearly complete. 99% of coverage of the Sun's photosphere by the Moon does not deliver 99% of the experience of a total eclipse—rather, it fails to differ much from a pretty cloudy day with an atmospherically obscured Sun. To assume the reverse is to risk missing out on the experience of totality when a short journey might award its unique and deep fulfilment. Once again we find the limitations of the quotidian everyday life interwoven with the great cosmic spectacle—regardless of whether we pay attention to them or are ignorant of their workings and influences.

---

[14] See Harrison, *The Territories of Science and Religion*. Chicago, IL, 2017.

[15] See Margaret Cavendish, *Experimental Philosophy*. London, 1666.

[16] 'The glory of God is to conceal a thing, but the glory of the king is to find it out; as if, according to the innocent play of children, the Divine Majesty took delight to hide his works, to the end to have them found out; and as if kings could not obtain a greater honour than to be God's playfellows in that game, considering the great commandment of wits and means, whereby nothing needeth to be hidden from them.' Francis Bacon, *Works*, Volume III, eds. J. Spedding, R. L. Ellis and D. D. Heath, Boston, MA, 1887.

[17] See, for example, Roger Penrose, *The Road to Reality: A Complete Guide to the Laws of the Universe*, New York, 2005, ch. 17, 'Spacetime.'

Revelation for Inspiration

True to its ever-ambivalent nature, the rare, fleeting, and illuminating appearance of total solar eclipses touches also upon the theme of eternity in the same paradoxical way in which Kate Russo has described total solar eclipses as a metaphor for life itself: 'A total eclipse lasts only a few minutes, but we are mesmerized, awestruck, and aware of every second. It is also precious and wonderful. And then it is over. A total eclipse is a perfect metaphor for life' (Figure C.1).[18]

In our work over these seven years and across the many disciplines represented in this volume, the total solar eclipse revealed itself to collapse and intensify all areas of experience, life, and knowledge. The struggle with its representation serves as a metaphor for the complexity of the phenomenon itself, as much as the tension between prediction/predictability and relative freedom of interpretation or invest-ment of emotional, religious, or magical thinking. Representation is an unavoidable core challenge in the anthropological history of total solar eclipses, since it relates

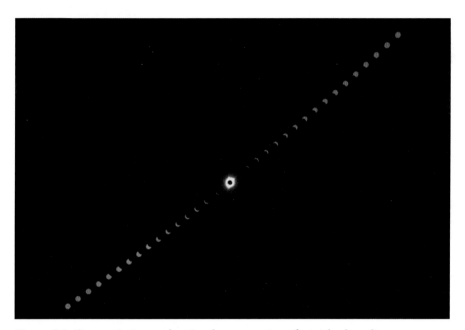

Figure C.1 Composite image showing the progression of a total solar eclipse over Madras, Oregon on 21 August 2017.
Source: NASA/Aubrey Gemignani.

---

[18] Russo contextualizes this quote as follows in relation to her loved one under the experience of severe illness: 'Each eclipse was a celebration of life, and a reminder that life is precious and wonderful. And ultimately life ends. We have one chance at it, and we ought not waste any moment as there will be a time where those moments no longer exist. Geordie's illness made me really appreciate how a total solar eclipse is a reminder of our lives.' Russo, *Total Addiction*, p. viii.

to communication and passing on the knowledge; its impossibility (as argued in the exploration of paradoxical strategies for phenomenal representation in Chapter 11); the merging of the scientific predictability, of a representation and interpretation of past eclipses, and of any new experience itself integrates all disciplines, human skills, and capabilities into a wider perspective across history, so that connections across time and space are interwoven with the interdisciplinary threads and patterns. Total solar eclipses strike a fine balance between the relative power of prediction and the total impossibility of final conditions to be controlled, speaking to both the pride and confidence of the human mind in its abilities of learning and calculation, memory-making, writing of history of cosmic phenomena as well as their interpretation on one hand, and the humility of recognizing its working under ultimately unpredictable conditions on the other. This, all in all, results in more gratefulness for whatever is possible, and with gratitude comes the appreciation of every moment.

For us, the corona has become a model of gathering, shared excitement, communication, and knowledge-production. The purpose of our collaboration was not to produce a definitive guide to the many diverse ways of predicting, perceiving, researching, or documenting an eclipse. Our work together rather aimed at further expanding the limits of perception or perceptibility while also understanding and sharing the experience of our lived scientific and scholarly practice in dialogue. We could only offer this as one possible attempt from our shared points of view; in fact, we would like to see a whole series of books, more perspectives from more disciplines, more expansive in chronological and geographical dimensions, on total solar eclipses and other spectacular phenomena. The collaboration represented in this volume has offered plenty of reasons for optimism about ever-new possibilities; just asking, 'what can be brought into my discipline from the observation of the methods, questions, and critical approaches of others?'—'how can my (artistic, writerly, ethical) practice be informed by that of others?'—starts to open ways towards new horizons of understanding.

In our final discussions, Anna Marie Roos suggested that there is a major benefit for the historian in considering the scientific importance of attention to empirical detail in balance with the pulling away from that detail to accomplish a kind of meta-analysis across disciplines. Not certain how describe the exact quality of light during a total solar eclipse (e.g. *not* dawn, *not* dusk, *not* day, *not* night, *not* blue hour, *not* golden hour), Roos determines that total solar eclipses represent a type of cognitive dissonance: 'Trying to describe it in prose or visually would seem to me to be as challenging as trying to portray the light of divinity. The closest I have seen this quality of light is in the twilight photography of Ansel Adams's gelatin silver print *Moonrise, Hernandez*.'[19] Finally, awe is associated with a deeper, enduring sense of appreciation. These are potent counterforces to the everyday indifference towards the world's cosmic context between chance and predictability. In constant flux, the vast cosmos includes the planet Earth, somehow also providing a unique place for each sentient being. Maybe there is a reason to be here after all—no matter how, from a religious,

---

[19] *Moonrise, Hernandez* was taken late in the afternoon on 1 November 1941, from a shoulder of highway US 84 / US 285 in the unincorporated community of Hernandez, New Mexico by the famous photographer of the American West, Ansel Adams.

philosophical, or personal viewpoint, one might interpret and act upon the conditions of this possibility. Awe brings a flash of attention into life; appreciation generates a lasting sense of meaning and connection.

Maybe what makes a total solar eclipse what it is for the observer-witness is the sense of partaking in the experience of the cooling shadow—of the big, all-governing light being switched off and on again. There is the drama of the temporal compression combined with the extraordinary aesthetics: The path of totality reveals more quickly and powerfully than any other phenomenon that everything around us is in constant circulation and movement; light, umbra, penumbra in constant flux, throughout the months (Moon around the Earth) and the years (Earth and Moon around the Sun). Being part of 'it' means to realize, by experience, not just that we are part of something—of nature; of community; of history; of personal and collective experience; of science history and technological progress; of any artistic, poetic, literary, political, or musical sphere; of the cosmic clockwork—but that we are part of all that *at once*, passive *and* active, paying attention to the gift of revelation that is at the core of inspiration and creativity.

The Eclipse-Chaser's Toolkit

Prepared by Mike Frost FRAS, FIET Director of the Historical Section of the British Astronomical Association (BAA)

A. IMPORTANT: Eye Safety Notice

The visible surface of the Sun, the photosphere, should never be looked at directly: If you look at it with the naked eye, a blink reflex is triggered and the eyes close automatically. Anyone looking at the Sun through any sort of magnification device risks irreparable damage to the eye before the blink reflex has time to trigger. This is why it is so important to consider eye safety before planning any other aspect of your total solar eclipse field trip.

During a total solar eclipse, the Moon moves in front of the Sun. The photosphere is still too bright to look at even during the partial phases of an eclipse, when the Sun is partly covered—so do not look directly at the Sun during a partial eclipse without suitable eye protection. If you want to monitor the progress of a partial eclipse, use certified eclipse glasses. Check the glasses are not scratched before you use them and discard any damaged pairs. Sunglasses are not sufficient. The Sun's image can also be safely projected, but if the projection involves magnification, make sure that no-one can accidentally look through the device doing the magnifying.

There is an important difference between the observation of partial solar eclipses and that of total solar eclipses: For the brief period of a *total* solar eclipse (which may only last seconds, at most minutes), the photosphere is completely covered by the Moon—and then, for the duration of totality, the eclipsed Sun is safe to look at. You can take off your eclipse glasses when totality begins. Be sure, however, that totality has commenced before using magnification—if you are not certain, it probably has not yet. For the period of totality, it is safe to observe with magnification—but the observation and experience of totality does not need magnification. Finally, if you use binoculars or other magnifying devices, make sure you put them down before third contact, the diamond ring, as it marks the end of totality. Visibility of the chromosphere, the lowest layer of the Sun's atmosphere, is a 'final warning' that the end of totality is imminent.

And if you are confused, err on the side of eye safety—just don't use any kind of magnification, and use certified eclipse glasses during the partial phases of the eclipse.

For further information, see: https://www.mreclipse.com/Totality3/TotalityCh11.html.

B. Eclipse Glossary

The Geometry of Eclipses:

> **total eclipse** An eclipse in which the whole of the eclipsed body is hidden.
> **partial eclipse** An eclipse in which only part of the eclipsed body is hidden.

solar eclipse An eclipse in which the Moon passes in front of the Sun.

annular eclipse A solar eclipse in which the Moon is too small to cover the Sun completely and a ring (annulus) of the Sun's photosphere remains uncovered.

lunar eclipse An eclipse in which the Moon passes into the shadow of the Earth.

saros A period of just over 18 years after which eclipses often repeat.

umbra The area of the shadow in which a total eclipse can be seen.

penumbra The area of the shadow in which a partial eclipse can be seen.

antumbra The area of the shadow in which an annular eclipse can be seen.

syzygy When three solar system bodies lie in a straight line (the case of a total solar eclipse is an example with the ordering Sun–Moon–Earth; a lunar eclipse aligns Sun–Earth–Moon).

totality The period of time when the eclipse is total from a given location: for a solar eclipse this is brief, minutes at most; for a lunar eclipse it can be longer.

track of totality The relatively small portion of the Earth's surface in which a total solar eclipse can be seen, generally sweeping from west to east across the Earth's surface.

centreline The centre of the track of totality, where (locally) the totality duration is longest.

Phases and Visual Phenomena of an Eclipse:

first contact The start of an eclipse (for a solar eclipse, the moment when the Moon's disc starts to cross the solar disk).

second contact The start of the total phase of an eclipse (for a solar eclipse, the moment when the Moon's disk completely covers the solar disk).

third contact The end of the total phase of an eclipse.

fourth contact The end of an eclipse (for a solar eclipse, the moment when the Moon's disc leaves the solar disk).

shadow bands Fleeting, low-contrast bands of darkness and brightness which can sometimes be seen on the ground, or projected onto high clouds, shortly before and shortly after a total solar eclipse (shadow bands are caused by atmospheric effects).

Baily's beads Beads of light seen around the surface of the Moon as the Sun is almost completely covered before and after a total eclipse (Baily's beads are caused by the rays of light passing through deeper valleys on the edge of the lunar disk). The last bead to disappear before totality, and the first to appear afterwards, are known as the diamond ring.

diamond ring The last ray of light from the solar disk prior to totality, and (more commonly) the first ray of light at the end of totality.

Parts of the Sun:

photosphere The dazzlingly bright visible solar surface, whose temperature is about 6,000 K.

chromosphere The lowest layer of the solar atmosphere, just above the photosphere.

corona The outer atmosphere of the Sun, whose temperature is about 2 million K.

sunspot A cooler area on the solar surface (associated with a concentration of magnetic field lines).

prominence An eruption of matter forming an arch above the solar surface (visible during totality and usually anchored in sunspots).

coronal mass ejection An eruption of matter from the Sun into the corona and out into interplanetary space.

coronal streamer (helmet streamer) Elongated cusp-like structures in the solar corona.

C. A Short Total Eclipse Reading List

Armitage, Geoff, *The Shadow of the Moon: British Solar Eclipse Mapping in the 18th Century.* London, 1997.

Brown, Elizabeth ('A Lady Astronomer'), *In Pursuit of a Shadow.* Gloucester, 1888.

Clark, David H. and F. Richard **Stephenson,** *The Historical Supernovae.* Oxford, 1977.

Espenak, Fred and Jay **Anderson,** *Eclipse Bulletin: Total Solar Eclipse of 2017 August 21.* Published 2015, available at MrEclipse.com: https://www.mreclipse.com/pubs/TSE2017.html.

Espenak, Fred and Jay **Anderson,** *Eclipse Bulletin: Total Solar Eclipse of 2024 April 08.* Published 2022, available at MrEclipse.com: https://www.mreclipse.com/pubs/EB2024.html.

Espenak, Fred and Jean **Meeus,** *Five Millennium Catalog of Lunar Eclipses: −1999 to +3000 (2000 BCE to 3000 CE).* Goddard Space Flight Center, MD, 2008 (NASA/TP-2008-214172 and 214173).

Green, Lucie, *15 Million Degrees: A Journey to the Centre of the Sun.* London, 2016.

Harris, Joel and Richard **Talcott,** *Chasing the Shadow: An Observer's Guide to Eclipses.* Waukesha, WI, 1994.

Kennefick, Daniel, *No Shadow of a Doubt: The 1919 Eclipse That Confirmed Einstein's Theory of Relativity.* Princeton, NJ, 2019.

Lomb, Nick and Toner **Stevenson,** *Eclipse Chasers.* Melbourne, 2023.

Russo, Kate, *Total Addiction: The Life of an Eclipse Chaser.* London, 2012.

Russo, Kate, *Being in the Shadow: Stories of the First-Time Total Eclipse Experience.* Belfast, 2017.

Steel, Duncan, *Eclipse: The Celestial Phenomenon that Changed the Course of History.* London, 1999.

Stephenson, F. Richard, *Historical Eclipses and Earth's Rotation.* Oxford, 2003.

Williams, Sheridan, *UK Solar Eclipses from Year 1 (an anthology of 3,000 years of solar eclipses).* Leighton Buzzard, Bedfordshire, 1996.

Zeiler, Michael and Michael E. **Bakich,** *Atlas of Solar Eclipses: 2020 to 2045.* Santa Fe, NM, 2020.

Zirker, J. B., *Total Eclipses of the Sun.* Princeton, NJ, 1995.

D. Upcoming Eclipses

For most of the time in most peoples' lives, a total solar eclipse is something very abstract, and for most of history it was rare to experience a total solar eclipse, with an average interval between total solar eclipses being about 380 years at a given location on Earth. Technology, logistics, travel, and the mobility of human beings has dramatically changed this—so much so in the past few decades that one can even experience the increased duration of totality when watching a total solar eclipse from the mobile location of an aeroplane in flight.

The book has been planned since the Great American Eclipse of 2017 to be published ahead of a sequence of total eclipses visible from some of the most easily accessible places on Earth—most prominently the Great American Eclipse of 2024, but there will be many more eclipses in the near future around the world. Many of the readers will live on or close to the tracks of totality; many more will be able to travel to view totality to experience first hand. (This volume,

Figure A.1 The eclipse of 8 April 2024.

Source: Fred Espenak, at NASA's GSFC—http://eclipse.gsfc.nasa.gov/.

of course, is making the case that one should never settle for a *partial* solar eclipse when a *total* solar eclipse is in a reachable distance.)

The next four total eclipses of the Sun are as follows:

8 April 2024 The track of totality crosses central Mexico, Texas, the American Midwest, the Great Lakes region of the USA and Canada, and the maritime provinces of Canada. Most of North and Central America will see a partial eclipse. The maximum duration of totality (on the centreline, over Mexico) will be 4 minutes 28 seconds.

12 August 2026 The track of totality crosses Greenland and western Iceland (including Reykjavik). It then heads south to cross Spain from north-west to east, also a tiny portion of Portugal. Many Spanish cities including Bilbao and Valencia see totality; the totality track is also close to Madrid and Barcelona. The eclipse ends with a sunset eclipse visible from the Balearic Islands (Majorca, Menorca, Ibiza, Formentera). Canada, New England, western Europe, and West Africa will see a partial eclipse. Maximum duration of totality is 2 minutes 18 seconds off Iceland.

SE2026Aug12T

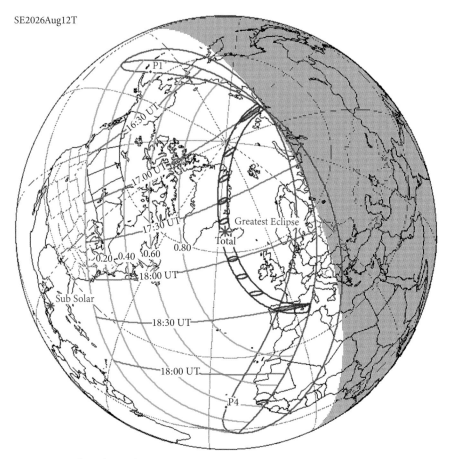

Figure A.2 The eclipse of 12 August 2026.

Source: Fred Espenak, at NASA's GSFC—http://eclipse.gsfc.nasa.gov/.

2 August 2027 The track of totality skims southern Spain and Gibraltar, and is total from several countries in North Africa—Morocco, Algeria, Tunisia, Libya, Egypt, and Sudan. It then crosses Saudi Arabia and the Horn of Africa. Europe, northern and central Africa, and southern Asia will see a partial eclipse. This eclipse has a particularly long duration of totality—6 minutes 23 seconds in Egypt.

22 July 2028 The track of totality crosses Australia from north-west to south-east, covering parts of the states of Western Australia, the Northern Territories, Queensland, and New South Wales. It is total from Sydney. The track continues across the Pacific, crossing the South Island of New Zealand, including Queenstown and Dunedin. South-East Asia and the whole of Australia and New Zealand will see a partial eclipse, as will a tiny portion of Antarctica. The maximum duration of totality is 5 minutes 10 seconds in Western Australia.

During the same period (2024–8) there are also several annular eclipses and partial eclipses. As we explain in Chapters 1 and 4, although these are worth viewing, they do not compare to the unique experience of a total solar eclipse.

SE2027Aug02T

Figure A.3 The eclipse of 2 August 2027.

Source: Fred Espenak, at NASA's GSFC—http://eclipse.gsfc.nasa.gov/.

2 October 2024 An annular eclipse visible across southern Chile and Argentina. The eastern Pacific, southern South America, and parts of Antarctica will see a partial eclipse.

29 March 2025 A partial eclipse (neither annular nor total, anywhere) visible from eastern North America, northern Russia, western Europe, and western Africa.

21 September 2025 A partial eclipse visible from the eastern coast of Australia, New Zealand, and parts of Antarctica.

17 February 2026 An annular eclipse visible across parts of Antarctica. South-east Africa and the tip of South America will see a partial eclipse.

6 February 2027 An annular eclipse visible across Chile, Argentina, Uruguay, and Brazil, ending with a sunset annular eclipse in West Africa. Most of South America and much of western Africa will see a partial eclipse.

26 January 2028 An annular eclipse visible from Ecuador, Peru, Colombia, Brazil, French Guyana, Suriname, and (at sunset) Morocco, Portugal, and Spain.

For further information, see *The Atlas of Solar Eclipses: 2020 to 2045* by Michael Zeiler and Michael E. Bakich.

SE2028Jul22T

Figure A.4 The eclipse of 22 July 2028.

Source: Fred Espenak, at NASA's GSFC—http://eclipse.gsfc.nasa.gov/.

Finally, there are also several of the less dramatic but still beautiful total *lunar* eclipses during this period. Fortunately, lunar eclipses are visible across an entire night-time hemisphere, so many of these will be visible from your back garden. Enjoy them!

13–14 March 2025 Best seen from the Americas.

7–8 September 2025 Best seen from Eastern Africa, Asia, and Australia.

2–3 March 2026 Best seen from the Pacific region.

31 December 2028–1 January 2029 Best seen from Asia and Western Australia.

(There are also a number of partial and penumbral lunar eclipses in this period.)

The NASA eclipse website http://eclipse.gsfc.nasa.gov/ gives detail on all these eclipses.

Select Bibliography

Armitage, Geoff, *The Shadow of the Moon: British Solar Eclipse Mapping in the 18th Century*. London, 1997.

Barfield, Owen, *Saving the Appearances: A Study in Idolatry*. London, 1957.

Baron, David, *American Eclipse: A Nation's Epic Race to Catch the Shadow of the Moon and Win the Glory of the World*. New York, 2017.

Baxandall, Michael, *Shadows and Enlightenment*. New Haven and London, 1995.

Bentley Hart, David, *The Beauty of the Infinite: The Aesthetics of Christian Truth*. Grand Rapids, MI, 2004.

Bierce, Ambrose, *The Devil's Dictionary*. London, 2011.

Brewer, Bryan, *Eclipse: History, Science, Awe*. Seattle, WA, 2017.

Brown, Elizabeth ('A Lady Astronomer'), *In Pursuit of a Shadow*. Gloucester, 1888.

Burke, Edmund, *Philosophical Enquiry into the Origin of Our Ideas of the Sublime and Beautiful*. London, 1757.

Cavendish, Margaret, *Experimental Philosophy*. London, 1666.

Clark, David H. and F. Richard Stephenson, *The Historical Supernovae*. Oxford, 1977.

Close, Frank, *Eclipse: Journeys to the Dark Side of the Moon*. Oxford, 2017.

Coleridge, Samuel Taylor, *The Works of Samuel Taylor Coleridge, Prose and Verse: Complete in One Volume*. Philadelphia, PA, 1852.

Cornish, Alison, *Reading Dante's Stars*. New Haven and London, 2000.

Cullen, Christopher, *Heavenly Numbers: Astronomy and Authority in Early Imperial China*. Oxford, 2017.

Cullen, Christopher, *The Foundations of Celestial Reckoning: Three Ancient Chinese Astronomical Systems*. Abingdon, 2017.

Dante Alighieri, *De vulgari eloquentia*. Trans. Steven Botterill. Cambridge, 1996.

Einstein, Albert and Leopold Infeld, *The Evolution of Physics*. Cambridge, 1938.

Espenak, Fred and Jean Meeus, *Five Millennium Catalog of Lunar Eclipses: −1999 to +3000 (2000 BCE to 3000 CE)*. Goddard Space Flight Center, MD, USA, 2008 (NASA/TP-2008-214172 and 214173).

Geneva, Ann, *Astrology and the Seventeenth-Century Mind: William Lilly and the Language of the Stars*. Manchester, 1995.

Golub, Leon and Jay M. Pasachoff, *Nearest Star: The Surprising Science of our Sun*. Cambridge, 2014.

Golub, Leon and Jay M. Pasachoff, *The Sun*. London, 2017.

Gombrich, Ernst, 'Mirror and Map: Theories of Pictorial Representation.' *Philosophical Transactions of the Royal Society of London B, Biological Sciences* 270, 1975, pp. 119–49.

Green, Lucie, *15 Million Degrees: A Journey to the Centre of the Sun*. London, 2016.

Grosseteste, Robert, *De Sphera*. Ed. C. Panti, trans. S. Sønnesyn, in *Mapping the Universe: Robert Grosseteste's* De sphera—On the Sphere. Eds. Giles E. M. Gasper et al. Oxford, 2023.

Guillermier, Pierre and Serge Koutchmy, *Total Eclipses: Science, Observations, Myths and Legends*. Chichester, 1998.

Guite, Malcolm, *Faith, Hope and Poetry*. London, 2012.

Harris, Joel and Richard Talcott, *Chasing the Shadow: An Observer's Guide to Eclipses*. Waukesha, WI, 1994.

Harrison, Peter, '"Science" and "Religion:" Constructing the Boundaries.' *The Journal of Religion* 86, 2006, pp. 81–106.

Harrison, Peter, *The Territories of Science and Religion*. Chicago, IL, 2015.

Hofstadter, Douglas, *Gödel, Escher, Bach: An Eternal Golden Braid*. New York, 1979.

Holland, Jocelyn, 'A Natural History of Disturbance: Time and the Solar Eclipse.' *Configurations* 23, 2015, pp. 215–33.

Kennefick, Daniel, *No Shadow of a Doubt: The 1919 Eclipse That Confirmed Einstein's Theory of Relativity*. Princeton, NJ, 2019.

Leatherbarrow, Bill, *The Moon*. London, 2018.

Littmann, Mark and Fred Espenak, *Totality: The Great American Eclipses of 2017 and 2024*. Oxford, 2017.

Lomb, Nick and Toner Stevenson, *Eclipse Chasers*. Melbourne, 2023.

Luck, Georg, *Arcana Mundi: Magic and the Occult in the Greek and Roman Worlds*. Baltimore, MD, 1985.

May, Brian and David J. Eicher, *Cosmic Clouds 3-D: Where Stars Are Born*. Cambridge, MA, 2020.

McCrea, W. H., 'Astronomer's Luck.' *Quarterly Journal of the Royal Astronomical Society* 13, 1972, pp. 506–19.

McLeish, Tom, *Faith and Wisdom in Science*. Oxford, 2014.

McLeish, Tom, *The Poetry and Music of Science: Comparing Creativity in Science and Art* (illustrated edition). Oxford, 2019.

McLeish, Tom, *The Poetry and Music of Science: Comparing Creativity in Science and Art* (new edition with an additional chapter on poetry). Oxford, 2022.

Moevs, Christian, *The Metaphysics of Dante's* Comedy. Oxford, 2005.

Moevs, Christian, '"Il punto che mi vinse": Incarnation, Revelation, and Self-Knowledge in Dante's *Commedia*.' In *Dante's Commedia: Theology as Poetry*. Eds. Vittorio Montemaggi and Matthew Treherne. Notre Dame, IN, 2010, pp. 267–85.

Montelle, Clemency, *Chasing Shadows: Mathematics, Astronomy, and the Early History of Eclipse Reckoning*. Baltimore, MD, 2011.

Moore, Donovan, *What Stars Are Made Of: The Life of Cecilia Payne-Gaposchkin*. Boston, MA, 2020.

Moore, Patrick, *On the Moon*. London, 2001.

Morgan, Daniel P., *Astral Sciences in Early Imperial China: Observation, Sagehood and the Individual*. Cambridge, 2017.

Newcomb, Simon, *The Coming Total Eclipse of the Sun*. New York, 1900.

Newcomb, Simon, *Astronomy for Everybody: A Popular Exposition of the Wonders of the Heavens*. New York, 1902.

Nordgren, Tyler, *Sun Moon Earth: The History of Solar Eclipses from Omens of Doom to Einstein and Exoplanets*. New York, 2016.

North, John, *Horoscopes and History*. London, 1986.

Nothaft, C. Philipp E., *Scandalous Error*. Oxford, 2019.

Ogilvie, Marilyn Bailey, 'Obligatory Amateurs: Annie Maunder (1868–1947) and British Women Astronomers at the Dawn of Professional Astronomy.' *The British Journal for the History of Science* 33, 2000, pp. 67–84.

Olson, Roberta J. M. and Jay M. Pasachoff, *Cosmos: The Art and Science of the Universe*. London, 2019.

Olson, Roberta J. M. and Jay M. Pasachoff, *Fire in the Sky: Comets and Meteors, the Decisive Centuries, in British Art and Science*. Cambridge, 1998.

Olson, Roberta J. M., *Fire and Ice: A History of Comets in Art*. New York, 1985.

Penrose, Roger, *The Road to Reality: A Complete Guide to the Laws of the Universe*. New York, 2005.

Ristoro d'Arezzo, *Il primo libro della composizione del mondo di Ristoro d'Arezzo, a cura di G. Amalfi*. Naples, 1888.

Roos, Anna Marie, *Luminaries in the Natural World: Perceptions of the Sun and Moon in England, 1400–1720*. Bern and Oxford, 2001.

Runge, Philipp Otto, *Die Farben-Kugel, oder Construction des Verhaeltnisses aller Farben zueinander*. Hamburg, 1810.

Russo, Kate, *Total Addiction: The Life of an Eclipse Chaser*. London, 2012.

Russo, Kate, *Being in the Shadow: Stories of the First-Time Total Eclipse Experience*. Belfast, 2017.

Sivin, Nathan, *Granting the Seasons: The Chinese Astronomical Reform of 1280, With a Study of Its Many Dimensions and an Annotated Translation of Its Records*. New York, 2009.

Somerville, Mary, *Personal Recollections, From Early Life to Old Age, of Mary Somerville*. London, 1873.

Sprat, Thomas, *The History of the Royal Society of London, for the Improving of Natural Knowledge*. London, 1667.

Steel, Duncan, *Eclipse: The Celestial Phenomenon which has changed the Course of History*. London, 1999.

Steele, John M., *Observations and Predictions of Eclipse Times by Early Astronomers*. Dordrecht, 2000.

Stephenson, F. Richard, *Historical Eclipses and Earth's Rotation*. Oxford, 2003.

Stoichita, Victor I., *A Short History of the Shadow*. London, 1997.

Thomann, Johannes and Matthias Vogel, *Schattenspur: Sonnenfinsternisse in Wissenschaft, Kunst und Mythos*. Basel, 1999.

Todd, Mabel Loomis, *Total Eclipses of the Sun*. Boston, MA, 1900.

Wallis, Faith and Robert Wisnovsky (eds), *Medieval Textual Cultures: Agents of Transmission, Translation and Transformation*. Berlin, 2016.

Westfall, John and William Sheehan, *Celestial Shadows: Eclipses, Transits, and Occultations*. New York, 2015.

White, Paul, 'Ministers of Culture: Arnold, Huxley and Liberal Anglican Reform of Learning.' *History of Science* 43, 2005, pp. 115–38.

Williams, Sheridan, *UK Solar Eclipses from Year 1 (an anthology of 3,000 years of solar eclipses)*. Leighton Buzzard, Bedfordshire, 1996.

Zeiler, Michael and Michael E. Bakich, *Atlas of Solar Eclipses: 2020 to 2045*. Santa Fe, NM, 2020.

Zirker, J. B., *Total Eclipses of the Sun*. Princeton, NJ, 1995.

Index